# PARTIAL DIFFERENTIAL EQUATIONS

# PARTIAL DIFFERENTIAL EQUATIONS

## E. T. COPSON

FORMERLY REGIUS PROFESSOR OF MATHEMATICS IN
THE UNIVERSITY OF ST ANDREWS

CAMBRIDGE UNIVERSITY PRESS

CAMBRIDGE

LONDON · NEW YORK · MELBOURNE

CAMBRIDGE UNIVERSITY PRESS
Cambridge, New York, Melbourne, Madrid, Cape Town, Singapore, São Paulo, Delhi

Cambridge University Press
The Edinburgh Building, Cambridge CB2 8RU, UK

Published in the United States of America by Cambridge University Press, New York

www.cambridge.org
Information on this title: www.cambridge.org/9780521205832

First published 1975
Re-issued in this digitally printed version 2008

*A catalogue record for this publication is available from the British Library*

*Library of Congress Cataloguing in Publication data*
Copson, Edward Thomas, 1901–
Partial differential equations.
Bibliography : p. 277
Includes index.
1. Differential equation, Partial. I. Title.
QA377.C77    515′.353    74–12965

ISBN 978-0-521-20583-2 hardback
ISBN 978-0-521-09893-9 paperback

# CONTENTS

# PREFACE

This book has been written in memory of my father-in-law the late Professor Sir Edmund Whittaker, F.R.S., in gratitude for all the help and encouragement he gave me for over thirty years. Today is the hundredth anniversary of his birth.

When I went to Edinburgh as a young lecturer in 1922, I was surprised to find how different the curriculum was from that in Oxford. It included such topics as Lebesgue integration, matrix theory, numerical analysis, Riemannian geometry, of which I knew nothing. I was particularly impressed by Whittaker's lectures on partial differential equations to undergraduate and postgraduate students, far different from the standard English textbooks of the time. This book is not based on Whittaker's lectures; yet without his inspiration it would never have been written.

I have frequently given courses of lectures on partial differential equations and have always regretted that there was no book to which I could refer my students. Friends told me that the remedy was to write one myself; and here it is, a presentation of *some* of the theory by the methods of classical analysis.

There are few references to original sources. After lecturing on the subject for so many years, I could not now say whence the material came. On page 277 will be found a list of the books which I have read with profit, many of them more advanced than this.

E.T.C.

St Andrews
24 October 1973

# 1

# PARTIAL DIFFERENTIAL EQUATIONS
# OF THE FIRST ORDER

## 1.1 Lagrange's equation

Lagrange's partial differential equation of the first order is of the form

$$Pp + Qq = R, \tag{1}$$

where $p = \partial u/\partial x$, $q = \partial u/\partial y$ and $P$, $Q$, $R$ are functions of $x$, $y$ and $u$; it is sometimes called a quasi-linear equation since it is linear in the derivatives. If $P$, $Q$ and $R$ do not involve $u$, Lagrange's equation is said to be linear; if only $R$ involves $u$ it is said to be semi-linear.

By a solution of (1), is meant a function $u(x, y)$ which satisfies the differential equation; but we often have to be content with a solution defined implicitly by a relation $f(x, y, u) = 0$. If we regard $(x, y, u)$ as rectangular Cartesian coordinates, $f(x, y, u) = 0$ is the equation of a surface; if $f = 0$ provides a solution of (1), the surface is called an integral surface. The fundamental problem is: given a regular arc†
$\Gamma$ in space, is there a unique integral surface through $\Gamma$? Alternatively, given a regular arc $\gamma$ in the $xy$-plane, is there a solution $u(x, y)$ of (1) which takes given values on $\gamma$?

Let the parametric equations of $\Gamma$ be

$$x = x_0(t), \quad y = y_0(t), \quad u = u_0(t).$$

On any surface, $du = p\,dx + q\,dy$. Hence, if there is an integral surface through $\Gamma$, the values $p_0(t)$, $q_0(t)$ of $p$ and $q$ on the integral surface at the point of parameter $t$ of $\Gamma$ satisfy

$$\dot{u}_0 = p_0\dot{x}_0 + q_0\dot{y}_0, \tag{2}$$

where dots denote differentiation with respect to $t$. If we denote by $P_0, Q_0, R_0$ the values of $P, Q, R$ at the point of $\Gamma$ of parameter $t$, we have

$$P_0 p_0 + Q_0 q_0 = R_0. \tag{3}$$

Hence if $\dot{x}_0 Q_0 - \dot{y}_0 P_0$ is not zero, $p_0$ and $q_0$ are determined.

It is conventional to denote the second derivatives $u_{xx}, u_{xy}, u_{yy}$ by $r, s, t$; the fact that we have also used $t$ to denote the parameter of $\Gamma$

---

† The term *regular arc* is defined in Note 3 of the Appendix.

will not cause any confusion. If we differentiate (1) with respect to $x$, we get

$$Pr + Qs = F(x, y, u, p, q),$$

so that, at the point of $\Gamma$ of parameter $t$,

$$P_0 r_0 + Q_0 s_0 = F_0.$$

Since $dp = r\, dx + s\, dy,$ $\qquad \dot{x}_0 r_0 + \dot{y}_0 s_0 = \dot{p}_0.$

If $\dot{x}_0 Q_0 - \dot{y}_0 P_0$ is not zero, $p_0$ and $q_0$ are determined on $\Gamma$, and hence so also are $r_0$ and $s_0$. Similarly we can find all the partial derivatives of $u$ on $\Gamma$. Thus we get a formal solution as a Taylor series

$$u = u_0 + \{p_0(x - x_0) + q_0(y - y_0)\}$$
$$+ \tfrac{1}{2}\{r_0(x - x_0)^2 + 2s_0(x - x_0)(y - y_0) + t_0(y - y_0)^2\} + \cdots.$$

Under suitable conditions, it can be shown that the series converges in a neighbourhood of $(x_0, y_0, u_0)$ of $\Gamma$, provided that $\dot{x}_0 Q_0 - \dot{y}_0 P_0$ is not zero.

Now drop the suffix zero which has served its purpose. At a point of $\Gamma$, an integral surface satisfies

$$Pp + Qq = R, \quad p\dot{x} + q\dot{y} = \dot{u}.$$

Hence $\qquad\qquad (Q\dot{x} - P\dot{y})\, q = R\dot{x} - P\dot{u},$

and similarly for $p$. If $Q\dot{x} - P\dot{y}$ vanishes at every point of $\Gamma$, this equation is impossible unless the transport equation

$$P\dot{u} = R\dot{x}$$

(or equivalently) $\qquad\qquad Q\dot{u} = R\dot{y}$

is satisfied. Hence, if $Q\dot{x} - P\dot{y} = 0$ on $\Gamma$, there is no integral surface through $\Gamma$ unless $u$ satisfies the transport equation; and then there are an infinite number of integral surfaces since $q$ can be chosen arbitrarily.

An arc $\Gamma$ which has this property is called a *characteristic*. There is one characteristic through each point of space at which $P, Q, R$ are not all zero; a characteristic satisfies

$$\frac{\dot{x}}{P} = \frac{\dot{y}}{Q} = \frac{\dot{u}}{R}.$$

A characteristic is the curve of intersection of two integral surfaces. For if $u = u_1(x, y)$, $u = u_2(x, y)$ are two intersecting integral surfaces,

$$p_1\, dx + q_1\, dy - du = 0,$$
$$p_2\, dx + q_2\, dy - du = 0,$$

in an obvious notation, and so

$$\frac{dx}{q_1 - q_2} = \frac{dy}{p_2 - p_1} = \frac{du}{p_2 q_1 - p_1 q_2}.$$

But
$$Pp_1 + Qq_1 = R, \quad Pp_2 + Qq_2 = R$$

so that
$$\frac{P}{q_1 - q_2} = \frac{Q}{p_2 - p_1} = \frac{R}{p_2 q_1 - p_1 q_2}.$$

Therefore on the curve of intersection of two integral surfaces,

$$\frac{dx}{P} = \frac{dy}{Q} = \frac{du}{R}.$$

The differential equations for a characteristic can be written as

$$\dot{x} = P, \quad \dot{y} = Q, \quad \dot{u} = R$$

by a change of parameter. The solution of these equations contains three constants of integration; two of these can be the coordinates of the point where the characteristic cuts, say, the plane $u = 0$, and the third can be fixed by measuring $t$ from that point – the differential equations are unaltered if we replace $t$ by $t + c$. The characteristics then form a two-parameter family. If $C$ is a non-characteristic arc, it can be shown that the unique integral surface through $C$ is generated by the one-parameter family of characteristics which intersect $C$. Again, if the two-parameter family of characteristics is given by

$$\phi(x, y, u) = a, \quad \psi(x, y, u) = b,$$

we can construct a one-parameter family by setting up a relation between $a$ and $b$, say $b = F(a)$. This one-parameter family generates the integral surface

$$\psi(x, y, u) = F\{\phi(x, y, u)\}.$$

The projection $\gamma$ of a characteristic $\Gamma$ on the plane $u = 0$ is called a *characteristic base-curve*. If $P$ and $Q$ do not involve $u$, the characteristic base-curves satisfy
$$\dot{x} = P, \quad \dot{y} = Q.$$

In order that there may be a solution which takes given values on $\gamma$, the data must satisfy the equation

$$P\dot{u} = R\dot{x}.$$

## 1.2 Two examples

We know that, if $u$ is a homogeneous function of $x$ and $y$ of degree $n$ then

$$x\frac{\partial u}{\partial x} + y\frac{\partial u}{\partial y} = nu.$$

We now prove the converse. The subsidiary equations of Lagrange are

$$\frac{dx}{x} = \frac{dy}{y} = \frac{du}{nu}.$$

From these equations we get

$$\frac{y}{x} = a, \quad \frac{u}{x^n} = b.$$

Hence the general solution is

$$\frac{u}{x^n} = f\left(\frac{y}{x}\right).$$

As a second example, let us find the integral surface of

$$(y-u)p + (u-x)q = x-y,$$

which goes through the curve $u = 0, xy = 1$.

The characteristics are given by

$$\dot{x} = y-u, \quad \dot{y} = u-x, \quad \dot{u} = x-y,$$

which give $\quad \dot{x}+\dot{y}+\dot{u} = 0, \quad x\dot{x}+y\dot{y}+u\dot{u} = 0.$

Hence the characteristics are circles,

$$x+y+u = a, \quad x^2+y^2+u^2 = b.$$

We have to choose the one-parameter family which goes through $u = 0, xy = 1$. When $u = 0, xy = 1$,

$$a^2 = (x+y)^2 = x^2+y^2+2xy = b+2.$$

The required integral surface is therefore

$$(x+y+u)^2 = x^2+y^2+u^2+2$$

or

$$u = \frac{1-xy}{x+y}.$$

## 1.3   The general first order equation

We now ask the same question concerning the general first order equation

$$F(x, y, u, p, q) = 0. \tag{1}$$

Does there exist an integral surface through a given regular arc $\Gamma$

$$x = x_0(t), \quad y = y_0(t), \quad u = u_0(t)?$$

The method is to try to construct a Taylor series which satisfies (1) and converges in a neighbourhood of an arbitrary point of $\Gamma$. This involves calculating at that point all the partial derivatives of $u$.

The first derivatives of $u$ satisfy the condition $du = p\,dx + q\,dy$, so their values $p_0$ and $q_0$ at the point of $\Gamma$ of parameter $t$ are given by

$$F(x_0, y_0, u_0, p_0, q_0) = 0,$$

$$\dot{x}_0 p_0 + \dot{y}_0 q_0 = \dot{u}_0.$$

We suppose that we can find a real pair $(p_0, q_0)$ which satisfies these equations; if we cannot, there is no real integral surface.

Next denote the partial derivatives of $F$ with respect to $x, y, u, p, q$ by $X, Y, U, P, Q$. Then, if we differentiate (1) partially with respect to $x$, the variables $u, p, q$ now being functions of $x$ and $y$, we get

$$Pr + Qs + X + Up = 0.$$

By hypothesis, $p$ is now known on $\Gamma$. Using the condition

$$dp = r\,dx + s\,dy,$$

the values of the second derivatives of $r$ and $s$ on $\Gamma$ satisfy

$$P_0 r_0 + Q_0 s_0 + X_0 + U_0 p_0 = 0,$$

$$\dot{x}_0 r_0 + \dot{y}_0 s_0 = \dot{p}_0.$$

Hence

$$(Q_0 \dot{x}_0 - P_0 \dot{y}_0) s_0 = -(X_0 + U_0 p_0)\dot{x}_0 - P_0 \dot{p}_0, \tag{2}$$

and similarly

$$(Q_0 \dot{x}_0 - P_0 \dot{y}_0) s_0 = (Y_0 + U_0 q_0)\dot{y}_0 + Q_0 \dot{q}_0. \tag{3}$$

Since

$$X\,dx + Y\,dy + U\,du + P\,dp + Q\,dq = 0,$$

where

$$du = p\,dx + q\,dy,$$

(2) and (3) are in fact the same equation.

If $Q_0 \dot{x}_0 - P_0 \dot{y}_0$ is not zero, the values $r_0, s_0, t_0$ of the second derivatives are determined on $\Gamma$, and similarly for the derivatives of higher orders. Thus we again get a formal solution as a double Taylor series which can, under suitable conditions, be shown to converge in some neighbourhood of the chosen point of $\Gamma$, provided that

$$Q_0 \dot{x}_0 - P_0 \dot{y}_0$$

does not vanish there.

Now drop the suffix zero. At a point of $\Gamma$, an integral surface satisfies

$$F(x, y, u, p, q) = 0,$$

$$\dot{u} = p\dot{x} + q\dot{y},$$

and

$$Pr + Qs = -X - pU,$$

$$Ps + Qt = -Y - qU,$$

where

$$r\dot{x} + s\dot{y} = \dot{p}, \quad s\dot{x} + t\dot{y} = \dot{q}.$$

Hence

$$(Q\dot{x} - P\dot{y})s = -(X + pU)\dot{x} - P\dot{p},$$

and

$$(Q\dot{x} - P\dot{y})s = (Y + qU)\dot{y} + Q\dot{q}.$$

If $Q\dot{x} - P\dot{y}$ vanishes at every point of $\Gamma$, there is no integral surface through $\Gamma$ unless the expressions on the right of the last two equations vanish. This means that there is no integral surface unless $u, p, q$ are appropriately chosen on $\Gamma$. Thus we have now not an arc but a strip; a sort of narrow ribbon formed by the arc $\Gamma$ and the associated surface elements specified by $p$ and $q$. Such a ribbon is called a *characteristic strip*. The arc carrying the strip may be called a characteristic.

The differential equations of a characteristic strip are

$$\frac{\dot{x}}{P} = \frac{\dot{y}}{Q} = \frac{\dot{u}}{pP + qQ} = -\frac{\dot{p}}{X + pU} = -\frac{\dot{q}}{Y + qU}$$

and also

$$F(x, y, u, p, q) = 0,$$

regarded, not as a differential equation, but as an equation in five variables. By a change in the parameter $t$, we can write the equations as

$$\dot{x} = P, \quad \dot{y} = Q, \quad \dot{u} = pP + qQ, \quad \dot{p} = -X - pU, \quad \dot{q} = -Y - qU,$$

where

$$F(x, y, u, p, q) = 0.$$

The characteristic strips form a three-parameter family. There are five constants of integration; one of these is fixed by the identity $F = 0$, the second by choice of the origin of $t$.

The unique integral surface which passes through a non-characteristic arc $C$ is generated by a one-parameter family of characteristic strips. The first step is to construct an initial integral strip by associating with each point of $C$ a surface element whose normal is in the direction $p:q:-1$. If the parametric equations of $C$ are

$$x = x_0(s), \quad y = y_0(s), \quad u = u_0(s),$$

where $s$ need not be the arc-length, we choose

$$p = p_0(s), \quad q = q_0(s)$$

so that

$$\frac{du_0}{ds} = p_0 \frac{dx_0}{ds} + q_0 \frac{\partial y_0}{ds}$$

and

$$F(x_0, y_0, u_0, p_0, q_0) = 0.$$

Through each surface element of the initial strip, there passes a unique characteristic strip. The one-parameter family of characteristic strips so formed generates the required integral surface, as illustrated by the example of the next section. This method is usually called the method of Lagrange and Charpit.

It will be noticed that although the quasi-linear equation

$$Pp + Qq = R$$

does possess characteristic strips no use is made of them in solving such an equation. This is because of an important geometrical difference between Lagrange's equation and the general equation

$$F(x, y, u, p, q) = 0.$$

If $(x_0, y_0, u_0)$ is a point on an integral surface of $F = 0$, the direction ratios $p_0:q_0:-1$ of the normal there satisfy $F(x_0, y_0, u_0, p_0, q_0) = 0$. Hence the normals to all possible integral surfaces through the point generate a cone $N$ whose equation is

$$F\left(x_0, y_0, u_0, -\frac{x-x_0}{u-u_0}, -\frac{y-y_0}{u-u_0}\right) = 0.$$

The tangent planes at $(x_0, y_0, u_0)$ to all possible integral surfaces through this point envelope another cone $T$ whose equation is obtained by eliminating $p_0$ and $q_0$ from the equations

$$u - u_0 = p_0(x - x_0) + q_0(y - y_0),$$

$$(x - x_0)Q_0 - (y - y_0)P_0 = 0,$$

$$F(x_0, y_0, u_0, p_0, q_0) = 0,$$

where $P_0$ and $Q_0$ denote the values of $\partial F/\partial p$ and $\partial F/\partial q$ at

$$(x_0, y_0, u_0, p_0, q_0).$$

The tangent plane to a particular integral surface at $(x_0, y_0, u_0)$ goes through a generator of the cone $T$; the normal there lies on the cone $N$.

In the case of Lagrange's equation, the cone $N$ degenerates into the plane

$$P_0(x - x_0) + Q_0(y - y_0) + R_0(u - u_0) = 0;$$

the cone $T$ becomes the straight line

$$\frac{x - x_0}{P_0} = \frac{y - y_0}{Q_0} = \frac{u - u_0}{R_0}.$$

## 1.4   An example of the Lagrange–Charpit method

We find by the method of characteristics the integral surface of $pq = xy$ which goes through the curve $u = x, y = 0$. The characteristic strips are given by the differential equations

$$\dot{x} = q, \quad \dot{y} = p, \quad \dot{u} = 2pq, \quad \dot{p} = y, \quad \dot{q} = x$$

and the relation $pq = xy$. It turns out that

$$x = Ae^t + Be^{-t}, \quad y = Ce^t + De^{-t}, \quad u = ACe^{2t} - BDe^{-2t} + E,$$

$$p = Ce^t - De^{-t}, \quad q = Ae^t - Be^{-t},$$

where the constants of integration are connected by

$$AD + BC = 0.$$

On the initial curve    $x = s, \quad y = 0, \quad u = s.$

On the initial integral strip, the equations

$$du = p\,dx + q\,dy, \quad pq = xy$$

give $p = 1, q = 0$. Let $t$ be measured from the initial curve. Then when $t = 0$, we have

$$A + B = s, \quad C + D = 0, \quad AC - BD + E = s,$$

$$C - D = 1, \quad A - B = 0$$

where    $AD + BC = 0.$

These give    $A = B = \tfrac{1}{2}s, \quad C = -D = \tfrac{1}{2}, \quad E = \tfrac{1}{2}s,$

the condition $AD + BC = 0$ being satisfied automatically since the initial strip is an integral strip.

The characteristics through the initial integral strip are therefore

$$x = s\cosh t, \quad y = \sinh t, \quad u = s\cosh^2 t,$$

$$p = \cosh t, \quad q = s\sinh t.$$

Eliminating $s$ and $t$ from the first three equations, we obtain

$$u^2 = x^2(1+y^2)$$

as the equation of the required integral surface.

## 1.5   An initial value problem†

In this section we prove that the equation

$$p = f(x, y, u, q) \qquad (1)$$

has, under certain conditions, a unique analytic solution which satisfies the initial conditions

$$u(x_0, y) = \phi(y), \quad q(x_0, y) = \phi'(y), \qquad (2)$$

where $\phi(y)$ is analytic. The result we obtain is a local result; we show, by the method of dominant functions, that there is a solution

$$u = u(x, y)$$

regular in a neighbourhood of any point $(x_0, y_0)$ of the initial line $x = x_0$. It is convenient to write $u_0$ for $\phi(y_0)$, $q_0$ for $\phi'(y_0)$. And we make the assumption that $f(x, y, u, q)$ is an analytic function of four independent variables, regular in a neighbourhood of $(x_0, y_0, u_0, q_0)$.

The problem can be transformed into one involving three quasi-linear equations with three dependent variables $u, p, q$. If there is an analytic solution, then $\partial q/\partial x = \partial p/\partial y$. From equation (1) we have

$$\frac{\partial p}{\partial x} = f_x + f_u p + f_q \frac{\partial q}{\partial x}.$$

Hence $u, p, q$ satisfy the equations

$$\left. \begin{aligned} \frac{\partial u}{\partial x} &= p, \\[1em] \frac{\partial p}{\partial x} &= f_x + f_u p + f_q \frac{\partial p}{\partial y}, \\[1em] \frac{\partial q}{\partial x} &= \frac{\partial p}{\partial y}, \end{aligned} \right\} \qquad (3)$$

under the initial conditions

$$u(x_0, y) = \phi(y), \quad p(x_0, y) = f(x_0, y, \phi(y), \phi'(y)), \quad q(x_0, y) = \phi'(y).$$

This system of three equations is equivalent to equation (1) under the initial conditions (2). From the first and last of equations (3),

$$\frac{\partial}{\partial x}\left(q - \frac{\partial u}{\partial y}\right) = 0$$

† See Notes 1 and 2 in the Appendix.

so that

$$q - \frac{\partial u}{\partial y} = \omega_1(y).$$

By the initial conditions, $\omega_1(y)$ is identically zero, so that $q = \partial u / \partial y$. The second equation of (3) gives

$$\frac{\partial p}{\partial x} = f_x + f_u p + f_q \frac{\partial q}{\partial x} = \frac{\partial f}{\partial x}$$

since

$$\frac{\partial q}{\partial x} = \frac{\partial^2 u}{\partial x \, \partial y} = \frac{\partial^2 u}{\partial y \, \partial x} = \frac{\partial p}{\partial y},$$

when $u$ is analytic. Hence

$$p = f(x, y, u, q) + \omega_2(y).$$

Again, by the initial conditions, $\omega_2(y)$ is identically zero, and so $p = f(x, y, u, q)$.

The coefficients in the second equation of (3) involve the independent variables $x$ and $y$. We can get rid of this restriction by introducing two additional dependent variables $\xi$ and $\eta$ defined by

$$\frac{\partial \xi}{\partial x} = \frac{\partial \eta}{\partial y}, \quad \frac{\partial \eta}{\partial x} = 0$$

under the initial conditions

$$\xi = 0, \quad \eta = y - y_0,$$

when $x = x_0$. Since $\eta$ is independent of $x$, $\eta = y - y_0$ for all $x$. Then $\partial \xi / \partial x = 1$ so that $\xi = x - x_0$. If we put $x = x_0 + \xi$, $y = y_0 + \eta$ in $f(x, y, u, p, q)$ we get an analytic function $g(\xi, \eta, u, p, q)$ of five variables regular in a neighbourhood of $(0, 0, u_0, p_0, q_0)$. We now have a system of five equations

$$\frac{\partial u}{\partial x} = p \frac{\partial \eta}{\partial y},$$

$$\frac{\partial p}{\partial x} = \{g_\xi + g_u p\} \frac{\partial \eta}{\partial y} + g_q \frac{\partial p}{\partial y},$$

$$\frac{\partial q}{\partial x} = \frac{\partial p}{\partial y},$$

$$\frac{\partial \xi}{\partial x} = \frac{\partial \eta}{\partial y}.$$

$$\frac{\partial \eta}{\partial x} = 0,$$

with the initial conditions

$$u = \phi(y), \quad p = f(x_0, y, \phi(y), \phi'(y)), \quad q = \phi'(y),$$

$$\xi = 0, \quad \eta = y - y_0$$

when $x = x_0$.

This system is of the form

$$\frac{\partial u_i}{\partial x} = \sum_{j=1}^{n} g_{ij} \frac{\partial u_j}{\partial y} \quad (i = 1, 2, ..., n)$$

with the initial conditions

$$u_i = \phi_i(y) \quad (i = 1, 2, ..., n)$$

when $x = x_0$. The coefficients $g_{ij}$ are analytic functions of the dependent variables only, regular in a neighbourhood of

$$(\phi_1(y_0), \phi_2(y_0), ..., \phi_n(y_0))$$

and the data $\phi_i(y)$ regular in a neighbourhood of $y_0$. Here $n = 5$; later, we shall need the case $n = 8$. The method of proof does not depend on the number of variables. The easiest case is when $n = 2$. The general case is discussed in Courant and Hilbert's book.

Consider, then, the simplest case of two equations

$$\frac{\partial u}{\partial x} = A \frac{\partial u}{\partial y} + B \frac{\partial v}{\partial y},$$

$$\frac{\partial v}{\partial x} = C \frac{\partial u}{\partial y} + D \frac{\partial v}{\partial y},$$

under the initial conditions

$$u = \phi(y), \quad v = \psi(y),$$

when $x = x_0$, where $\phi$ and $\psi$ are analytic functions regular in a neighbourhood of $y_0$. For simplicity in writing, shift the origin so that $x_0$ and $y_0$ are zero. The coefficients $A, B, C, D$ are analytic functions of $u$ and $v$, regular in a neighbourhood of $(u_0, v_0)$, where $u_0 = \phi(0), v_0 = \psi(0)$. Again, by a shift of origin in the $uv$-plane, we can take $u_0$ and $v_0$ to be zero.

Since $\phi, \psi$ and the coefficients are analytic, we can find constants $M$ and $R$ so that $\phi(y)$ and $\psi(y)$ are dominated by $My/(1-y/R)$ in $|y| < R$ and $A, B, C, D$ by $M/\{1 - (u+v)/R\}$ in $|u+v| < R$. We can use the same constants $M$ and $R$ by taking $M$ large enough, $R$ small enough. And, by changes of scale, we may take $R = 1$ which simplifies writing.

Formal solutions of the initial value problem as double power series

$$u = \sum_{m,\,n=0}^{\infty} \frac{u_{mn}}{m!\,n!}\, x^m y^n,$$

$$v = \sum_{m,\,n=0}^{\infty} \frac{v_{mn}}{m!\,n!}\, x^m y^n,$$

can be found by calculating the coefficients

$$u_{mn} = \left(\frac{\partial^{m+n} u}{\partial x^m \,\partial y^n}\right)_0, \quad v_{mn} = \left(\frac{\partial^{m+n} v}{\partial x^m \,\partial y^n}\right)_0,$$

where the suffix zero denotes that the derivatives are evaluated at the origin.

Since $\phi$ and $\psi$ are analytic, they have convergent expansions

$$\phi(y) = \sum_0^{\infty} \frac{\phi_n}{n!}\, y^n, \quad \psi(y) = \sum_0^{\infty} \frac{\psi_n}{n!}\, y^n,$$

where $\phi_0 = \psi_0 = 0$. Hence

$$u_{0n} = \phi_n, \quad v_{0n} = \psi_n.$$

From the differential equations we have

$$u_{10} = A(0,0)\,\phi_1 + B(0,0)\,\psi_1, \quad v_{10} = C(0,0)\,\phi_1 + D(0,0)\,\psi_1.$$

Next,

$$\frac{\partial^2 u}{\partial x\,\partial y} = A\,\frac{\partial^2 u}{\partial y^2} + V\,\frac{\partial^2 v}{\partial y^2} + A_u \left(\frac{\partial u}{\partial y}\right)^2 + (A_v + B_u)\,\frac{\partial u}{\partial y}\,\frac{\partial v}{\partial y} + B_v \left(\frac{\partial v}{\partial y}\right)^2$$

so that

$$u_{11} = A(0,0)\,\phi_2 + B(0,0)\,\psi_2 + A_u(0,0)\,\phi_1^2$$
$$\qquad + \{A_v(0,0) + B_u(0,0)\}\,\phi_1 \psi_1 + B_v(0,0)\,\psi_1^2.$$

Similarly we can calculate successively the coefficients $u_{1n}$ and $v_{1n}$. Next, by considering $\partial^2 u/\partial x^2$, $\partial^3 u/\partial x^2 \partial y$, ... we can calculate $u_{20}, u_{21}, \ldots$. In a similar way, the coefficients $u_{mn}$ and $v_{mn}$ in the formal expansions for $u$ and $v$ can all be determined.

If we replace $A, B, C, D, \phi$ and $\psi$ by their majorants, we obtain the system of equations

$$\frac{\partial U}{\partial x} = \frac{M}{1 - U - V}\left(\frac{\partial U}{\partial y} + \frac{\partial V}{\partial y}\right),$$

$$\frac{\partial V}{\partial x} = \frac{M}{1 - U - V}\left(\frac{\partial U}{\partial y} + \frac{\partial V}{\partial y}\right),$$

under the conditions    $U = V = My/(1-y)$,

when $x = 0$, $|y| < 1$, the factor $y$ occurring because $\phi$ and $\psi$ vanish when $y = 0$. We show that this initial value problem has a unique solution regular near the origin, and the functions $U$ and $V$ will be majorants of $u$ and $v$.

Since

$$\frac{\partial}{\partial x}(U - V) = 0,$$

$U - V$ is a function of $y$ alone, and by the initial condition is identically zero. Hence $U = V$ and

$$\frac{\partial U}{\partial x} = \frac{2M}{1 - 2U}\frac{\partial U}{\partial y}.$$

Since the Jacobian of $U$ and $2Mx + (1-2U)y$ vanishes identically, the general solution of this equation is

$$2Mx + (1 - 2U)y = F(U),$$

and, by the initial condition, $F(U) = U(1 - 2U)/(M + U)$. Hence the required solution satisfies the quadratic equation

$$2Mx(M + U) + (1 - 2U)(M + U)y = U(1 - 2U).$$

This equation has two solutions; one takes the value 0 at the origin, the other the value $\frac{1}{2}$. It is the former which is needed; it can be written down explicitly, and is evidently an analytic function, regular in a neighbourhood of the origin.

If we write down the expansion of $U$ near the origin as

$$U = \sum_{m,n=0}^{\infty} \frac{U_{mn}}{m!\,n!}x^m y^n,$$

the coefficients can be calculated successively just as in the original problem. Putting $x = 0$, we get

$$U_{00} = 0, \qquad \frac{U_{0n}}{n!} = M.$$

From the differential equation,

$$U_{10} = 2M^2.$$

Next,          $$\frac{\partial^2 U}{\partial x\,\partial y} = \frac{2M}{1 - 2U}\frac{\partial^2 U}{\partial y^2} + \frac{4M}{(1 - 2U)^2}\left(\frac{\partial U}{\partial y}\right)^2$$

so that          $$U_{11} = 4M^2 + 4M^3.$$

By repeated differentiations and respect to $y$, we can calculate all the coefficients $U_{1n}$; each is positive. Again

$$\frac{\partial^2 U}{\partial x^2} = \frac{2M}{1-2U}\frac{\partial^2 U}{\partial x\,\partial y} + \frac{4M}{(1-2U)^2}\frac{\partial U}{\partial x}\frac{\partial U}{\partial y}$$

so that $\quad U_{20} = 2MU_{11} + 4MU_{10}U_{01} = 4M^3 + 8M^4.$

By repeated differentiations with respect to $y$, we can calculate all the coefficients $U_{2n}$; again, each is positive. Similarly all the coefficients $U_{mn}$ can be found, and each is positive.

Lastly, by this iterative process, we see that

$$|u_{mn}| < U_{mn}, \quad |v_{mn}| < U_{mn}$$

so that the formal series $u$ and $v$ are both dominated by $U$. Therefore there is a unique solution, regular near $(x_0, y_0)$, of the initial value problem with which we started.

The result proved in this section is not as general as might appear at first sight. We do not need to assume the data are analytic. For example, if we assume that $\phi(y)$ is everywhere differentiable, the equation $xp - yq + u = 0$, under the initial conditions

$$u(x_0, y) = \phi(y), \quad q(x_0, y) = \phi'(y),$$

where $x_0 \neq 0$, has the solution

$$u = \frac{x_0}{x}\,\phi\left(\frac{xy}{x_0}\right),$$

which is differentiable but not necessarily regular in the neighbourhood of any point $(x_0, y_0)$.

## 1.6   Systems of semi-linear equations of the first order

A system of semi-linear equations of the first order is of the form

$$\left.\begin{array}{l} a_{11}u_x + a_{12}v_x + b_{11}u_y + b_{12}v_y = h_1, \\ a_{21}u_x + a_{22}v_x + b_{21}u_y + b_{22}v_y = h_2, \end{array}\right\} \tag{1}$$

where the coefficients $a_{ij}, b_{ij}$ depend only on $x$ and $y$, but $h_1$ and $h_2$ may also involve $u$ and $v$. The coefficients $a_{ij}, b_{ij}$ are analytic functions, regular in some domain (open connected set) $D$ and $h_1$ and $h_2$ are regular when $(x, y)$ belongs to $D$ and $|u| < K, |v| < K$ for some constant $K$. The problem is to see whether there exists a unique analytic solution, regular in a neighbourhood of a regular arc $\gamma$, which takes given values on $\gamma$. The first step in a solution by Taylor series is to show that, at any point of $\gamma$, the partial derivatives of $u$ and $v$ of all orders are determinate.

Let the parametric equations of $\gamma$ be $x = x_0(t)$, $y = y_0(t)$; on $\gamma$, we are given that $u = u_0(t)$, $v = v_0(t)$. As we shall need derivatives of these four functions of all orders, we assume that they are analytic functions of $t$, regular near any given value of $t$.

Dropping the suffix zero, we have

$$\dot{u} = u_x \dot{x} + u_y \dot{y}, \quad \dot{v} = v_x \dot{x} + v_y \dot{y},$$

on $\gamma$, where dots denote derivatives with respect to $t$. By (1) we then have

$$\left. \begin{array}{l} (b_{11}\dot{x} - a_{11}\dot{y})\,u_y + (b_{12}\dot{x} - a_{12}\dot{y})\,v_y = h_1\dot{x} - a_{11}\dot{u} - a_{12}\dot{v}, \\ (b_{21}\dot{x} - a_{21}\dot{y})\,u_y + (b_{22}\dot{x} - a_{22}\dot{y})\,v_y = h_2\dot{x} - a_{21}\dot{u} - a_{22}\dot{v}. \end{array} \right\} \quad (2)$$

Usually these equations determine the first derivatives of $u$ and $v$ at any point of $\gamma$.

The six second derivatives satisfy four similar equations, namely

$$\left. \begin{array}{l} a_{11}u_{xx} + a_{12}v_{xx} + b_{11}u_{xy} + b_{12}v_{xy} = k_1, \\ a_{21}u_{xx} + a_{22}v_{xx} + b_{21}u_{xy} + b_{22}v_{xy} = k_2, \\ a_{11}u_{xy} + a_{12}v_{xy} + b_{11}u_{yy} + b_{12}v_{yy} = k_3, \\ a_{21}u_{xy} + a_{22}v_{xy} + b_{21}u_{yy} + b_{22}v_{yy} = k_4, \end{array} \right\} \quad (3)$$

where $k_1, k_2, k_3, k_4$ involve, not only $x, y, u, v$, but also the first derivatives now supposed known on $\gamma$. Since $\ddot{x}, \ddot{y}, \ddot{u}, \ddot{v}$ are assumed to exist, we have two other equations

$$\ddot{u} = u_{xx}\dot{x}^2 + 2u_{xy}\dot{x}\dot{y} + u_{yy}\dot{y}^2 + u_x\ddot{x} + u_y\ddot{y},$$

$$\ddot{v} = v_{xx}\dot{x}^2 + 2v_{xy}\dot{x}\dot{y} + v_{yy}\dot{y}^2 + v_x\ddot{x} + v_y\ddot{y}$$

on $\gamma$. These six equations usually determine the six second derivatives on $\gamma$. A similar method determines the partial derivatives of all orders.

We can simplify the calculation by using only the first two of equations (3) and the relations

$$\dot{u}_x = u_{xx}\dot{x} + u_{xy}\dot{y}, \quad \dot{v}_x = v_{xx}\dot{x} + v_{xy}\dot{y}.$$

We then obtain a pair of equations similar to (2), namely

$$\left. \begin{array}{l} (b_{11}\dot{x} - a_{11}\dot{y})\,u_{xy} + (b_{12}\dot{x} - a_{12}\dot{y})\,v_{xy} = k_1\dot{x} - a_{11}\dot{u}_x - a_{12}\dot{v}_x, \\ (b_{21}\dot{x} - a_{21}\dot{y})\,u_{xy} + (b_{22}\dot{x} - a_{22}\dot{y})\,v_{xy} = k_2\dot{x} - a_{21}\dot{u}_x - a_{22}\dot{v}_x. \end{array} \right\} \quad (4)$$

Again, these usually determine $u_{xy}$ and $v_{xy}$ on $\gamma$, and hence the other second derivatives there.

Equations (2) determine $u_y$ and $v_y$ uniquely only if the determinant

$$\Delta = \begin{vmatrix} b_{11}\dot{x} - a_{11}\dot{y} & b_{12}\dot{x} - a_{12}\dot{y} \\ b_{21}\dot{x} - a_{21}\dot{y} & b_{22}\dot{x} - a_{22}\dot{y} \end{vmatrix}$$

is not zero. If the determinant is zero, the equations are usually inconsistent. If $\Delta$ is zero, and if the determinants

$$\Delta_1 = \begin{vmatrix} b_{11}\dot{x} - a_{11}\dot{y} & h_1\dot{x} - a_{11}\dot{u} - a_{12}\dot{v} \\ b_{21}\dot{x} - a_{21}\dot{y} & h_2\dot{x} - a_{21}\dot{u} - a_{22}\dot{v} \end{vmatrix},$$

$$\Delta_2 = \begin{vmatrix} b_{21}\dot{x} - a_{12}\dot{y} & h_1\dot{x} - a_{11}\dot{u} - a_{12}\dot{v} \\ b_{22}\dot{x} - a_{22}\dot{y} & h_2\dot{x} - a_{21}\dot{u} - a_{22}\dot{v} \end{vmatrix}$$

are both zero, equations (2) are identical; we can assign, say, $u_y$ as we please and calculate $v_y$, but the equations do not determine $u_y$ and $v_y$ uniquely.

The equation $\Delta = 0$ is a quadratic equation in $\dot{x}:\dot{y}$. At any point $(x, y)$ it determines two directions, called the characteristic directions. If the characteristic directions at $(x, y)$ are real and distinct, the system is said to be of *hyperbolic type* there. If the quadratic equation has equal roots, the system is said to be of *parabolic type*. But if the roots of $\Delta = 0$ are complex at $(x, y)$, the system is said to be of *elliptic type* there.

A real arc whose tangent at every point is in a characteristic direction is called a characteristic base curve, or, briefly, a *characteristic*. It does not depend on $u$ or $v$. If the system (1) is of elliptic type everywhere in $D$, there are no characteristics in $D$. If it is of hyperbolic type there, there are two distinct families of characteristics. But if it is everywhere of parabolic type, there is only one family.

If $\gamma$ is a characteristic of (1) and equations (2) are consistent, there exists a function $f(t)$ such that the second equation of (2) becomes identical with the first if we multiply through by $f(t)$. Hence, if there is a solution of our problem, the values of $u$ and $v$ on $\gamma$ satisfy the ordinary differential equation

$$h_1\dot{x} - a_{11}\dot{u} - a_{12}\dot{v} = f\{h_2\dot{x} - a_{21}\dot{u} - a_{22}\dot{v}\}.$$

This is called a *transport equation*.

A hyperbolic system has two transport equations, one corresponding to each system of characteristics. To illustrate this, consider the simple example
$$u_x - v_y = h_1, \quad v_x - u_y = h_2.$$

Equations (2) then become

$$\dot{y}u_y + \dot{x}v_y = \dot{u} - h_1\dot{x},$$
$$\dot{x}u_y + \dot{y}v_y = \dot{v} - h_2\dot{x}.$$

The characteristic directions are given by $\dot{x}^2 - \dot{y}^2 = 0$. The system is of hyperbolic type, the two families of characteristics being $x \pm y = $ constant. On a characteristic $x + y = $ constant, the transport equation is

$$\dot{u} + \dot{v} = \dot{x}(h_1 + h_2);$$

on $x - y = $ constant,     $\dot{u} - \dot{v} = \dot{x}(h_1 - h_2).$

In the simplest case when $h_1$ and $h_2$ are identically zero, the transport equations can be solved immediately. On each characteristic $x + y = $ constant, $u + v$ is constant; on $x - y = $ constant, $u - v$ is constant. Therefore

$$u + v = 2F(x+y), \quad u - v = 2G(x-y),$$

where $F$ and $G$ are arbitrary functions. Hence

$$u = F(x+y) + G(x-y), \quad v = F(x+y) - G(x-y).$$

Again, for the Cauchy–Riemann equations

$$u_x - v_y = 0, \quad v_x + u_y = 0,$$

the characteristic directions are given by $\dot{x}^2 + \dot{y}^2 = 0$, so that the system is of elliptic type. Lastly, the system

$$u_x = v, \quad v_x - u_y = 0$$

is of parabolic type, with one system of characteristics, $y = $ constant.

A system of semi-linear equations with two independent variables $x$ and $y$ and $n$ dependent variables $u_1, u_2, \ldots, u_n$ can be written in vector form

$$A\frac{\partial u}{\partial x} + B\frac{\partial u}{\partial y} = h, \tag{5}$$

where $A = \{a_{ij}\}$ and $B = \{b_{ij}\}$ are non-singular $n \times n$ matrices and $u$ and $h$ are column vectors

$$\begin{bmatrix} u_1 \\ u_2 \\ \vdots \\ u_n \end{bmatrix}, \begin{bmatrix} h_1 \\ h_2 \\ \vdots \\ h_n \end{bmatrix}.$$

Each coefficient $a_{ij}, b_{ij}$ is a function of $x$ and $y$ alone, regular in some domain $D$. Each function $h_i$ depends on $x, y$ and the components of $u$; each is assumed regular when $(x, y)$ is in $D$ and $|u_i| < K$ for all $i$.

Let $\gamma$ be a regular arc in $D$, with parametric equations $x = x(t)$, $y = y(t)$. If we wish to show that there is a unique vector $u$ which satisfies (5) and takes given values on $\gamma$, then we must first show that we can calculate uniquely the values of all the partial derivatives of each component $u_i$ at any point of $\gamma$. The procedure is just the same as in the case of two dependent variables.

On $\gamma$, $u$ is known and

$$\dot{u} = \dot{x}\frac{\partial u}{\partial x} + \dot{y}\frac{\partial u}{\partial y}.$$

Hence

$$(\dot{x}B - \dot{y}A)\frac{\partial u}{\partial y} = \dot{x}h - A\dot{u}. \tag{6}$$

The first derivatives are thus determined provided that the direction of $\gamma$ at the point considered does not make $\dot{x}B - \dot{y}A$ singular.

If $u_{mn}$ denote $\partial^{m+n}u/\partial x^m \partial y^n$, it follows from (5) that

$$Au_{20} + Bu_{11} = k_1,$$

$$Au_{11} + Bu_{02} = k_2,$$

where $k_1$ and $k_2$ depend on $x, y, u$ and the first derivatives of $u$, now known on $\gamma$. But

$$\dot{u} = \dot{x}^2 u_{20} + 2\dot{x}\dot{y}u_{11} + \dot{y}^2 u_{02} + \ddot{x}u_{10} + \ddot{y}u_{01},$$

so that we have three equations to determine the three partial derivatives. Since we are assuming that $A$ and $B$ are not singular,

$$u_{20} = A^{-1}k_1 - A^{-1}Bu_{11}, \quad u_{02} = B^{-1}k_2 - B^{-1}Au_{11}.$$

Therefore $u_{11}$ satisfies on $\gamma$ the equation

$$(\dot{x}^2 A^{-1}B - 2\dot{x}\dot{y} + \dot{y}^2 B^{-1}A)u_{11} = k_3,$$

where the vector $k_3$ is known on $\gamma$. This equation can be written as

$$B^{-1}(\dot{x}B - \dot{y}A)A^{-1}(\dot{x}B - \dot{y}A)u_{11} = k_3.$$

Hence, if $\dot{x}B - \dot{y}A$ is not singular,

$$u_{11} = (\dot{x}B - \dot{y}A)^{-1}A(\dot{x}B - \dot{y}A)^{-1}Bk_3.$$

on $\gamma$. The second derivatives are thus determined provided that the direction of $\gamma$, at the point considered, does not make $\dot{x}B - \dot{y}A$ singular.

A shorter proof can be given which depends on the equation

$$A\frac{\partial v}{\partial x} + B\frac{\partial v}{\partial y} = k_1,$$

where $v = \partial u/\partial x$. It follows that $u_{20} = \partial v/\partial x$, and $u_{11} = \partial v/\partial y$ are determined on $\gamma$ if $\dot{x}B - \dot{y}A$ is not singular. Similarly from

$$A\frac{\partial w}{\partial x} + B\frac{\partial w}{\partial y} = k_2,$$

where $w = \partial u/\partial y$, we can get the values of $u_{11}$ and $u_{02}$ when $\dot{x}B - \dot{y}A$ is not singular. And similarly for the derivatives of all orders. We shall not discuss the convergence of the resulting series.

If however the matrix $\dot{x}B - \dot{y}A$ is singular at the point of $\gamma$ under consideration, equation (6) either has no solution or an infinity of solutions, depending on the rank of the extended matrix

$$[\dot{x}B - \dot{y}A, \dot{x}h - A\dot{u}].$$

In the latter case, the components of the vector $\dot{x}h - A\dot{u}$ are not linearly independent; there then exists a transport equation

$$\sum_{i=1}^{n} \lambda_i \dot{x}h_i - \sum_{i,j} \lambda_i a_{ij}\dot{u}_j = 0$$

satisfied by the data on $\gamma$.

Since
$$\det(\dot{x}B - \dot{y}A) = \dot{x}^n \det B + \ldots + (-1)^n \dot{y}^n \det A,$$

the equation $\det(\dot{x}B - \dot{y}A) = 0$ is of degree $n$ in $\dot{x}:\dot{y}$. At any point $(x, y)$ it determines $n$ directions, called the characteristic directions. If the characteristic directions at $(x, y)$ are all real and distinct, the system (5) is said to be of hyperbolic type there; if they are all complex, the system is said to be of elliptic type there. These are the two extreme cases.

A regular arc $\gamma$ is called a characteristic if its tangent at every point is in a characteristic direction. If there are $n$ distinct families of real characteristics, the system is said to be of hyperbolic type; if there are no real characteristics, it is of elliptic type.

If $\gamma$ is a characteristic, there is usually no solution which takes given values on $\gamma$. But if $u$ satisfies the transport equation on $\gamma$, there are an infinity of solutions.

It will be recalled that we assumed that $A$ and $B$ are not singular. We can always rotate the axes about the point of $\gamma$ considered so that $\dot{x} = 0$ and $\dot{y} = 0$ are not characteristic directions; after the rotation, $\det A$ and $\det B$ are not zero. Hence the assumption that $A$ and $B$ are not singular at the point considered is no restriction.

We could have dealt in much the same way with the quasi-linear case in which the matrices $A$ and $B$ also depend on the components of $u$. An example is given at the end of the chapter.

## 1.7 An application of the method of characteristics

The linear system
$$u_x - v_y = 0, \quad v_x - u_y = 0$$
is of hyperbolic type, with characteristics $x \pm y =$ constant. On each characteristic $x - y =$ constant, $u - v$ is constant; on each characteristic $x + y =$ constant, $u + v$ is constant. Let us consider the problem of finding solutions $u$ and $v$ in the strip $0 \leqslant x \leqslant l$ given that $u = 0$ when $x = 0$ and when $x = l$, and that
$$u = \phi(x), \quad v = \psi(x)$$
when $y = 0$, $0 \leqslant x \leqslant l$. We assume that the data are continuous and continuously differentiable; in particular, $\phi(0) = \phi(l) = 0$.

Consider the rectangle with vertices
$$P(x, y), \quad Q(l, l + y - x), \quad R(l - x, l + y), \quad S(0, y + x),$$
where $0 \leqslant x \leqslant l$; the sides are characteristics. Then
$$u_P - v_P = u_Q - v_Q, \quad u_P + v_P = u_S + v_S,$$
$$u_R + v_R = u_Q + v_Q, \quad u_R - v_R = u_S - v_S.$$
Since $u_Q$ and $u_S$ are zero,
$$u_R = -u_P, \quad v_R = v_P,$$
that is    $u(l - x, l + y) = -u(x, y), \quad v(l - x, l + y) = v(x, y).$

From this it follows that
$$u(x, y) = u(x, y + 2l), \quad v(x, y) = v(x, y + 2l)$$
so that $u$ and $v$ are periodic functions of period $2l$. Hence we only need to solve the boundary value problem for $0 \leqslant y \leqslant l$.

In fig. 1, $AB$ is $x = l$, and the inclined lines are characteristics. There are four regions to consider, the triangles $OAK$, $ABK$, $OKC$, and the square $KBLC$.

In the triangle $OAK$,
$$u(x, y) - v(x, y) = u(x - y, 0) - v(x - y, 0)$$
$$= \phi(x - y) - \psi(x - y) = 2F(x - y),$$
$$u(x, y) + v(x, y) = u(x + y, 0) + v(x + y, 0)$$
$$= \phi(x + y) + \psi(x + y) = 2G(x + y),$$

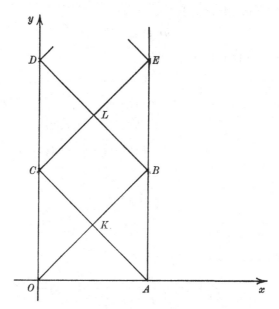

Fig. 1

since $u+v$ and $u-v$ are constant on characteristics. Therefore

$$u(x,y) = F(x-y)+G(x+y), \quad v(x,y) = -F(x-y)+G(x+y),$$

where $\phi-\psi = 2F$, $\phi+\psi = 2G$.

Next, in the triangle $OKC$,

$$u(x,y)-v(x,y) = u(0,y-x)-v(0,y-x) = -v(0,y-x),$$

$$u(x,y)+v(x,y) = u(x+y,0)+v(x+y,0) = 2G(x+y).$$

But

$$u(0,y-x)+v(0,y-x) = u(y-x,0)+v(y-x,0) = 2G(y-x),$$

so that $$v(0,y-x) = 2G(y-x).$$

Hence, in the triangle $OKC$,

$$u(x,y) = -G(y-x)+G(y+x), \quad v(x,y) = G(y-x)+G(y+x).$$

In particular, $$v(0,y) = 2G(y).$$

Note, that in the triangle $OKC$, $u$ and $v$ do not depend on $F$.

Similarly, in the triangle $ABK$,

$$u(x,y) = F(x-y)-F(2l-x-y), \quad v(x,y) = -F(x-y)-F(2l-x-y)$$

are independent of $G$. In particular

$$v(l, y) = -2F(l-y).$$

Lastly, in the square $KBLC$,

$$u(x, y) - v(x, y) = u(0, y-x) - v(0, y-x) = -2G(y-x),$$
$$u(x, y) + v(x, y) = u(l, x+y-l) + v(l, x+y-l) = -2F(2l-x-y).$$

Therefore
$$u(x, y) = -F(2l-x-y) - G(y-x),$$
$$v(x, y) = -F(2l-x-y) + G(y-x).$$

## Exercises

**1.** Solve the problem of §1.7 when $u = kv$ on $x = l$, $u = -kv$ when $x = 0$, where $k \neq 1$.

**2.** Of what type is the system

$$\frac{\partial u}{\partial x} + f(x, y)\frac{\partial v}{\partial y} = 0, \quad \frac{\partial u}{\partial y} + f(x, y)\frac{\partial v}{\partial x} = 0?$$

Find the characteristics and the corresponding transport equations.

**3.** Find the characteristics and corresponding transport equations of the system

$$xyu_x - v_y = 0, \quad xu_y - yv_x = 0.$$

**4.** Show that the system

$$xyu_x + v_y = 0, \quad u_y - v_x = 0$$

is of elliptic type when $xy > 0$, of hyperbolic type when $xy < 0$ and of parabolic type on the axes.

**5.** A semi-linear system has two distinct families of characteristics $\xi(x, y) = $ constant, $\eta(x, y) = $ constant. Prove that, if $\xi$ and $\eta$ are taken as independent variables, the equations of the system are

$$a_1\frac{\partial u}{\partial \xi} + a_2\frac{\partial v}{\partial \xi} = h_1, \quad b_1\frac{\partial u}{\partial \eta} + b_2\frac{\partial v}{\partial \eta} = h_2.$$

($\xi$ and $\eta$ are called *characteristic variables*.)

**6.** In the equation $aw_{xx} + 2bw_{xy} + cw_{yy} = h$, the coefficients $a, b, c$ depend only on $x$ and $y$, but $h$ is a function of $x, y, w_x, w_y$. Show that it is equivalent to the half-linear system

$$au_x + bv_x + bu_y + cv_y = h(x, y, u, v),$$
$$v_x - u_y = 0$$

with characteristics given by

$$a\dot{y}^2 - 2b\dot{x}\dot{y} + c\dot{x}^2 = 0.$$

**7.** Find the integral surface of $yp-xq = 1+u^2$ which goes through the circle $u = 0, x^2+y^2 = 1$.

**8.** Find the integral surface of $yp+xq = u$ which goes through the curve $u = x^3, y = 0$.

**9.** Find the integral surface of $xpq+yq^2 = 1$ which goes through the curve $u = x, y = 0$.

**10.** Find the characteristic strips of $xp+yq = pq$. Deduce the equation of the integral surface through $u = \frac{1}{2}x, y = 0$.

**11.** $u$ and $v$ satisfy the system of quasi-linear equations

$$\frac{\partial u}{\partial x}+u\frac{\partial u}{\partial y}+2v\frac{\partial v}{\partial y} = 0,$$

$$2\frac{\partial v}{\partial x}+2u\frac{\partial v}{\partial y}+v\frac{\partial u}{\partial y} = 0.$$

If $u$ and $v$ are given on a regular arc $\gamma$, prove that the partial derivatives of $u$ and $v$ on $\gamma$ can be found uniquely if $\dot{y} \neq (u\pm v)\dot{x}$.

Show that, if $\dot{y} = (u+v)\dot{x}$ on $\gamma$, values can be found for the first derivatives, though not uniquely, if $u+2v$ is constant on $\gamma$. Find the corresponding result if $\dot{x} = (u-v)\dot{y}$ on $\gamma$.

If there is a family of straight lines $x = at, y = bt$ which are characteristic base curves on which $\dot{y} = (u+v)\dot{x}$, prove that $u$ and $v$ are functions of $y/x$ and hence that

$$u = \frac{2y}{3x}+A, \quad v = \frac{y}{3x}-A,$$

where $A$ is an arbitrary constant.

**12.** If, in Ex. 11, $u$ and $v$ are taken as independent variables and $x$ and $y$ as dependent variables, show that the equations become

$$2v\frac{\partial x}{\partial u}-u\frac{\partial x}{\partial v}+\frac{\partial y}{\partial v} = 0, \quad 2u\frac{\partial x}{\partial u}-v\frac{\partial x}{\partial v}-2\frac{\partial y}{\partial u} = 0.$$

Prove that this system has characteristics $u\pm 2v = $ constant. Show that, if $\xi = u+2v, \eta = u-2v$,

$$(\xi+3\eta)\frac{\partial x}{\partial \xi} = 4\frac{\partial y}{\partial \xi}, \quad (3\xi+\eta)\frac{\partial x}{\partial \eta} = 4\frac{\partial y}{\partial \eta}.$$

# 2

# CHARACTERISTICS OF EQUATIONS OF THE SECOND ORDER

## 2.1 The general equation of the second order

The general equation of the second order with two independent variables is

$$F(x, y, u, p, q, r, s, t) = 0, \tag{1}$$

where, in the usual notation, $p = u_x$, $q = u_y$, $r = u_{xx}$, $s = u_{xy}$, $t = u_{yy}$. Given a regular arc $\Gamma$ in space, $(x, y, u)$ being rectangular Cartesian coordinates, does there exist a unique solution of (1), for which $p$ and $q$ take given values on $\Gamma$? This is called the *Problem of Cauchy*; the existence theorem under appropriate conditions is the theorem of Cauchy and Kowalewsky. There are two other ways of formulating the problem. If $\Delta$ is a developable surface containing $\Gamma$, does there exist a unique integral surface which touches $\Delta$ along $\Gamma$? Alternatively, if $\gamma$ is a regular arc in the $x, y$-plane, does there exist a unique solution of (1) for which $u, p, q$ take given values on $\gamma$? The data must satisfy the strip condition $du = p\,dx + q\,dy$ on $\gamma$.

The Cauchy–Kowalewsky theorem is a local theorem. It is assumed that $F$ is an analytic function of the eight variables, and that the data $u, p, q$ are analytic functions of the parameter, say the arc, which defines position on $\gamma$. The result is that, if $(x_0, y_0)$ is any point of $\gamma$, there exists a unique solution of (1) in the form of a convergent Taylor series

$$u = u_{00} + \sum_1^\infty u_{mn} \frac{(x - x_0)^m (y - y_0)^n}{m! \, n!},$$

where $u_{mn}$ is the value of $\partial^{m+n} u / \partial x^m \, \partial y^n$ at $(x_0, y_0)$. The first step is to show that $u$ has unique partial derivatives of all orders at every point of $\gamma$. The result is that the problem has a unique solution in a domain containing $\gamma$.

On $\gamma$, we have $u$, $p$ and $q$, analytic functions of the arc length $l$ say. Denote differentiation with respect to the arc by dashes. Then on $\gamma$

$$p' = rx' + sy', \quad q' = sx' + ty'. \tag{2}$$

We have three equations (1) and (2) to determine $r, s, t$ on $\gamma$. There are three possibilities. Firstly, there may be no set of real values of $(r, s, t)$ which satisfies (1) and (2); in this case the problem has no

[ 24 ]

solution. Secondly, there may be just one real set $(r, s, t)$ which satisfies (1) and (2); in this case, if there is a solution, it is unique. Thirdly, there may be several real sets, in which case we may get a solution corresponding to each set.

Suppose that there is just one set. The next step is to calculate the third derivatives. If we differentiate (1) partially with respect to $x$, we obtain
$$Rr_x + Ss_x + Tt_x = -X - Up - Pr - Qs \tag{3}$$
in the obvious notation $(R = \partial F/\partial r, \ldots)$. In this equation, everything is known on $\gamma$ except $r_x, s_x, t_x$. Since $dr = r_x dx + s_x dy, ds = s_x dx + t_x dy$, because $r_y = s_x, s_y = t_x$, we have on $\gamma$
$$x'r_x + y's_x = r', \quad x's_x + y't_x = s'. \tag{4}$$
We now have three equations to determine $r_x$, $s_x$ and $t_x$ on $\gamma$, namely (3) and (4). The three derivatives are determined uniquely provided that the determinant of (3) and (4) does not vanish. That is, provided that
$$Ry'^2 - Sx'y' + Tx'^2$$
does not vanish. Similarly we can calculate all the partial derivatives of $u$ on $\gamma$ provided that this condition is satisfied. If $Ry'^2 - Sx'y' + Tx'^2$ vanishes everywhere on $\gamma$, $\gamma$ is called a characteristic base curve.

The condition
$$Ry'^2 - Sx'y' + Tx'^2 = 0$$
for a curve $\gamma$ to be a characteristic usually depends on the values of $u$, $p$ and $q$ given on $\gamma$. The following examples make this clearer. The curve $\gamma$ is a characteristic of
$$p^2 r - 3pqs + 2q^2 t = 0$$
if
$$py'^2 - 3pqx'y' + 2q^2 x'^2 = 0.$$
If $y = x$ is a characteristic, $p^2 - 3pq + 2q^2 = 0$. The data must satisfy $p = q$ or $p = 2q$ on the line; otherwise it is not a characteristic. The curve $\gamma$ is a characteristic of
$$r - 3q + 2t = 0$$
if
$$y'^2 - 3x'y' + 2x'^2 = 0,$$
a condition which does not depend on the data. Either $y' = x'$ or $y' = 2x'$ so that there are two real families of characteristics $y = x + \alpha$ and $y = 2x + \beta$ where $\alpha$ and $\beta$ are constants. Lastly, the curve $\gamma$ is a characteristic of
$$pr - qs = 0$$
if
$$py'^2 - qx'y' = 0.$$
There is one family, $y = $ constant, of characteristics independent of the data; the other family, given by $py' - qx' = 0$, depends on the data.

## 2.2 The Cauchy–Kowalewsky theorem

If $F(x, y, u, p, q, r, s, t) = 0$ involves $r$ explicitly, we can solve for $r$, getting

$$r = f(x, y, u, p, q, s, t). \tag{1}$$

This is no restriction; for if it does not involve $r$ explicitly, we can always make a change of independent variables to $(x', y')$ say, so that $\partial^2 u / \partial x'^2$ does occur explicitly. The Cauchy–Kowalewsky Theorem asserts that, if $f$ is an analytic function of the seven variables, there is a unique analytic solution of (1) which is regular in a neighbourhood of $(x_0, y_0)$ and which satisfies the conditions

$$u = \phi(y), \quad p = \psi(y), \tag{2}$$

when $x = x_0$, provided that $\phi(y)$ and $\psi(y)$ are regular in a neighbourhood of $y_0$. If these conditions are satisfied at every point $(x_0, y_0)$ of a finite interval $\gamma$ of the initial line $x = x_0$, the Problem of Cauchy posed in (1) and (2) has a unique solution regular near $\gamma$. The case when $\gamma$ is a regular arc $x = \omega(y)$ can be reduced to this case by a change of independent variable.

The values of $u$ and its first and second derivatives at $(x_0, y_0)$ are

$$u_0 = \phi(y_0), \quad p_0 = \psi(y_0), \quad q_0 = \phi'(y_0),$$

$$s_0 = \psi'(y_0), \quad t_0 = \phi''(y_0),$$

$$r_0 = f(x_0, y_0, u_0, p_0, q_0, s_0, t_0).$$

And all the other partial derivatives are uniquely determined at $(x_0, y_0)$. The existence of a unique analytic solution of (1) is proved by transforming the problem into one of solving a system of six quasi-linear equations with six dependent variables $u, p, q, r, s, t$. The argument is like that of §1.5. We do not now assume that $p, q, r, s, t$ are partial derivatives of $u$; that they are, will be a consequence of the system of equations. The system is

$$\left. \begin{aligned} u_x = p, \quad p_x = r, \quad q_x = p_y, \quad s_x = r_y, \quad t_x = s_y, \\ r_x = X + Up + Pr + Qp_y + Sr_y + Ts_y, \end{aligned} \right\} \tag{3}$$

where $X = \partial f / \partial x$, etc., $f(x, y, u, p, q, s, t)$ being regarded as a function of seven independent variables. Note that in each of equations (3), the partial derivatives with respect to $x$ occur on the left but not on the right. The conditions when $x = x_0$ are

$$u = \phi(y), \quad p = \psi(y), \quad q = \phi'(y), \quad s = \psi'(y), \quad t = \phi''(y), \tag{4}$$

and $\quad r = f(x_0, y, \phi(y), \psi(y), \phi'(y), \psi'(y), \phi''(y)).$

The first and second equations of (3) ensure that $p$ and $r$ have their usual meanings. The first and third give

$$\frac{\partial}{\partial x}\left(q - \frac{\partial u}{\partial y}\right) = 0,$$

so that $q - u_y$ is independent of $x$. Using (4), we find that $q = u_y$. The second and fourth equations give

$$\frac{\partial}{\partial x}\left(s - \frac{\partial p}{\partial y}\right) = 0,$$

so that $s - p_y$ is independent of $x$. Using (4), we find that $s = p_y = u_{xy}$. Similarly for $t$ and $r$. Thus system (3) with conditions (4) is equivalent to equation (1) with conditions (2).

The coefficients in the last equation of (3) involve $x$ and $y$ as well as the five dependent variables $u, p, q, s, t$. By introducing two more variables $\xi$ and $\eta$, we get a system in which the coefficients do not involve $x$ and $y$. We define $\eta$ by $\eta_x = 0$ with the condition that $\eta = y - y_0$ when $x = x_0$; then $\eta = y - y_0$ for all $x$. We define $\xi$ by $\xi_x = \eta_y$, where $\xi = 0$ when $x = x_0$. This gives $\xi_x = 1$ and so $\xi = x - x_0$.

Now write

$$f(x_0 + \xi, y_0 + \eta, u, p, q, s, t) \equiv g(\xi, \eta, u, p, q, s, t).$$

We now have the system

$$u_x = p\eta_y, \quad p_x = r\eta_y, \quad q_x = p_y,$$

$$s_x = r_y, \quad t_x = s_y,$$

$$r_x = (g_\xi + g_u p + g_p r)\,\eta_y + g_q p_y + g_s r_y + g_t s_y,$$

$$\xi_x = \eta_y, \quad \eta_x = 0$$

under the given conditions when $x = x_0$. This system is of the form

$$\frac{\partial u_i}{\partial x} = \sum_{j=1}^{n} g_{ij}\frac{\partial u_j}{\partial y}, \quad (i = 1, 2, ..., n)$$

with initial conditions $\qquad u_i = \phi_i(y),$

when $x = x_0$, which was discussed in §1.5. The coefficients $g_{ij}$ are functions of the $u_l$ only, and are regular in a neighbourhood of $\{\phi_i(y_0)\}$; the functions $\phi_i(y)$ are regular in a neighbourhood of $y_0$. In the present case, $n = 8$. The system then has a solution regular in a neighbourhood of $(x_0, y_0)$. The Cauchy–Kowalewsky theorem follows.

## 2.3  The linear equation

The simplest second order equation with two independent variables
is the linear equation

$$ar + 2hs + bt + 2gp + 2fq + cu + d = 0,$$

where the coefficients are all analytic functions of $x$ and $y$ alone,
regular in some open connected set (domain) $D$. Write this equation as

$$ar + 2hs + bt + F = 0. \tag{1}$$

Let $\gamma$ be a regular arc in the $xy$-plane. Then, by the Cauchy–
Kowalewsky theorem, there exists a unique solution of (1) for which
$u, p, q$ assume given analytic values on $\gamma$, and this solution is regular
in a neighbourhood of $\gamma$, provided always that $\gamma$ is not a characteristic base curve.

On $\gamma$, the data satisfy the strip condition $u' = px' + qy'$ where
dashes denote differentiation with respect to the arc length. In solving
(1) by a Taylor series, we have to find all the derivatives of $u$ on $\gamma$.
In particular, we have

$$p' = rx' + sy', \quad q' = sx' + ty',$$

so that the value of $s$ on $\gamma$ satisfies

$$(ay'^2 - 2hx'y' + bx'^2)s = ap'y' + bq'x' + Fx'y'. \tag{2}$$

Thus $s$, and hence $r$ and $t$, are determined uniquely on $\gamma$ provided
that $\gamma$ is not a characteristic base curve.

The characteristic base curves are given by

$$ay'^2 - 2hx'y' + bx'^2 = 0.$$

As $a, h, b$ depend only on $x$ and $y$, the characteristic base curves do
not depend on the data; the characteristics on an integral surface do
not depend on the integral surface. The characteristic base curves are
therefore frequently called the characteristics.

If $ab - h^2 > 0$ in $D$, there are no real characteristics, and the equation
is of elliptic type. If $ab - h^2 < 0$ in $D$, there are two distinct families
of real characteristics, and the equation is of hyperbolic type. If
$ab - h^2 = 0$ everywhere in $D$, there is but one family of characteristics and the equation is of parabolic type. Laplace's equation
$r + t = 0$ is of elliptic type; the equation of wave motions $r - t = 0$ is
of hyperbolic type, and the equation of conduction of heat $r - q = 0$
is of parabolic type. There are also equations of mixed type; $yr + t = 0$
is of elliptic type when $y > 0$, but of hyperbolic type when $y < 0$.

Return now to equation (2). If $\gamma$ is a characteristic, (2) gives, in general, an infinite value for $s$, and so there is no integral surface satisfying the prescribed conditions. But if the data on $\gamma$ satisfy

$$ap'y' + bq'x' + Fx'y' = 0 \tag{3}$$

we can assign $s$ arbitrarily on $\gamma$. This corresponds to the transport relation we found in the theory of linear systems of the first order. In fact, if (1) does not involve $u$ explicitly, we can replace (1) by such a system, viz.

$$av_x + hw_x + hv_y + bw_y + F = 0,$$

$$w_x - v_y = 0,$$

where $F$ is linear in $v$ and $w$, the second equation being the condition that $v\,dx + w\,dy$ is an exact differential $du$.

If $\gamma$ is a characteristic, there is no solution unless the data on $\gamma$ satisfy the transport condition (3); and, if the condition is satisfied, the solution is not unique.

The simplest equation of hyperbolic type is the equation of wave motions $r - t = 0$. The characteristics are given by $y' = \pm x'$, and so are straight lines $x - y = $ constant and $x + y = $ constant. The data on a characteristic must satisfy $p'y' - q'x' = 0$. On $x - y = $ constant, this shows that $p - q$ is constant. Hence

$$p - q = 2\phi(x - y).$$

On $x + y = $ constant, $p + q$ is constant, and so

$$p + q = 2\psi(x + y).$$

Therefore

$$p = \phi(x - y) + \psi(x + y), \quad q = -\phi(x - y) + \psi(x + y),$$

and so

$$u = \Phi(x - y) + \Psi(x + y),$$

the well-known solution. From this, the solution of the initial value problem $u = f(x), u_y = g(x)$ when $y = 0$, viz.

$$u = \tfrac{1}{2}\{f(x - y) + f(x + y)\} + \tfrac{1}{2}\int_{x-y}^{x+y} g(\tau)\,d\tau$$

follows.

The problem of the vibrations of a string with fixed ends is to find the solution of $r - t = 0$ in $0 \leqslant x \leqslant l, y \geqslant 0$ given that $u = f(x), u_y = g(x)$ when $y = 0, 0 \leqslant x \leqslant l$ and $u = 0$ when $x = 0$ or $l$ and $y \geqslant 0$. It can be discussed by the method of characteristics as in §1.7.

The problem of Cauchy is of little importance for equations of elliptic type. The fundamental problem for Laplace's equation $r+t = 0$ is the problem of Dirichlet, to find a solution which takes given values on a simple closed regular curve $C$ and which is analytic inside $C$. For example, if $C$ is the circle $x^2 + y^2 = a^2$, the solution which takes values $\cos\theta$ in polar coordinates on $C$ and is analytic inside $C$ is $u = x/a$; and the values of $p$ and $q$ on $C$ follow. It is possible to ask whether there is a solution which satisfies Cauchy conditions on $C$. For example, if $u = \cos\theta$, $\partial u/\partial r = 0$ on $C$, the solution is

$$u = \tfrac{1}{2}\frac{r}{a}\cos\theta + \tfrac{1}{2}\frac{a}{r}\cos\theta.$$

This is regular except at the origin. This Cauchy problem has a solution regular near $C$, but no solution regular everywhere inside $C$.

## 2.4 The quasi-linear equation

The quasi-linear equation of the second order is of the form

$$Ar + 2Hs + Bt + F = 0, \tag{1}$$

where $A, H, B, F$ are analytic functions of five variables $(x, y, u, p, q)$ regular in some appropriate domain. Given a regular arc $\Gamma$ in $xyu$-space, the theorem of Cauchy and Kowalewsky shows that there is a unique analytic solution, regular near $\Gamma$, on which $p$ and $q$ take given analytic values satisfying the strip condition $du = p\,dx + q\,dy$, provided that $\Gamma$ is not a characteristic.

Since the coefficients involve $u$, $p$ and $q$, there is no unique family of characteristic base curves, so that the theory is more complicated than for the linear equation. On $\Gamma$, $p' = rx' + sy'$, $q' = sx' + ty'$, where dashes denote derivatives with respect to the arc length. Eliminating $r$ and $t$ from (1) we have

$$(Ay'^2 - 2Hx'y' + Bx'^2)s = Ap'y' + Bq'x' + Fx'y'$$

just as for the linear equation. Usually this determines $s$ and so also $r$ and $t$ on $\Gamma$. But if

$$Ay'^2 - 2Hx'y' + Bx'^2 = 0, \tag{2}$$

there is no solution unless

$$Ap'y' + Bq'x' + Fx'y' = 0. \tag{3}$$

If this last condition is satisfied, we can choose $s$ arbitrarily and get an infinite member of solutions. A strip which satisfies equations (2) and (3) is called a *characteristic strip*.

Suppose that $A$ and $B$ are not both zero. If they are, we make a linear change of variable. If $A$ is zero, we interchange the parts played by $x$ and $y$ in what follows. Equation (2), regarded as a quadratic in $y'/x'$, has two roots $\rho$ and $\sigma$. If $\rho$ and $\sigma$ are complex, the equation is of elliptic type, and no characteristic strips exist. If $\rho$ and $\sigma$ are real and distinct, the equation is of hyperbolic type, and there are two families of characteristic strips.

On the first family,
$$y' = \rho x',$$
$$u' = px' + qy',$$
$$Ap'y' + Bq'x' + Fx'y' = 0.$$

Since $B = \rho\sigma A$, these equations become

$$\left.\begin{aligned} \frac{dy}{dx} &= \rho, \\ \frac{du}{dx} &= p + q\rho, \\ A\frac{dp}{dx} + A\sigma\frac{dq}{dx} + F &= 0. \end{aligned}\right\} \tag{4}$$

The equations of the second system are obtained by interchanging $\rho$ and $\sigma$.

In equations (4), $A$, $F$, $\rho$ and $\sigma$ are functions of $x, y, u, p$ and $q$. So we have three equations to determine $y, u, p, q$ on a characteristic strip. If we put one of these variables equal to some arbitrary function of $x$, we have three equations to determine the other three. So the equations of a characteristic strip involve an arbitrary function.

There are, in general, two families of characteristic strips. It can be shown that a surface generated by a one-parameter family of characteristic strips is an integral surface, and conversely. In practice, the method is complicated and of little use. We sometimes can make use of the differential equations of characteristic strips in a different way to find *intermediate integrals*, which are partial differential equations of the first order.

The differential equations of one family of characteristic strips are
$$dy - \rho\, dx = 0,$$
$$du - p\, dx - q\, dy = 0,$$
$$\rho A\, dp + B\, dq + F\rho\, dx = 0.$$

If we can find functions $\alpha, \beta, \gamma$ such that
$$\alpha(dy - \rho\, dx) + \beta(du - p\, dx - q\, dy) + \gamma(A\rho\, dp + B\, dq + F\rho\, dx)$$

is the exact differential $dV$ of a function of five independent valuables $x, y, u, p, q$, the equation $V = $ constant is called a first integral. It is a first order partial differential equation, and it can be shown that every solution of it satisfies $Ar + 2Hs + Bt + F = 0$.

If we can find two first integrals $V_1 = $ constant and $V_2 = $ constant, then $\Phi(V_1, V_2) = 0$ is also a first integral for any arbitrary function $\Phi$. $\Phi = 0$ is usually called an intermediate integral. If, as is most unlikely, we can solve $\Phi = 0$, the general solution will involve two arbitrary functions, one from $\Phi$, the other, the arbitrary function occurring in the general solution of a first order partial differential equation.

As an example of the method, we find two intermediate integrals of

$$x^2 r - y^2 t = 0.$$

On any regular arc,

$$p' = rx' + sy', \quad q' = sx' + ty',$$

and so $\quad (x^2 y'^2 - y^2 x'^2) s = x^2 y' p' - y^2 x' q'.$

The characteristics are given by $xy' = \pm yx'$, and on any characteristic strip

$$x^2 y' p' - y^2 x' q' = 0.$$

If $xy' = yx'$, $y = kx$ is the equation of one family of characteristics, $k$ being constant. The transport equation becomes

$$xp' - yq' = 0$$

or $\quad p' - kq' = 0.$

Therefore $p - kq$ is constant on $y = kx$, and so an intermediate integral is

$$xp - yq = xf(y/x),$$

where $f$ is an arbitrary function. This is an equation of Lagrange's type; its general solution is

$$u = F(xy) + xG(y/x)$$

where $F$ is an arbitrary function, and $G(\theta)$ is connected with $f(\theta)$ by

$$G(\theta) - 2\theta G'(\theta) = f(\theta).$$

Thus, from a knowledge of an intermediate integral, we have found a general solution of

$$x^2 r - y^2 t = 0$$

involving two arbitrary functions.

The second family of characteristics, given by $xy' = -yx'$, is $xy = k$, for any constant $k$. The transport equation is

$$xp' + yq' = 0,$$

satisfied when $xy = k$. But

$$d(u - xp - yq) = (du - p\,dx - q\,dy) - (x\,dp + y\,dq)$$

The first term on the right vanishes by the strip condition; the second term vanishes on a characteristic $xy = k$. Therefore $u - xp - yq$ is constant on a characteristic $xy = k$; and so we have another intermediate integral

$$u - xp - yq = g(xy),$$

where $q$ is an arbitrary function. This again is a first order equation of Lagrange's type.

A better method of solving $x^2 r - y^2 t = 0$ is to reduce the equation to normal form by a change of independent variables, as in the next section.

## 2.5   The normal form of a half-linear equation

By a half-linear equation is meant an equation of the form

$$L(u) \equiv au_{xx} + 2hu_{xy} + bu_{yy} + F = 0, \qquad (1)$$

where $F$ is a function of $x, y, u$ and the first derivatives of $u$, but $a, h, b$ depend only on $x, y$. An equation of this type can be reduced to a simple normal form by the use of characteristic variables. The transformation may be only a local one; for example $yu_{xx} + u_{yy} = 0$, which is of elliptic type in $y > 0$ but hyperbolic type in $y < 0$, has different normal forms in the elliptic and hyperbolic half-planes.

We suppose that $a, h, b$ are analytic functions of $x, y$ regular in a domain $D$ and that $L(u) = 0$ is of fixed type in $D$. Let

$$\xi = \xi(x, y), \quad \eta = \eta(x, y)$$

be a bijection which maps $D$ onto a domain $\Delta$. Every point of $D$ has a unique image in $\Delta$; every point of $\Delta$ has a unique inverse image in $D$. We suppose also that $\xi$ and $\eta$ are real and have continuous second derivatives. The Jacobian

$$J = \frac{\partial(\xi, \eta)}{\partial(x, y)}$$

does not vanish on $D$.

This mapping gives

$$L(u) \equiv \alpha u_{\xi\xi} + 2\kappa u_{\xi\eta} + \beta u_{\eta\eta} + 2G, \qquad (2)$$

where $G$ does not involve the second derivatives of $u$. A straightforward calculation shows that

$$\alpha = Q(\xi_x, \xi_y), \quad \beta = Q(\eta_x, \eta_y), \quad \kappa = \Pi(\xi_x, \xi_y; \eta_x, \eta_y),$$

where $Q$ is the quadratic form

$$Q(X, Y) = aX^2 + 2hXY + bY^2,$$

and $\Pi$ is the polar form of $Q$, namely

$$\Pi(X, Y; X', Y') = aXX' + h(XY' + X'Y) + bYY'.$$

At any fixed point $(x, y)$ there, are three cases. If $ab - h^2 > 0$, the equation $Q(X, Y) = \pm 1$, where the sign is $+$ or $-$ according as $a > 0$ or $a < 0$, represents an ellipse in the $XY$-plane; the equation $L(u) = 0$ is then of elliptic type at $(x, y)$. If $ab - h^2 < 0$, $Q(X, Y) = 1$ is the equation of a hyperbola; and $L(u) = 0$ is of hyperbolic type at $(x, y)$. Lastly, if $ab - h^2 = 0$, $aQ(X, Y) = 1$ is the equation of a pair of parallel lines; and $L(u) = 0$ is of parabolic type at $(x, y)$.

Suppose that $L(u) = 0$ is of hyperbolic type at every point of some domain $D$. We can choose $\xi$ and $\eta$ so that $\alpha$ and $\beta$ both vanish everywhere. Then $\xi$ and $\eta$ both satisfy the equation

$$a\phi_x^2 + 2h\phi_x\phi_y + b\phi_y^2 = 0.$$

The curves $\xi = $ constant and $\eta = $ constant are therefore the characteristics of $L(u) = 0$ since on them

$$a\,dy^2 - 2h\,dy\,dx + b\,dx^2 = 0. \tag{3}$$

Since $a, h, b$ are analytic in $D$, so also are $\xi$ and $\eta$. When $\xi$ and $\eta$ are chosen in this way, we obtain the normal form

$$\kappa\frac{\partial^2 u}{\partial\xi\,\partial\eta} + G = 0, \tag{4}$$

the form of the original equation when $a$ and $b$ are both zero. If $a$ is not zero, (3) gives

$$\frac{dy}{dx} = \frac{h \pm \sqrt{(h^2 - ab)}}{a},$$

so that the curves $\xi = $ constant and $\eta = $ constant are nowhere tangent, and the Jacobian $J$ does not vanish. If $a$ is zero, $\xi$ is equal to $x$, and the changes are straightforward.

A different normal form is obtained by putting $\xi + \eta = \lambda$, $\xi - \eta = \mu$ in (4), namely,

$$\kappa\left\{\frac{\partial^2 u}{\partial\lambda^2} - \frac{\partial^2 u}{\partial\mu^2}\right\} + G = 0.$$

If $L(u)$ is of elliptic type in $D$, the functions $\xi$ and $\eta$ are conjugate complex functions. In equation (4), we put $\xi + \eta = \lambda$, $\xi - \eta = i\nu$, where

$\lambda$ and $\nu$ are real. This gives the normal form

$$\kappa\left\{\frac{\partial^2 u}{\partial\lambda^2}+\frac{\partial^2 u}{\partial\nu^2}\right\}+G = 0$$

for an equation of elliptic type with two independent variables.

Lastly, if $ab-h^2 = 0$, we can have $a = 0$, $h = 0$, and $L(u) = 0$ becomes

$$bu_{yy}+F = 0,$$

which is already of normal form. A similar result follows if $b = 0$, $h = 0$. If $a,b,h$ are not all zero, equation (3) becomes

$$a\,dy - h\,dx = 0.$$

If this has solution $\eta = $ constant, the coefficient $\beta$ vanishes. If we take any analytic function $\eta(x,y)$ such that the Jacobian $J$ is not zero, we have

$$a\Pi(\xi_x,\xi_y;\eta_x,\eta_y) = a\xi_x(a\eta_x+h\eta_y)+h\xi_y(a\eta_x+h\eta_y) = 0,$$

so that $\kappa = 0$. The normal form for an equation of parabolic type is then

$$\alpha u_{\xi\xi}+G = 0.$$

The linear equation with constant coefficients can be still further simplified. The transformation from $(x,y)$ to $(\lambda,\mu)$ in the hyperbolic case is a non-singular linear transformation, and the transformed equation becomes

$$u_{\lambda\lambda} - u_{\mu\mu}+ 2gu_\lambda + 2fu_\mu+cu = 0,$$

where $g,f,c$ are constants. If we put $u = v\exp(-g\lambda+f\mu)$, we obtain

$$v_{\lambda\lambda} - v_{\mu\mu}+(c+f^2-g^2)\,v = 0.$$

And similarly for the equation of elliptic type.

## 2.6   The half-linear equation with three independent variables

The half-linear equation of the second order with independent variables $x,y,z$ is of the form

$$au_{xx}+bu_{yy}+cu_{zz}+2fu_{yz}+2gu_{zx}+2hu_{xy}+F = 0, \tag{1}$$

where $F$ is a function of $x,y,z,u$ and the first derivatives of $u$, but $a,b,c,f,g,h$ depend only on $x$, $y$ and $z$. The problem of Cauchy for this equation is to show that, under appropriate conditions, there is a unique solution of (1) for which $u$ and its normal derivative $\partial u/\partial N$ take given values on a given regular cap $C$.† If the coefficients in (1)

† See Note 6.

and the data on $C$ are analytic functions regular in a neighbourhood of every point of $C$, the proof of the existence of the solution goes very like that for two independent variables; we do not go into this in detail.

The proof depends on showing that, in a neighbourhood of any point of $C$, a formal Taylor series satisfying the equation can usually be constructed and its convergence proved by the method of dominant functions. Let $N$ be the unit vector normal to $C$. Since $u$ and $\partial u/\partial N$ are given on $C$, all the first derivatives of $u$ are known on $C$. Now, if $w$ is a scalar, $\operatorname{grad} w$ can be resolved into a component $N.\operatorname{grad} w$ in the direction of $N$ and $N \times \operatorname{grad} w$ in a direction perpendicular to $N$. Hence if $w$ is known on $C$, $N \times \operatorname{grad} w$, being a tangential derivative, is known on $C$. In particular, the three vectors

$$N \times \operatorname{grad} u_x, \quad N \times \operatorname{grad} u_y, \quad N \times \operatorname{grad} u_z$$

are known on $C$. They are not independent, since each is perpendicular to $N$. In this way, we obtain nine expressions, each linear in the second derivatives of $u$, which are known on $C$.

If $(l, m, n)$ are the components of $N$, we can choose five of these expressions which are linearly independent. Such a set is

$$\left. \begin{aligned} nu_{xx} - lu_{xz} &= \Phi_1, \\ lu_{yy} - mu_{xy} &= \Phi_2, \\ mu_{zz} - nu_{yz} &= \Phi_3, \\ mu_{xz} - nu_{xy} &= \Phi_4, \\ lu_{yz} - mu_{zx} &= \Phi_5, \end{aligned} \right\} \tag{2}$$

where $\quad au_{xx} + bu_{yy} + cu_{zz} + 2fu_{yz} + 2gu_{zx} + 2hu_{xy} = -F.$

The expressions on the right are all known on $C$. Hence we can find all the second derivatives on $C$, provided that the determinant

$$\begin{vmatrix} a & b & c & 2f & 2g & 2h \\ n & 0 & 0 & 0 & -l & 0 \\ 0 & l & 0 & 0 & 0 & -m \\ 0 & 0 & m & -n & 0 & 0 \\ 0 & 0 & 0 & 0 & m & -n \\ 0 & 0 & 0 & l & -m & 0 \end{vmatrix}$$

is not zero. This determinant is equal to

$$lmn(al^2 + bm^2 + cn^2 + 2fmn + 2gnl + 2hlm).$$

If $lmn$ is not zero, all the second derivatives of $u$ are determined on the cap $C$, provided that the quadratic form

$$Q = al^2 + bm^2 + cn^2 + 2fmn + 2gml + 2hlm$$

does not vanish. This condition is also sufficient in the exceptional case $lmn = 0$.

If $l = m = 0$, $n = 1$, $C$ lies on a plane $z = $ constant. The data determine all the second derivatives except $u_{zz}$; the differential equation then gives $u_{zz}$ provided that $c \neq 0$.

If $n = 0$, $lm \neq 0$, $u_{xz}, u_{yz}, u_{zz}$ are all determined on $C$. We then have

$$au_{xx} + bu_{yy} + 2hu_{xy} = -F_1,$$

$$lu_{yy} - mu_{xy} = \Phi_2,$$

$$lu_{xy} - mu_{xx} = \Phi_6,$$

which determine $u_{xx}, u_{yy}, u_{xy}$ provided that $al^2 + bm^2 + 2hlm \neq 0$. Thus, in any case, unless $Q = 0$, the second derivatives of $u$ are determined on $C$.

We could go on to discuss higher derivatives, and prove the Cauchy–Kowalewsky theorem for the half-linear equation. But what happens when $Q = 0$ is more interesting; for if $Q = 0$ everywhere on $C$, the problem proposed does not have a unique solution. $C$ is said to be a characteristic.

Suppose that the equation of $C$ is $\phi(x, y, z) = 0$. Since

$$l : m : n = \phi_x : \phi_y : \phi_z,$$

we have the partial differential equation

$$a\phi_x^2 + b\phi_y^2 + c\phi_z^2 + 2f\phi_y\phi_z + 2g\phi_z\phi_x + 2h\phi_x\phi_y = 0, \tag{3}$$

satisfied by the characteristic surfaces of (1). Since the coefficients were assumed to be functions of $x, y, z$ alone in the semi-linear equation (1), the family of characteristics is independent of the data in the Cauchy problem.

We can arrive at the condition (3) in a somewhat different way. We can ask what conditions $\phi$ must satisfy if the problem of Cauchy does not have a unique solution with data on the surface $\phi(x, y, z) = 0$. In other words, can there exist two solutions $u = u_1(x, y, z)$ and $u = u_2(x, y, z)$ of (1) with the properties

$$u_1 = u_2, \quad \frac{\partial u_1}{\partial x} = \frac{\partial u_2}{\partial x}, \quad \frac{\partial u_1}{\partial y} = \frac{\partial u_2}{\partial y}, \quad \frac{\partial u_1}{\partial z} = \frac{\partial u_2}{\partial z}$$

on $\phi = 0$?

Write $u_1 - u_2 = v$, so that $v$ and its first derivatives vanish on $\phi = 0$. Since $F$ has the same values for $u_1$ and $u_2$ on $\phi = 0$,

$$av_{xx} + bv_{yy} + cv_{zz} + 2fv_{yz} + 2gv_{zx} + 2hv_{xy} = 0 \qquad (4)$$

there. Let $\gamma$ be a regular arc $x = x(t)$, $y = y(t)$, $z = z(t)$ on $\phi = 0$; then

$$\frac{\partial \phi}{\partial x}\dot{x} + \frac{\partial \phi}{\partial y}\dot{y} + \frac{\partial \phi}{\partial z}\dot{z} = 0.$$

Since $\partial v/\partial x$ vanishes on $\phi = 0$,

$$v_{xx}\dot{x} + v_{xy}\dot{y} + v_{xz}\dot{z} = 0.$$

Eliminating $\dot{z}$, we have

$$(\phi_x v_{xz} - \phi_z v_{xx})\dot{x} + (\phi_y v_{xz} - \phi_z v_{xy})\dot{y} = 0.$$

But since $\gamma$ is an arbitrary regular arc, we can choose $\dot{x}, \dot{y}$ as we please and so
$$\phi_x v_{xz} - \phi_z v_{xx} = 0, \quad \phi_y v_{xz} - \phi_z v_{xy} = 0.$$

Hence $\qquad\qquad v_{xx}:v_{xy}:v_{xz} = \phi_x:\phi_y:\phi_z$

on $\phi = 0$. Similarly $\quad v_{yx}:v_{yy}:v_{yz} = \phi_x:\phi_y:\phi_z,$

$$v_{zx}:v_{zy}:v_{zz} = \phi_x:\phi_y:\phi_z.$$

Therefore on $\phi = 0$,

$$v_{xx}:v_{yy}:v_{zz}:v_{yz}:v_{zx}:v_{xy} = \phi_x^2:\phi_y^2:\phi_z^2:\phi_y\phi_z:\phi_z\phi_x:\phi_x\phi_y.$$

Hence, by (4)

$$a\phi_x^2 + b\phi_y^2 + c\phi_z^2 + 2f\phi_y\phi_z + 2g\phi_z\phi_x + 2h\phi_x\phi_y = 0,$$

so that $\phi = 0$ is a characteristic.

If we solve $\phi(x, y, z) = 0$ for $z$, and if we denote the partial derivatives of $z$ by $p$ and $q$, the differential equation for the characteristics becomes
$$ap^2 + bq^2 + c - 2fq - 2gp + 2hpq = 0. \qquad (5)$$

The characteristics can be built up from the characteristics of this first order equation, as we saw in Ch. 1. The characteristics of (5) are called the *bicharacteristics* of equation (1).

## 2.7   The half-linear equation in general

The half-linear equation with $n$ independent variables $(x_1, x_2, ..., x_n)$, is of the form
$$\sum_1^n a_{ij}\frac{\partial^2 u}{\partial x_i \partial x_j} + F = 0, \qquad (1)$$

where $a_{ij}$ are functions of the independent variables alone, but $F$ can also involve $u$ and its first derivatives; it can be treated by the method of §2.6. Its characteristics satisfy the first-order equation

$$\sum_1^n a_{ij} \frac{\partial\phi}{\partial x_i} \frac{\partial\phi}{\partial x_j} = 0. \tag{2}$$

Consider the quadratic form

$$Q = \sum_1^n a_{ij}\xi_i\xi_2,$$

where the coefficients are evaluated at any fixed point $P_0$. By a suitable non-singular linear transformation, we can express $Q$ as a sum of squares

$$Q = \sum_{i=1}^m \mu_i \eta_i^2, \tag{3}$$

where $m \leqslant n$. The number $m$ does not depend on the particular linear transformation. If $m$ is less than $n$, the equation (1) is said to be of parabolic type at $P_0$.

If $m = n$, the number of positive coefficients $\mu_i$ does not depend on the particular linear transformation. If the transformation is orthogonal, the $\mu_i$ are the latent roots (eigenvalues) of the matrix $(a_{ij})$.

If all the coefficients $\mu_i$ are of the same sign, (1) is said to be of elliptic type at $P_0$. An equation which is everywhere of elliptic type has no characteristics.

If all but one of the $\mu_i$ are of the same sign, the equation is said to be of hyperbolic type at $P_0$.

There are intermediate cases, such as

$$u_{xx} + u_{yy} - u_{zz} - u_{tt} = 0,$$

which are of neither type; about these, little is known.

## 2.8   The half-linear equation with constant coefficients

The half-linear equation with $n$ independent variables $(x_1, x_2, ..., x_n)$ is of the form

$$\sum_1^n a_{ij} \frac{\partial^2 u}{\partial x_i \partial x_j} + F = 0,$$

where $F$ is a function of the independent variables and possibly of $u$ and its first derivatives. If we introduce new independent variables

$$\xi_i = \xi_i(x_1, x_2, ..., x_n) \quad (i = 1, 2, ..., n)$$

and try to choose the functions $\xi_i$ so that the resulting equation is of the form

$$\sum_1^n b_i \frac{\partial^2 u}{\partial \xi_i^2} + G = 0,$$

we find that the $n$ functions $\xi_i$ have to satisfy $\frac{1}{2}n(n-1)$ differential equations of the first order. This is, in general, impossible if $n > 3$; if $n = 3$, there are just enough conditions to determine the functions $\xi_i$. There are special cases when it is possible to make a transformation of this kind when $n > 3$; a particular case is when the coefficients $a_{ij}$ are constants.

When the coefficients are constants, the differential operator

$$L(u) = \Sigma a_{ij} \frac{\partial^2 u}{\partial x_i \partial x_j}$$

can be written in vector form

$$L(u) = \delta^T A \delta u,$$

where $A$ is the real symmetric matrix $(a_{ij})$, $\delta^T$ is the row-vector operator

$$\delta^T = \left( \frac{\partial}{\partial x_1}, \frac{\partial}{\partial x_2}, \ldots, \frac{\partial}{\partial x_n} \right)$$

and $\delta$ is the transpose of $\delta^T$.

Let $x' = Mx$ be a real non-singular linear transformation from the variables $(x_1, x_2, \ldots, x_n)$ to $(x'_1, x'_2, \ldots, x'_n)$, using the obvious vector notation. Then

$$\delta = M^T \delta',$$

where $\delta'$ is the transpose of the row-vector operator

$$(\delta')^T = \left( \frac{\partial}{\partial x'_1}, \frac{\partial}{\partial x'_2}, \ldots, \frac{\partial}{\partial x'_n} \right),$$

and $M^T$ is the transpose of $M$.

Then

$$L(u) = (\delta')^T M A M^T \delta' u = (\delta')^T B \delta' u,$$

where

$$B = M A M^T.$$

Since $A$ is a real symmetric matrix, we can find an orthogonal matrix $M$ so that $M A M^T$ is a diagonal matrix. The diagonal elements of the resulting matrix $B$ will be the non-zero latent roots $\mu_1, \mu_2, \ldots, \mu_r$ of $A$, where $r$ is the rank of $A$, and $n - r$ zero elements if $r < n$; all the latent roots are real. We have thus shown that

$$L(u) = \sum_{i=1}^{r} \mu_i \frac{\partial^2 u}{\partial x_i'^2}.$$

If we put $x'_i = \xi_i \sqrt{|\mu_i|}$, we get

$$L(u) = \sum_{i=1}^{r} \nu_i \frac{\partial^2 u}{\partial \xi_i^2},$$

where each $\nu_i$ is $\pm 1$.

This reduction of a half-linear equation with constant coefficients can be done in an elementary way by using Hermite's reduction of a quadratic form to a sum of squares, which is merely the process of 'completing the square', instead of an orthogonal transformation. This is illustrated below.

Suppose that

$$L(u) = u_{xx} + 3u_{yy} + 84u_{zz} + 28u_{yz} + 16u_{zx} + 2u_{xy}$$

$$= (\delta_1^2 + 3\delta_2^2 + 84\delta_3^2 + 28\delta_2\delta_3 + 16\delta_3\delta_1 + 2\delta_1\delta_2)\, u,$$

where $\delta_1$, $\delta_2$, $\delta_3$ are the three components of the column-vector operator $\boldsymbol{\delta}$. Now

$$\delta_1^2 + 3\delta_2^2 + 84\delta_3^2 + 28\delta_2\delta_3 + 16\delta_3\delta_1 + 2\delta_1\delta_2$$

$$= (\delta_1 + \delta_2 + 8\delta_3)^2 + 2\delta_2^2 + 12\delta_2\delta_3 + 20\delta_3^2$$

$$= (\delta_1 + \delta_2 + 8\delta_3)^2 + 2(\delta_2 + 3\delta_3)^2 + 2\delta_3^2$$

$$= \delta_1'^2 + 2\delta_2'^2 + 2\delta_3'^2,$$

where
$$\delta_1' = \frac{\partial}{\partial x_1'} = \delta_1 + \delta_2 + 8\delta_3,$$

$$\delta_2' = \frac{\partial}{\partial x_2'} = \delta_2 + 3\delta_3,$$

$$\delta_3' = \delta_3.$$

If the linear transformation is $x = Nx'$, then $\boldsymbol{\delta}' = N^T\boldsymbol{\delta}$ and so

$$N^T = \begin{bmatrix} 1 & 1 & 8 \\ 0 & 1 & 3 \\ 0 & 0 & 1 \end{bmatrix}.$$

Therefore      $x = x'$,   $y = x' + y'$,   $z = 8x' + 3y' + z'$,

or            $x' = x$,   $y' = y - x$,   $z' = -5x - 3y + z$.

With this transformation

$$L(u) = \frac{\partial^2 u}{\partial x_1'^2} + 2\frac{\partial^2 u}{\partial x_2'^2} + 2\frac{\partial^2 u}{\partial x_3'^2}.$$

The process has to be modified slightly if the associated quadratic form contains no squares. For example, if

$$(Lu) = 4u_{xy} + 2u_{yz} + 2u_{zx},$$

the quadratic form is      $4\delta_1\delta_2 + 2\delta_2\delta_3 + 2\delta_3\delta_1.$

If we put $\delta_1 = \alpha - \beta$, $\delta_2 = \alpha + \beta$, this becomes

$$4\alpha^2 - 4\beta^2 + 4\alpha\delta_3$$
$$= (2\alpha + \delta_3)^2 - 4\beta^2 - \delta_3^2$$
$$= (\delta_1 + \delta_2 + \delta_3)^2 - (\delta_1 - \delta_2)^2 - \delta_3^2$$
$$= \delta_1'^2 - \delta_2'^2 - \delta_3'^2,$$

where $\quad\quad \delta_1' = \delta_1 + \delta_2 + \delta_3, \quad \delta_2' = \delta_1 - \delta_2, \quad \delta_3' = \delta_3.$

Then $\quad\quad\quad x = x' + y', \quad y = x' - y', \quad z = x' + z'.$

In terms of these new variables,

$$L(u) = \frac{\partial^2 u}{\partial x'^2} - \frac{\partial^2 u}{\partial y'^2} - \frac{\partial^2 u}{\partial z'^2}.$$

## Exercises

**1.** Find the types of the following differential equations, and reduce them to normal form:

(i) $y^2 r - t = 0$,
(ii) $y^2 r + 2ys + t = p$,
(iii) $y^2 r + t = 0$,
(iv) $r - 2s + 3t - q - u = 0$,
(v) $(1 + x^2)^2 r - t = 0$,
(vi) $x^2 r - 2xys + y^2 t = 0$,
(vii) $x^2 r + 2xys + y^2 t = 0$.

**2.** Reduce the following equations to normal form:

(i) $\dfrac{\partial^2 u}{\partial x^2} + 2\dfrac{\partial^2 u}{\partial y^2} + 3\dfrac{\partial^2 u}{\partial z^2} + 4\dfrac{\partial^2 u}{\partial t^2} + 2\dfrac{\partial^2 u}{\partial x \partial y} + 2\dfrac{\partial^2 u}{\partial x \partial z} + 2\dfrac{\partial^2 u}{\partial x \partial t}$

$$+ 4\dfrac{\partial^2 u}{\partial y \partial z} + 4\dfrac{\partial^2 u}{\partial y \partial t} + 6\dfrac{\partial^2 u}{\partial x \partial t} = 0.$$

(ii) $\dfrac{\partial x^2}{\partial x^2} + \dfrac{\partial^2 u}{\partial z^2} + 2\dfrac{\partial^2 u}{\partial x \partial y} + 2\dfrac{\partial^2 u}{\partial x \partial z} + 2\dfrac{\partial^2 u}{\partial x \partial t} + 2\dfrac{\partial^2 u}{\partial z \partial t} = 0.$

(iii) $\dfrac{\partial^2 u}{\partial x^2} + \dfrac{\partial^2 u}{\partial x \partial y} + \dfrac{\partial^2 u}{\partial x \partial z} + \dfrac{\partial^2 u}{\partial x \partial t} = 0.$

**3.** Prove that $\quad\quad\quad q^2 r - 2pqs + p^2 t = 0$

has an intermediate integral $p = qf(u)$, where $f$ is an arbitrary function. Hence show that

$$y + xf(u) = g(u),$$

where $g$ is a second arbitrary function.

**4.** Transform the equation

$$x^2 \frac{\partial^2 u}{\partial x^2} = y^2 \frac{\partial^2 u}{\partial y^2}$$

into one with characteristic variables. Hence show that

$$u = f(xy) + xg\left(\frac{y}{x}\right),$$

where $f$ and $g$ are arbitrary functions.

**5.** Find the characteristics of $r - yt = 0$. Reduce this equation to normal form when $y > 0$.

**6.** Find the characteristics of $(1+x^2)r - (1+y^2)t = 0$, and reduce the equation to normal form.

**7.** Prove that $x^2 r + 2xys + y^2 t = 0$ has an intermediate integral

$$px + qy - u + f\left(\frac{y}{x}\right) = 0,$$

where $f$ is an arbitrary function. Hence show that

$$u = f\left(\frac{y}{x}\right) + xg\left(\frac{y}{x}\right).$$

where $g$ is also an arbitrary function.

**8.** Show that $qr + (uq - p)s - upt = 0$ has an intermediate integral

$$p + qu = f(u),$$

where $f$ is an arbitrary function. Hence solve the equation.

**9.** Prove that the two families of characteristic strips of

$$s = F(x, y, u, p, q)$$

are given by

(i) $dx = 0$, $du - qdy = 0$, $dp - Fdy = 0$,

and

(ii) $dy = 0$, $du - pdx = 0$, $dq - Fdx = 0$.

Hence solve the equation $s = pq$.

# 3

# BOUNDARY VALUE AND INITIAL VALUE PROBLEMS

## 3.1 Laplace's equation

The theorem of Cauchy and Kowalewsky states that, under certain conditions, a second-order partial differential equation has a unique solution. But this result is of little importance in many problems of mathematical physics. Firstly, the solution in the Cauchy–Kowalewsky theorem is only a local solution, valid near the manifold carrying the Cauchy type data; whereas a global solution is often needed. Secondly, the Cauchy type data are analytic functions, not necessarily the case in practice. Thirdly, in some problems, we have too much data for the existence of a global solution. In short, are we asking the right question?

The simplest equation of elliptic type is Laplace's equation

$$\nabla^2 u \equiv \frac{\partial^2 u}{\partial x^2} + \frac{\partial^2 u}{\partial y^2} + \frac{\partial^2 u}{\partial z^2} = 0.$$

It can be shown that there is but one solution which is regular in the sphere $r \leqslant a$, in spherical polar coordinates, and which takes the value $\cos \theta$ on $r = a$, namely $(r/a) \cos \theta$. But there is no solution, regular in $r \leqslant a$, which satisfies the Cauchy conditions

$$u = \cos \theta, \quad \frac{\partial u}{\partial r} = 0,$$

on $r = a$. There is the solution

$$u = \frac{2}{3} \frac{a}{r} \cos \theta + \frac{1}{3} \frac{a^2}{r^2} \cos \theta,$$

which is regular when $r > 0$, but has a singularity at the origin. Thus, when we ask for a solution regular in $r \leqslant a$ and satisfying Cauchy conditions on $r = a$, we are asking the wrong question.

The right questions are suggested by physical problems. Let $D$ be a spherical cavity inside an earthed perfectly conducting body; let $S$ be the boundary of $D$. If there is a point charge $e$ at a point $P_0$ inside $S$, a charge is induced on $S$, and the total electrostatic potential in $D$ is

$$\phi = u + \frac{e}{R_0},$$

[ 44 ]

where $R_0$ is the distance from $P_0$. The function $u$, which is the potential of the induced charge, has no singularities in $D$ and satisfies Laplace's equation. On $S$, $u = -e/R_0$. It is obvious to the physicist that such a potential $u$ is uniquely determined; and a knowledge of $u$ determines the normal derivative of $u$ on $S$. We cannot assign the normal derivative arbitrarily. The problem of finding $u$ is a particular case of the problem of Dirichlet, a boundary value problem.

Next let $D$ be an earthed perfectly conducting spherical conductor with surface $S$. If there is a point charge at a point $P_0$ outside $S$, a charge is induced on $S$, and the total electrostatic potential outside $S$ is

$$\phi = u + \frac{e}{R_0},$$

where $R_0$ is the distance from $P_0$. The potential $u$ of the induced charge satisfies Laplace's equation, has no singularities outside $S$ and has a gradient which vanishes at infinity. On $S$, $u = -e/R_0$. Again it is 'obvious' that $u$ is uniquely determined. Finding $u$ is a particular case of the external problem of Dirichlet.

Laplace's equation is also satisfied by the velocity potential defining the irrotational motion of an incompressible perfect fluid. Suppose that the motion of such a fluid in a spherical cavity $D$ with rigid wall $S$ is caused for a source $m$ at $P_0$ and an equal sink at $P_1$. If the distances of a point of the cavity from $P_0$ and $P_1$ are $R_0$ and $R_1$ respectively, the velocity potential is

$$\phi = u + \frac{m}{R_0} - \frac{m}{R_1},$$

where $u$ is a solution of Laplace's equation with no singularity in $D$. Since there is no flow across $S$, the normal velocity $\partial\phi/\partial N$ vanishes on $S$. Hence $u$ satisfies the condition

$$\frac{\partial u}{\partial N} = \frac{\partial}{\partial N}\left(\frac{m}{R_1} - \frac{m}{R_0}\right)$$

on $S$. Obviously $u$ is uniquely determined, up to an additive constant. This is a particular case of the problem of Neumann.

Lastly suppose that we have a rigid spherical obstacle $D$ fixed in a perfect fluid whose flow at infinity is uniform with velocity $V$. The velocity potential outside $D$ is

$$\phi = u + Vx.$$

The function $u$ is a solution of Laplace's equation with no singularity outside $D$. On the boundary of $D$,

$$\frac{\partial u}{\partial N} = -\frac{\partial}{\partial N}(Vx)$$

and at infinity grad $u$ vanishes. This is a particular case of the external problem of Neumann. Naturally, in all these problems, there is no special merit in considering only spherical boundaries except that it saves a lengthy explanation of the nature of the boundary.

The important problems for Laplace's equations are thus boundary value problems, in which only one condition is satisfied on the boundary.

## 3.2   The equation of wave motions

The simplest equation of hyperbolic type is the equation of wave motions

$$\nabla^2 u = \frac{1}{c^2}\frac{\partial^2 u}{\partial t^2},$$

where $c$ is a constant.

For example, suppose that we have an infinitely long uniform string of density $\rho$, which is taut with tension $T$. For small vibrations, the displacement $u$ at the point of abscissa $x$ at time $t$ satisfies

$$\frac{\partial^2 u}{\partial x^2} = \frac{1}{c^2}\frac{\partial^2 u}{\partial t^2}, \tag{1}$$

where
$$c^2 = T/\rho.$$

The displacement at any instant is determined if we are given the initial displacement and initial velocity of each point of the string, that is, if we are given

$$u(x, 0) = \Phi(x), \quad u_t(x, 0) = \Psi(x).$$

In this case, the data are of Cauchy type, and the Cauchy–Kowalewsky theorem shows that there is a unique solution analytic near every point of the $x$-axis if the data are analytic. But there is no need to assume analytic data.

In terms of the characteristic variables $\xi = x + ct$, $\eta = x - ct$, equation (1) is

$$\frac{\partial^2 u}{\partial \xi \, \partial \eta} = 0.$$

Hence we have

$$u = F(\xi) + G(\eta) = F(x + ct) + G(x - ct)$$

so that the general solution represents the sum of two disturbances propagated with velocities $\pm c$. If we use the initial conditions, we get the well-known solution

$$u = \tfrac{1}{2}\Phi(x - ct) + \tfrac{1}{2}\Phi(x + ct) + \frac{1}{2c}\int_{x-ct}^{x+ct}\Psi(\tau)\,d\tau$$

when $t \geqslant 0$. This solution has continuous derivatives of the second order, provided that $\Phi''(\tau)$ and $\Psi'(\tau)$ are continuous.

In some physical problems, the data are not everywhere continuously differentiable. For example, if

$$\Phi(x) = \max\,(l - |x|, 0), \quad \Psi(x) = 0,$$

the graph of $u$ for any fixed value of $t\,(\geqslant 0)$ consists of straight segments. At the corners, the partial differential equation does not hold; $u$ is not differentiable.

Suppose next that the string is 'stopped' at $x = \pm l$. The motion of the portion of the string when $|x| \leqslant l$ then satisfies the conditions

$$u(x, 0) = \Phi(x), \quad u_t(x, 0) = \Psi(x),$$
$$u(l, t) = 0, \quad u(-l, t) = 0,$$

where $\Phi(\pm l) = 0$, $\Psi(\pm l) = 0$. This is a mixed problem; initial conditions have to be satisfied on $t = 0$, boundary conditions on $x = \pm l$.

The equation

$$\frac{\partial^2 u}{\partial x^2} + \frac{\partial^2 u}{\partial y^2} = \frac{1}{c^2}\frac{\partial^2 u}{\partial t^2}$$

occurs in the theory of small vibrations of a uniform stretched membrane. If the membrane covers the whole $xy$-plane, the relevant problem is an initial value problem. But if the membrane is like the top of a drum, we have to solve a mixed problem – an initial value problem and a boundary value problem resulting from the fact that the membrane is fixed at its rim.

The equation

$$\frac{\partial^2 u}{\partial x^2} + \frac{\partial^2 u}{\partial y^2} + \frac{\partial^2 u}{\partial z^2} = \frac{1}{c^2}\frac{\partial^2 u}{\partial t^2}$$

appears in the theory of sound waves of small amplitude, and in other branches of applied mathematics.

## 3.3   Characteristics as wave fronts

The characteristics of

$$\frac{\partial^2 u}{\partial x^2} + \frac{\partial^2 u}{\partial y^2} = \frac{1}{c^2}\frac{\partial^2 u}{\partial t^2} \tag{1}$$

satisfy the first-order equation

$$p^2 + q^2 = \frac{1}{c^2}, \tag{2}$$

where $p$ and $q$ denote $\partial t/\partial x$ and $\partial t/\partial y$ respectively. The characteristics of (2) are called the bicharacteristics of (1) and satisfy the Lagrange–Charpit equations

$$\frac{dx}{ds} = pc, \quad \frac{dy}{ds} = qc, \quad \frac{dt}{ds} = \frac{1}{c}, \quad \frac{dp}{ds} = 0, \quad \frac{dq}{ds} = 0,$$

where $s$ is a parametric variable. Hence on a bicharacteristic strip,

$$p = \frac{1}{c}\cos\alpha, \quad q = \frac{1}{c}\sin\alpha,$$

where $\alpha$ is a constant, and

$$x = x_0 + s\cos\alpha, \quad y = y_0 + s\sin\alpha, \quad ct = ct_0 + s.$$

The bicharacteristic lines are straight lines making a constant angle $\cos^{-1} 1/\sqrt{(1+c^2)}$ with $Ot$; the bicharacteristic strips lie on planes

$$(x-x_0)\cos\alpha + (y-y_0)\sin\alpha = c(t-t_0). \tag{3}$$

Any characteristic of (1) is the envelope of a one-parametric family of such planes

$$x\cos\alpha + y\sin\alpha = ct + f(\alpha).$$

If $x_0, y_0, t_0$ are constants, the envelope of (3) is the characteristic cone

$$(x-x_0)^2 + (y-y_0)^2 = c^2(t-t_0)^2.$$

Similarly, for
$$\frac{\partial^2 u}{\partial x^2} + \frac{\partial^2 u}{\partial y^2} + \frac{\partial^2 u}{\partial z^2} = \frac{1}{c^2}\frac{\partial^2 u}{\partial t^2}, \tag{4}$$

the characteristic cone with vertex $(x_0, y_0, z_0, t_0)$ is

$$(x-x_0)^2 + (y-y_0)^2 + (z-z_0)^2 = c^2(t-t_0)^2,$$

the light-cone of special relativity.

If we transform (4) to spherical polar coordinates $(r, \theta, \phi)$, a solution independent of $\theta$ and $\phi$ satisfies

$$\frac{\partial^2 u}{\partial r^2} + \frac{2}{r}\frac{\partial u}{\partial r} = \frac{1}{c^2}\frac{\partial^2 u}{\partial t^2}:$$

and, if we put $u = v/r$, we obtain the one-dimensional wave equation

$$\frac{\partial^2 v}{\partial r^2} = \frac{1}{c^2}\frac{\partial^2 v}{\partial t^2}.$$

Hence a solution of (4) which represents expanding waves with spherical symmetry is of the form

$$u = \frac{1}{r}f(ct-r).$$

3.3]     BOUNDARY AND INITIAL VALUE PROBLEMS     [49

Let us suppose that $f(r)$ and $f'(r)$ vanish when $r \leqslant 0$. Then

$$u = \frac{1}{r} f(ct - r), \quad r \leqslant ct;$$

$$= 0, \quad r \geqslant ct,$$

is the velocity potential of sound waves of small amplitude due to a source at the origin expanding into the gas at rest. The disturbance at a point distance $r$ from the origin is zero until the instant $t = r/c$. The plane sections $t = $ constant of the characteristic cone $x^2 + y^2 + z^2 = c^2 t^2$ give the successive positions of the moving wave front.

More generally, a disturbance with wave front $f(x, y, z) = t$ propagated into the gas initially at rest is specified by a velocity potential $u$ which is zero on one side of the wave front and non-zero on the other. If the motion is not a shock wave, $u$ and its first derivatives vanish on the wave front. Hence we have two solutions, $u = u_1$ representing the propagated waves and $u = 0$ representing the state of rest, which are equal and have equal first derivatives on the wave front, which is therefore a characteristic.

Suppose that the normal at $(x, y, z)$ on $f(x, y, z) = t$ cuts $f(x, y, z) = t'$ where $t' > t$ at $(x', y', z')$. If the distance from $(x, y, z)$ to $(x', y', z')$ is $\delta$,

$$f(x + l\delta, y + m\delta, z + n\delta) = t',$$

where $(l, m, n)$ are the direction cosines of the normal. Then

$$t' - t = f(x + l\delta, y + m\delta, z + n\delta) - f(x, y, z)$$

$$= \delta\{lf_x + mf_y + nf_z\},$$

where $f_x, f_y, f_z$ are evaluated at $(x + \theta l\delta, y + \theta m\delta, z + \theta n\delta)$ where

$$0 < \theta < 1.$$

Hence     $$\lim_{t' \to t} \frac{\delta}{t' - t} = \frac{1}{lf_x(x, y, z) + mf_y(x, y, z) + nf_z(x, y, z)}.$$

But     $$l : m : n = f_x : f_y : f_z,$$

where     $$f_x^2 + f_y^2 + f_z^2 = \frac{1}{c^2},$$

since $f(x, y, z) = t$ is a characteristic of (4). Therefore $f_x = l/c$, etc. and so

$$\lim_{t' \to t} \frac{\delta}{t' - t} = c.$$

The wave front is therefore advancing with normal velocity $c$.

### 3.4 The equation of telegraphy

If a uniform telegraph cable has self-inductance $L$, resistance $R$ and capacity $C$, all per unit length, the electric potential $V$ and the current $i$ at a distance $x$ from an end satisfy the equations

$$\frac{\partial V}{\partial x} = -L\frac{\partial i}{\partial t} - Ri,$$

$$C\frac{\partial V}{\partial t} = -\frac{\partial i}{\partial x}.$$

Eliminating $i$, we have Heaviside's equation of telegraphy

$$\frac{1}{C}\frac{\partial^2 V}{\partial x^2} = L\frac{\partial^2 V}{\partial t^2} + R\frac{\partial V}{\partial t}.$$

The physical problem to be solved for this equation is a mixed one. We are given $V$ and $\partial V/\partial t$ when $t = 0$, $x \geqslant 0$ and also we are given $V$ at the end point $x = 0$ for all $t$.

### 3.5 The equation of heat

The best known example of an equation of parabolic type is the diffusion equation, the equation governing the conduction of heat in a uniform solid of conductivity $k$, specific heat $c$ and density $\rho$. The temperature $u$ satisfies

$$\frac{\partial^2 u}{\partial x^2} + \frac{\partial^2 u}{\partial y^2} + \frac{\partial^2 u}{\partial z^2} = \frac{1}{a^2}\frac{\partial u}{\partial t} \quad \left(a^2 = \frac{k}{c\rho}\right),$$

when there is no internal source of heat. If the body fills all space, a knowledge of the initial temperature distribution determines the temperature everywhere at any subsequent instant, assuming that the temperature at infinity remains finite as it obviously must.

If, however, the body in which heat is flowing is bounded, there are also boundary conditions to be satisfied. There are three forms of boundary condition:

(i) The temperature on the boundary may be given. This may be constant or a given function of the time.

(ii) The rate of flow of heat, $k\,\partial u/\partial N$ may be given. If there is no flow of heat, $\partial u/\partial N = 0$.

(iii) The rate of flow of heat at the surface may be proportional to the difference between the temperature at the boundary and the temperature $u_0$ (not necessarily constant) of the medium in which the body is immersed. Then on the boundary

$$k\frac{\partial u}{\partial N} = -h(u - u_0),$$

when $\partial/\partial N$ denotes differentiation along the outward normal and $h$ is non-negative.

There are various particular cases. For a uniform straight rod whose surface is thermally insulated, the equation of heat reduces to

$$a^2\frac{\partial^2 u}{\partial x^2} = \frac{\partial u}{\partial t},$$

provided that the temperature is uniform over every cross section. But if the rod loses heat from its surface at a rate proportional to the temperature, the conduction of heat is governed by

$$a^2\frac{\partial^2 u}{\partial x^2} = \frac{\partial u}{\partial t} + \nu u,$$

where $\nu$ is a positive constant depending on the size, shape and material of the rod.

If the telegraph cable of §3.4 has negligible self-inductance the equation of telegraphy reduces to the diffusion equation

$$\frac{\partial^2 V}{\partial x^2} = RC\frac{\partial V}{\partial t},$$

but now the data are simply that $V$ is known when $x \geqslant 0, t = 0$ and when $x = 0, t \geqslant 0$. We do not need to assign $\partial V/\partial t$ on $t = 0$; it is determined from the differential equation.

## 3.6  Well-posed problems

It is essential to know whether a boundary value or initial value problem has a solution and whether the solution is unique. But more than this is needed in mathematical physics. The data in a physical problem, being derived from experiment, are necessarily approximate. If the problem is to have any physical meaning, a small change in the data must give rise to a small change in the solution.

Again, the numerical analyst, solving such a problem numerically, is forced to approximate his data in order to apply numerical methods. It is not enough to know that the problem has a unique solution. It must also be known that the process of making an approximation to the data produces only a slight change in the solution.

The same difficulty arises in connexion with the Cauchy–Kowalew-sky theorem, in which the initial data are analytic. We can approximate to any continuous function as accurately as we please by polynomials. So we could apply the Cauchy–Kowalewsky theorem with continuous data by using polynomial approximations only if a small change in the analytic data produces a small change in the solution.

An initial value problem, boundary value problem or mixed problem which has a unique solution is said to be *well posed* if a small change in the data produces a small change in the solution.

If $\Phi(x)$, $\Phi'(x)$, $\Psi'(x)$ are differentiable, the initial value problem

$$\frac{\partial^2 u}{\partial x^2} = \frac{\partial^2 u}{\partial t^2},$$

where        $u(x, 0) = \Phi(x), \quad u_t(x, 0) = \Psi'(x)$

has the d'Alembert solution

$$u = \tfrac{1}{2}\{\Phi(x+t) + \Phi(x-t)\} + \frac{1}{2} \int_{x-t}^{x+t} \Psi'(\tau) \, d\tau.$$

If we replace $\Phi$ by $\Phi + \delta\Phi$, $\Psi'$ by $\Psi' + \delta\Psi'$ on an interval $(a, b)$ where $|\delta\Phi| < \epsilon$, $|\delta\Psi'| < \epsilon$, the absolute value of the change in $u$ is less than $\epsilon + \tfrac{1}{2}\epsilon(b - a)$. Hence the problem is well posed.

Next suppose that $f(\theta)$ is continuous. Then the problem of Dirichlet for

$$\frac{\partial^2 u}{\partial x^2} + \frac{\partial^2 u}{\partial y^2} = 0,$$

with        $u(R\cos\theta, R\sin\theta) = f(\theta) \quad (0 \leqslant \theta \leqslant 2\pi),$

has a unique solution, regular in $x^2 + y^2 < R^2$, given by Poisson's formula

$$u(r\cos\theta, r\sin\theta) = \frac{1}{2\pi} \int_0^{2\pi} \frac{R^2 - r^2}{R^2 - 2Rr\cos(\theta - \phi) + r^2} f(\phi) \, d\phi.$$

If we replace $f$ by $f + \delta f$, the absolute value of the change in $u$ is less than or equal to

$$\frac{1}{2\pi} \sup |\delta f(\phi)| \int_0^{2\pi} \frac{R^2 - r^2}{R^2 - 2Rr\cos(\theta - \phi) + r^2} \, d\phi$$

$$= \sup |\delta f(\phi)|.$$

Hence if $|\delta f(\phi)| < \epsilon$, the change in the absolute value of the solution $u$ is less than $\epsilon$. This problem is thus well posed.

The problem of Cauchy for

$$\frac{\partial^2 u}{\partial x^2} + \frac{\partial^2 u}{\partial y^2} = 0$$

with data on $y = 0$ is not well posed. If

$$u(x, 0) = 0, \quad u_y(x, 0) = \frac{1}{n} \sin nx,$$

then
$$u(x, y) = \frac{1}{n^2} \sin nx \sinh ny.$$

The data can be made as small as we please by choosing $n$ large enough. But if $y \neq 0$, the solution is unbounded as $n \to \infty$.

Now consider the two Cauchy initial value problems:

(i) $\quad\quad u_1(x, 0) = x^2, \quad \partial u_1(x, 0)/\partial y = 0;$

(ii) $\quad\quad u_2(x, 0) = x^2, \quad \partial u_2(x, 0)/\partial y = \frac{1}{n} \sin nx.$

Then
$$u_1 = x^2 - y^2,$$
$$u_2 = x^2 - y^2 + \frac{1}{n^2} \sin nx \sinh ny.$$

Although the data for $u_2$ tend to the data for $u_1$ as $n \to \infty$, $u_2$ does not tend to $u_1$.

**4**

# EQUATIONS OF HYPERBOLIC TYPE

## 4.1 The half-linear equation of hyperbolic type

The general half-linear equation with two independent variables is of the form
$$ar + 2hs + bt = f(x, y, u, p, q),$$
where $a, h, b$ are functions of $x$ and $y$ alone. As we saw in §2.5, if this equation is of hyperbolic type, it can be reduced to the normal form
$$s = f(x, y, u, p, q) \tag{1}$$
by using characteristic variables, though the transformation to characteristic variables may be only a local one. The characteristics are $x =$ constant on which
$$\frac{du}{dy} = q, \quad \frac{dp}{dy} = f,$$
and $y =$ constant on which
$$\frac{du}{dx} = p, \quad \frac{dq}{dx} = f.$$

In particular, if $f$ is identically zero, $p$ is constant on each characteristic $x =$ constant and $q$ is constant on each characteristic $y =$ constant.

If we rotate the axes in (1) so that the characteristics are
$$x \pm y = \text{constant},$$
we obtain a second normal form
$$r - t = f(x, y, u, p, q). \tag{2}$$

If $f$ is identically zero, $p + q$ is constant on every characteristic $x + y =$ constant, $p - q$ is constant on every characteristic $x - y =$ constant.

## 4.2 The equation $u_{xy} = 0$

The general solution of $u_{xy} = 0$ is $u = F(x) + G(y)$, so that $p$ is constant on every characteristic $x =$ constant, $q$ is constant on every characteristic $y =$ constant. If $u$ has continuous second derivatives, $F''$ and $G''$ are continuous.

The problem of Cauchy cannot be solved with Cauchy data on a characteristic, for if we try to solve the Cauchy problem with data

$$u = f(x), \quad p = g(x), \quad q = h(x),$$

on a characteristic $y = b$, where $f' = g$ by the strip condition, we have to find $F$ and $G$ to satisfy

$$F(x) + G(b) = f(x), \quad F'(x) = g(x) = f'(x), \quad G'(b) = h(x).$$

The third equation is impossible for arbitrary $h(x)$; it can only be satisfied if $h(x)$ is a constant $C$ and $G(y) = Cy$. The solution then is

$$u = f(x) + C(y - b).$$

In what is called the characteristic boundary value problem, $u$ is given on two characteristics $x = a$ and $y = b$. On $x = a$, $u = \Phi(y)$; on $y = b$, $u = \Psi(x)$ where $\Psi(a) = \Phi(b)$ because the data are assumed continuous. The solution is

$$u = \Psi(x) + \Phi(y) - \Psi(a).$$

The most interesting case of the problem of Cauchy for $u_{xy} = 0$ is that which corresponds to the initial value problem for the wave equation; the Cauchy data

$$u = f(x), \quad p = g(x), \quad q = h(x),$$

are given on the line $x + y = 0$. The strip condition is then

$$f'(x) = g(x) - h(x).$$

It is readily verified that the solution is

$$u = f(x) + \int_{-y}^{x} h(\tau)\, d\tau.$$

In order that this solution may have continuous derivatives of the second order, it is necessary and sufficient that $f''$, $g'$ and $h'$ be continuous; and

$$p = g(x), \quad q = h(-y)$$

everywhere.

If, however, $h(x)$ has a finite jump at $x = a$, so that the limits $h(a + 0)$ and $h(a - 0)$ exist but are not equal, then

$$q(x, -a+\epsilon) - q(x, -a-\epsilon) = h(a-\epsilon) - h(a+\epsilon)$$

does not tend to zero as $\epsilon \to 0$. Hence $q(x, y)$ is discontinuous across $y = -a$. A discontinuity in the given value of $q$ at $(a, -a)$ persists across the whole characteristic $y = -a$. Similarly a discontinuity in the given value of $p$ at $(a, -a)$ is propagated along the whole character-istic $x = a$.

Next consider the case when

$$u = f(x), \quad p = g(x), \quad q = h(x)$$

on $x+y = 0$, $x > 0$, where $f' = g - h$, and

$$u = F(x), \quad p = G(x), \quad q = H(x)$$

on $x+y = 0$, $x < 0$, where $F' = G - H$. Suppose that $f'', g', h'$ are continuous in $x \geqslant 0$, and that $F'', G', H'$ are continuous in $x \leqslant 0$. We assume in addition that $f(+0) = F(-0)$; for, if not, the solution $u$, if it existed, would be discontinuous at the origin.

The characteristics $x = 0, y = 0$ divide the half-plane $x+y \geqslant 0$ into three regions $D_1, D_2, D_3$ where

$$(x \geqslant 0, y \leqslant 0), \quad (x \leqslant 0, y \geqslant 0), \quad (x \geqslant 0, y \geqslant 0)$$

respectively. In $D_1$, $\qquad u = f(x) + \int_{-y}^{x} h(\tau)\,d\tau.$

In $D_2$, $\qquad\qquad\qquad u = F(x) + \int_{-y}^{x} H(\tau)\,d\tau.$

Hence when $x \geqslant 0, y = 0$,

$$u(x, 0) = f(x) + \int_{0}^{x} h(\tau)\,d\tau,$$

but when $x = 0, y > 0$,

$$u(0, y) = F(0) + \int_{-y}^{0} H(\tau)\,d\tau.$$

To get $u$ in $D_3$, we have to solve a characteristic boundary value problem, the data being continuous since $f(0) = F(0)$. The solution is

$$u(x, y) = f(x) + \int_{0}^{x} h(\tau)\,d\tau + \int_{-y}^{0} H(\tau)\,d\tau,$$

which is just what we should have got if we had forgotten that the given values of $q$ on $x+y = 0$ are not necessarily continuous at the origin.

The first derivatives are given by

$$p = f'(x) + h(x), \qquad q = h(-y), \qquad \text{in } D_1,$$
$$p = F'(x) + H(x), \qquad q = H(-y), \qquad \text{in } D_2,$$
$$p = f'(x) + h(x), \qquad q = H(-y), \qquad \text{in } D_3.$$

In $D_1 \cup D_3$, $p$ is continuous, but $q$ is discontinuous across the axis of $x$, the jump being $H(0) - h(0)$. In $D_2 \cup D_3$, $q$ is continuous, but $p$ is discontinuous across the axis of $y$, the jump being $g(0) - G(0)$. If, however, $G(0) = g(0)$, $H(0) = h(0)$, the first derivatives are continuous in the half plane $x + y > 0$.

If the first derivatives are continuous in $x + y > 0$, we have

$$r = f''(x) + h'(x) = g'(x) \quad \text{in} \quad D_1 \cup D_3$$
$$r = F''(x) + H'(x) = G'(x) \quad \text{in} \quad D_2.$$

Hence, if $g'(0) \neq G'(0)$, $r$ is discontinuous across $Oy$ in the half-plane $x + y > 0$. A discontinuity in the data for $p$ produces a discontinuity in $r$ along $Oy$. Similarly a discontinuity in the data for $q$ produces a discontinuity in $t$ along $Ox$.

The argument can be extended by a change of variable to the case when the Cauchy data are given on a curve $y = \phi(x)$, where $\phi(x)$ is strictly monotonic and is such that no characteristic touches $y = \phi(x)$ or cuts it in more than one point. Such a curve is said to be *duly inclined*.

## 4.3    The uniqueness theorem for $u_{xy} = 0$

In § 4.2 we found a solution of $u_{xy} = 0$ satisfying the Cauchy conditions

$$u = f(x), \quad p = g(x), \quad q = h(x)$$

on $x + y = 0$, where $f' = g - h$. If $AB$ is a closed segment of $x + y = 0$ on which $f''$, $g'$, $h'$ are continuous, the solution we found has continuous second derivatives on the square with $AB$ as diagonal. If there were two such solutions $u_1$ and $u_2$, $u = u_1 - u_2$ would also be a solution of $u_{xy} = 0$ satisfying the conditions $u = p = q = 0$ on $AB$. To prove uniqueness, we have to show that $u_1 - u_2$ is identically zero on the square.

Suppose, then, that $u$ is a solution of $u_{xy} = 0$ with zero Cauchy data on the segment $AB$ of $x + y = 0$, and that $u$ has continuous second derivatives on the square with $AB$ as diagonal. Let the characteristics through any point $P(x, y)$ of the square cut $AB$ in $Q(-y, y)$ and $R(x, -x)$. If $D$ is the triangle $PQR$, $\Gamma$ its perimeter, we have, by Green's theorem given in Note 5 of the Appendix,

$$\int_\Gamma (l u_\eta^2 + m u_\xi^2) \, ds = 2 \iint_D (u_\eta u_{\eta\xi} + u_\xi u_{\xi\eta}) \, d\xi \, d\eta = 0,$$

$l$ and $m$ being the direction cosines of the outward normal to $\Gamma$. Therefore

$$\int_{-y}^{x} u_\xi^2(\xi, y) \, d\xi + \int_{-x}^{y} u_\eta^2(x, \eta) \, d\eta = 0$$

3

since the first derivatives of $u$ vanish on $QR$. Hence

$$u_\xi(\xi, y) = 0, (-y < \xi < x); \quad u_\eta(x, y) = 0, (-x < \eta < y).$$

It follows that $p$ and $q$ are zero everywhere on the square, and so $u$ is constant. But $u$ vanishes on $AB$. Hence $u$ is identically zero on the square.

## 4.4 The Cauchy problem for the half-linear equation of hyperbolic type

The Cauchy–Kowalewsky theorem asserts that the half-linear equation

$$s = F(x, y, u, p, q) \tag{1}$$

has a unique solution analytic in a neighbourhood of a duly-inclined arc carrying the analytic Cauchy data. We now turn to a case when the data are not analytic; suppose that on the line $x + y = 0$ we are given that

$$u = f(x), \quad p = g(x), \quad q = h(x),$$

where the derivatives $f''$, $g'$ and $h'$ are continuous, and $f' = g - h$ by the strip condition.

If $F \equiv 0$, so that the equation is $s = 0$, the solution is

$$u_0 = f(x) + \int_{-y}^{x} h(\tau)\, d\tau.$$

If we put $u = u_0 + U$ in (1) we obtain an equation of the same form, but $U, U_x, U_y$ vanish on $x + y = 0$. Thus we need only consider equation (1) in the case when

$$u = 0, \quad p = 0, \quad q = 0$$

on $x + y = 0$. We show that this problem has a unique solution on any square $|x| \leqslant \gamma, |y| \leqslant \gamma$ provided that $F$ is a continuous function of $(x, y, u, p, q)$ when $|x| \leqslant \gamma, |y| \leqslant \gamma, u^2 + p^2 + q^2 \leqslant \delta^2$, where it satisfies a Lipschitz condition,

$$|F(x, y, u_1, p_1, q_1) - F(x, y, u_2, p_2, q_2)|$$
$$\leqslant M\{|u_1 - u_2| + |p_1 - p_2| + |q_1 - q_2|\},$$

$M$ being a constant. Such a condition is certainly satisfied when $F$ is linear in $u, p, q$ and has continuous coefficients.

Let $P(x, y)$ be any point of the square $|\xi| < \gamma, |\eta| < \gamma$ and let $D$ be the triangle bounded by $\xi + \eta = 0, \xi = x, \eta = y$. If $u_{xy}$ is continuous,

$$\iint_D u_{\xi\eta}(\xi, \eta)\, d\xi\, d\eta = \int_{-x}^{y} d\eta \int_{-\eta}^{x} \frac{\partial u_\eta}{\partial \xi}\, d\xi = \int_{-x}^{y} u_\eta(x, \eta)\, d\eta$$

since $u_\eta$ vanishes when $\xi = -\eta$. Integrating again, we have

$$u(x,y) = \iint_D u_{\xi\eta}(\xi,\eta)\,d\xi\,d\eta,$$

and so

$$u(x,y) = \iint_D F(\xi,\eta,u,p,q)\,d\xi\,d\eta, \tag{2}$$

it being understood that $u, p, q$ are functions of the coordinates $(\xi,\eta)$ of the integration point. Thus, if there is a solution with continuous second derivatives, it satisfies the integro-differential equation (2).

Conversely, suppose that (2) has a continuously differentiable solution. Evidently it vanishes on $x+y = 0$. Since the integrand is continuous, we may write the integral as a repeated integral in two ways and differentiate to obtain

$$\left.\begin{aligned}
p(x,y) &= \int_{-x}^{y} F(x,\eta,u(x,\eta),p(x,\eta),q(x,\eta))\,d\eta, \\
q(x,y) &= \int_{-y}^{x} F(\xi,y,u(\xi,y),p(\xi,y),q(\xi,y))\,d\xi,
\end{aligned}\right\} \tag{3}$$

so that $p$ and $q$ vanish on $x+y = 0$. Since the integrands in (3) are also continuous, we see that $p$ and $q$ are also differentiable with respect to $y$ and $x$ respectively, giving

$$\frac{\partial p}{\partial y} = \frac{\partial q}{\partial x} = F(x,y,u,p,q).$$

Therefore $s$ is continuous, and $u$ satisfies the partial differential equation.

The integro-differential equation can be solved by Picard's method of successive approximations, by constructing a sequence of functions $\{u_n(x,y)\}$ connected by the recurrence formula

$$u_{n+1}(x,y) = \iint_D F(\xi,\eta,u_n(\xi,\eta),p_n(\xi,\eta),q_n(\xi,\eta))\,d\xi\,d\eta,$$

the first term $u_0(x,y)$ being any continuously differential function which, together with its first derivatives, vanishes on $x+y = 0$. By induction, all the functions of the sequence are continuously differentiable and

$$p_{n+1}(x,y) = \int_{-x}^{y} F(x,\eta,u_n(x,\eta),p_n(x,\eta),q_n(x,\eta))\,d\eta$$

$$q_{n+1}(x,y) = \int_{-y}^{x} F(\xi,y,u_n(\xi,y),p_n(\xi,y),q_n(\xi,y))\,d\xi.$$

$u_n, p_n, q_n$ all vanish on $x+y = 0$.

We show that $u_n, p_n, q_n$ tend to limits uniformly on $|x| \leqslant \gamma, |y| \leqslant \gamma$ by proving that the series

$$\sum_0^\infty (u_{n+1} - u_n), \quad \sum_0^\infty (p_{n+1} - p_n), \quad \sum_0^\infty (q_{n+1} - q_n),$$

are uniformly and absolutely convergent. We may suppose $x + y \geqslant 0$; for, if not, we have merely to replace $x$ and $y$ by $-x$ and $-y$.

By the Lipschitz condition,

$$|u_{n+1} - u_n| \leqslant M \iint_D (|u_n - u_{n-1}| + |p_n - p_{n-1}| + |q_n - q_{n-1}|) \, d\xi \, d\eta,$$

$$|p_{n+1} - p_n| \leqslant M \int_{-x}^y (|u_n - u_{n-1}| + |p_n - p_{n-1}| + |q_n - q_{n-1}|)_{\xi = x} \, d\eta,$$

with a similar expression for $|q_{n+1} - q_n|$.

Since $u_0$ and $u_1$ are continuously differentiable, they and their first derivatives are bounded on $|x| \leqslant \gamma, |y| \leqslant \gamma$. Hence there exists a constant $K$ such that

$$|u_2(x, y) - u_1(x, y)| \leqslant MK \iint_D d\xi \, d\eta = \tfrac{1}{2} MK(x + y)^2,$$

$$|p_2(x, y) - p_1(x, y)| \leqslant MK \int_{-x}^y d\eta = MK(x + y),$$

$$|q_2(x, y) - q_1(x, y)| \leqslant MK(x + y).$$

Then     $|u_2 - u_1| + |p_2 - p_1| + |q_2 - q_1| \leqslant MK(\gamma + 2)(x + y),$

since $0 \leqslant x + y \leqslant 2\gamma$.

Next

$$|u_3(x, y) - u_2(x, y)| \leqslant M^2 K(\gamma + 2) \iint_D (\xi + \eta) \, d\xi \, d\eta$$

$$= \frac{M^2 K(\gamma + 2)}{3!} (x + y)^3,$$

$$|p_3(x, y) - p_2(x, y)| \leqslant M^2 K(\gamma + 2) \int_{-x}^y (x + \eta) \, d\eta = \frac{M^2 K(\gamma + 2)}{2!} (x + y)^2,$$

$$|q_3(x, y) - q_3(x, y)| \leqslant \frac{M^2 K(\gamma + 2)}{2!} (x + y)^2,$$

and so     $|u_3 - u_2| + |p_3 - p_2| + |q_3 - q_2| \leqslant \dfrac{M^2 K(\gamma + 2)^2}{2!} (x + y)^2.$

By induction,

$$|u_{n+1}-u_n| \leqslant \frac{M^n K(\gamma+2)^{n-1}(x+y)^{n+1}}{(n+1)!},$$

$$|p_{p+1}-p_n| \leqslant \frac{M^n K(\gamma+2)^{n-1}(x+y)^{n}}{n!},$$

$$|q_{n+1}-q_n| \leqslant \frac{M^n K(\gamma+2)^{n-1}(x+y)^{n}}{n!}.$$

By comparison with the exponential series, the series $\Sigma(u_{n+1}-u_n)$ and the series obtained by term-by-term differentiation are uniformly and absolutely convergent on $|x| \leqslant \gamma, |y| \leqslant \gamma$. Hence the integro-differential equation has a continuously differentiable solution. Since $u_{xy} = F(x,y,u,p,q)$, $u_{xy}$ is also continuous.

To prove the solution is unique, we have to show that the limit does not depend on the initial member of the sequence. If we construct two sequences $\{u_n(x,y)\}$ and $\{u_n'(x,y)\}$, we have

$$|u_n(x,y)-u_n'(x,y)| \leqslant \iint_D |F(\xi,\eta,u_n(\xi,\eta),p_n(\xi,\eta),q_n(\xi,\eta))$$

$$- F(\xi,\eta,u_n'(\xi,\eta),p_n'(\xi,\eta),q_n'(\xi,\eta))|\,d\xi d\eta$$

$$\leqslant M \iint_D (|u_n-u_n'|+|p_n-p_n'|+|q_n-q_n'|)\,d\xi d\eta,$$

with similar formulae for $|p_n(x,y)-p_n'(x,y)|$ and $|q_n(x,y)-q_n'(x,y)|$. By induction, we can show that, with some constant $K_1$,

$$|u_n(x,y)-u_n'(x,y)| \leqslant \frac{M^n K_1(\gamma+2)^{n-1}}{(n+1)!}(x+y)^n$$

so that $u_n-u_n'$ tends to zero as $n\to\infty$. Therefore the sequences $\{u_n(x,y)\}$ and $\{u_n'(x,y)\}$ converge to the same limit whatever $u_0(x,y)$ and $u_0'(x,y)$ may be.

Lastly, to show that the problem is well posed, we revert to the original form. The unique solution of $s = F(x,y,u,p,q)$ which satisfies on a segment $AB$ of $x+y = 0$ the Cauchy conditions

$$u = f(x), \quad p = g(x), \quad q = h(x),$$

where $f' = g-h$, satisfies

$$u(x,y) = f(x)+\int_{-y}^{x} h(\tau)\,d\tau+\iint_D F(\xi,\eta,u,p,q)\,d\xi d\eta.$$

Let $u_1$ be another unique solution which satisfies the conditions $u_1 = f_1$, $p_1 = g_1$, $q_1 = h_1$ where $f_1' = g_1 - h_1$. We have to show that if $|f_1 - f| < \epsilon$, $|g_1 - g| < \epsilon$, $|h_1 - h| < \epsilon$, where $\epsilon$ is small, then $u_1 - u$ is small, at any rate near the segment $AB$.

Consider the strip $0 \leqslant x + y \leqslant \delta$ in the square with diagonal $AB$. Let the least upper bound of $|u_1 - u| + |p_1 - p| + |q_1 - q|$ in this strip be $K$. Then

$$|u_1(x,y) - u(x,y)| \leqslant |f_1(x) - f(x)| + \int_{-y}^{x} |h_1(\tau) - h(\tau)|\, d\tau$$
$$+ \iint_D |F(\xi, \eta, u_1, p_1, q_1) - F(\xi, \eta, u, p, q)|\, d\xi\, d\eta$$
$$< \epsilon + \epsilon(x+y) + MK \iint_D d\xi\, d\eta$$
$$\leqslant \epsilon + \epsilon\delta + \tfrac{1}{2}MK\delta^2.$$

Next

$$|p_1(x,y) - p(x,y)| \leqslant |g_1(x) - g(x)| + \int_{-y}^{x} |F(x, \eta, u_1, p_1, q_1)$$
$$- F(x, \eta, u, p, q)|\, d\eta$$
$$< \epsilon + MK \int_{-y}^{x} d\eta \leqslant \epsilon + MK\delta;$$

and similarly for $|q_1 - q|$.

Hence we have, on $0 \leqslant x + y \leqslant \delta$,

$$|u_1 - u| + |p_1 - p| + |q_1 - q| < 3\epsilon + \epsilon\delta + 2MK\delta + \tfrac{1}{2}MK\delta^2$$
$$= 3\epsilon + \epsilon\delta + \tfrac{1}{2}K$$

if we choose $\delta\ (>0)$ so that $M\delta^2 + 4M\delta = 1$. Since the least upper bound of the expression on the left is $K$, we have

$$\tfrac{1}{2}K \leqslant 3\epsilon + \epsilon\delta,$$

and so $u_1 - u$, $p_1 - p$, $q_1 - q$ are all of the order of $\epsilon$ in the strip $0 \leqslant x + y \leqslant \delta$, where $\delta$ does not depend on $\epsilon$.

## 4.5 Two other applications of Picard's method

In the characteristic boundary value problem for

$$s = F(x, y, u, p, q) \tag{1}$$

$u$ is given on two intersecting characteristics which we may take to be the axes. We are given that

$$u = \phi(x)\ (y = 0, x \geqslant 0), \quad u = \psi(y)\ (x = 0, y \geqslant 0),$$

where $\phi(0) = \psi(0)$, and $\phi$ and $\psi$ have continuous second derivatives.

The problem is to show that there is a unique solution in $x \geqslant 0, y \geqslant 0$ which has continuous second derivatives. If $F$ is identically zero, the solution is

$$u_0 = \phi(x) + \psi(y) - \phi(0).$$

If we put $u = u_0 + v$, we find that $v$ satisfies an equation

$$v_{xy} = F_1(x, y, v, v_x, v_y)$$

and vanishes on the positive parts of the axes. Thus it suffices to deal with (1) when $\phi$ and $\psi$ are identically zero. It can be shown that this simplified problem is equivalent to solving the integro-differential equation

$$u(x, y) = \iint_D F(\xi, \eta, u(\xi, \eta), p(\xi, \eta), q(\xi, \eta)) \, d\xi \, d\eta \quad (x > 0, y > 0),$$

where $D$ is now the rectangle with the origin and the point $(x, y)$ as opposite corners. The existence and uniqueness of a solution of this problem can be demonstrated by Picard's method when $F$ satisfies a Lipschitz condition.

The second problem is the problem of Goursat when $u$ is given on a characteristic and on a duly-inclined curve, which we may take to be $y = x$. The data are

$$u = \phi(x) \, (y = 0, x \geqslant 0), \quad u = \psi(x) \, (y = x, x \geqslant 0)$$

when $\phi(0) = \psi(0)$, and $\phi$ and $\psi$ have continuous second derivatives. We wish to prove that the problem has a unique solution in the angle $x \geqslant y \geqslant 0$.

If $F$ is identically zero, the problem has a unique solution

$$u_0 = \phi(x) - \phi(y) + \psi(y).$$

If we put $u = u_0 + v$, we find that $v$ satisfies a half-linear equation of the form (1) and vanishes on $y = 0, x \geqslant 0$ and on $y = x, x \geqslant 0$. Thus it suffices to deal with (1) when $\phi$ and $\psi$ are identically zero. This simplified problem is equivalent to showing that

$$u(x, y) = \iint_D F(\xi, \eta, u(\xi, \eta), p(\xi, \eta), q(\xi, \eta)) \, d\xi,$$

where $D$ is the rectangle with sides $\xi = x, \xi = y, \eta = 0, \eta = y \, (x \geqslant y \geqslant 0)$, has a unique solution with continuous second derivatives. This can be done by Picard's method when $F$ satisfies a Lipschitz condition.

## 4.6   Duly inclined initial lines

A simple example shows why the curve carrying the Cauchy data for a hyperbolic equation with two independent variables must be duly inclined. Consider the Cauchy problem for $u_{xy} = 0$ with data

$$u = f(x), \quad p = g(x), \quad q = h(x)$$

on the parabola $y = x^2$; we assume that $f'', g', h'$ are continuous and that the strip condition $f' = g + 2xh$ is satisfied.

At a point $(x, y)$ in the region $y > x^2$, a unique solution with continuous second derivatives is determined by the Cauchy data on the arc from $(x, x^2)$ to $(\sqrt{y}, y)$. Another such solution is determined by the data on the arc from $(-\sqrt{y}, y)$ to $(x, x^2)$. These two solutions are not necessarily the same, as $f, g, h$ are arbitrary functions which merely have to satisfy the condition that $f'', g', h'$ are continuous. The Cauchy problem in this case does not have a unique solution. The curve $y = x^2$ is not duly inclined; it touches the characteristic $y = 0$ and cuts the characteristic $y = a^2$ in two points $(\pm a, a^2)$.

If $y = \phi(x)$ is a duly inclined curve, the change of variable $x_1 = \phi(x)$ does not alter the form of the equation

$$s = F(x, y, u, p, q),$$

and the Cauchy data, originally carried by $y = \phi(x)$, are now carried by the straight line $y = x_1$.

## 4.7   The equation of wave motions

The equation of wave motions in one dimension

$$\frac{\partial^2 u}{\partial x^2} - \frac{1}{c^2} \frac{\partial^2 u}{\partial t^2} = 0,$$

where $c$ is a constant becomes

$$\frac{\partial^2 u}{\partial x^2} - \frac{\partial^2 u}{\partial y^2} = 0 \qquad (1)$$

if we put $y$ for $ct$. This is a hyperbolic equation with characteristics $x \pm y = \text{constant}$. It is equivalent to the system

$$p_x - q_y = 0, \quad q_x - p_y = 0$$

discussed in §1.7, since the second equation shows that $p\,dx + q\,dy$ is an exact differential $du$.

The general solution of (1) is

$$u = F(x + y) + G(x - y),$$

from which it follows that $p + q$ is constant on each characteristic $x + y = \text{constant}$, $p - q$ is constant on each characteristic $x - y = \text{constant}$.

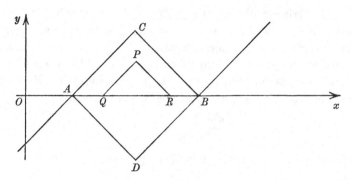

Fig. 2.

The solution of the initial value problem

$$u = f(x), \quad q = h(x)$$

when $y = 0$, where $f'', g'$ are continuous, is

$$u = \tfrac{1}{2}\{f(x+y)+f(x-y)\} + \tfrac{1}{2}\int_{x-y}^{x+y} h(\tau)\,d\tau. \tag{2}$$

We can relax the conditions to continuity of $f$ and piece-wise continuity of $f', f'', h, h'$, that is, these four functions are continuous on closed segments. A discontinuity of $f'$ or $h$ at $x = \alpha$ produces discontinuities of $p$ and $q$ across the characteristics $x \pm y = \alpha$. A discontinuity of $f''$ or $h'$ at $x = \beta$ produces discontinuities of $r$, $s$ and $t$ across $x \pm y = \beta$. The discontinuities in the data are propagated along characteristics.

If $f$ and $g$ are given only along the segment from $A(a, 0)$ to $B(b, 0)$, equation (2) determines the solution only if $x \pm y$ lie between $a$ and $b$. Let $C$ be $(\tfrac{1}{2}(a+b), \tfrac{1}{2}(b-a))$, $D(\tfrac{1}{2}(a+b), \tfrac{1}{2}(a-b))$ as in fig. 2; the characteristics through $A$ and $B$ meet at $C$ and $D$. Then $u$ is determined at $P(x, y)$ by the data on $AB$ if and only if $P$ lies in the square $ADBC$. And we can then assign whatever data we please outside $AB$ without altering the value of $u$ at $P$. The square $ADBC$ is called the *domain of influence* of the data on $AB$.

Let $P$ be any point of the square, and let the characteristics through $P$ cut $Ox$ in $Q$ and $R$. The value of $u$ at $P$ depends only on the data on $QR$. $QR$ is called the *domain of dependence* associated with the point $P$. The ideas of domains of influence and dependence occur in the theory of hyperbolic equations because data on an initial line or curve are propagated along characteristics.

The violin string problem, the problem of determining the small transverse vibrations of a taut uniform string with fixed end points, depends on finding the solution of a mixed initial value and boundary value problem for the equation of wave motion. We are given the initial displacement and velocity of every point of the string. Replacing $ct$ by $y$ in the equation of wave motions, we have to solve

$$\frac{\partial^2 u}{\partial x^2} - \frac{\xi^2 u}{\partial y^2} = 0,$$

given that $\qquad u = f(x), \quad q = h(x) \quad (y = 0),$

where $f$ vanishes when $x = 0$ or $l$, and

$$u = 0 \quad (x = 0, l).$$

We assume that $f''$ and $h'$ are continuous; if we merely assumed piece-wise continuity of $f', f'', h, h'$ the solution would have discontinuities propagated along characteristics.

We could solve this problem by translating the results for the system $\qquad p_x - q_y = 0, \quad q_x - p_y = 0$

obtained in §1.7, but prefer to base the discussion on the use of Green's theorem, preparing for a discussion of Riemann's method.

Let $C$ be a regular closed curve bounding a domain $D$ whose closure is $\bar{D}$. Then if $u, p, q$ are continuous in $\bar{D}$ and $r$ and $t$ are continuous in $D$,

$$\iint_D (u_{xx} - u_{yy})\, dx\, dy = \int_C u_y\, dx + u_x\, dy$$

where $C$ is described in the positive sense. Hence if $u$ satisfies the wave equation

$$\int_C u_y\, dx + u_x\, dy = 0.$$

Take $C$ to be the rectangle $PQRS$ of fig. 3, where $AQ$ is the line $x = l$. If $P$ is $(x, y)$, then $R$ is $(l - x, l + y)$.

On $PQ$ and $RS$, $dy = dx$; on $PS$ and $QR$, $dy = -dx$. Then

$$0 = \int_C u_y\, dx + u_x\, dy$$

$$= \int_{PQ} u_x\, dx + u_y\, dy - \int_{QR} u_x\, dx + u_y\, dy + \int_{RS} u_x\, dx + u_y\, dy$$

$$- \int_{SP} u_x\, dx + u_y\, dy$$

$$= [u]_P^Q - [u]_Q^R + [u]_R^S - [u]_S^P$$

$$= -2u_P - 2u_R,$$

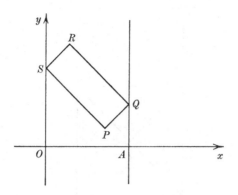

Fig. 3.

since $u$ vanishes at $Q$ and $S$. Therefore $u_R = -u_P$, or

$$u(l-x, l+y) = -u(x, y)$$

If we can find $u$ for $0 \leqslant y \leqslant l$, this formula gives $u$ for $l \leqslant y \leqslant 2l$. Moreover, the equation gives

$$u(x, y+2l) = u(x, y)$$

so that $u$ is periodic in $y$ of period $2l$. It suffices to find $u$ in the square $0 \leqslant x \leqslant l, 0 \leqslant y \leqslant l$. Divide the square into four triangles $D_1, D_2, D_3, D_4$ by the characteristics through $O$ and $A$, as in Fig. 4.

When $P$ is in $D_1$, let the characteristics through $P$ cut $Ox$ in $Q$ and $R$. Then, taking $C$ to be $PQR$,

$$0 = \int_{QR} u_y \, dx - \int_{RP} (u_x \, dx + u_y \, dy) + \int_{PQ} (u_x \, dx + u_y \, dy)$$

$$= \int_{QR} h(x) \, dx + u_R - u_P + u_Q - u_P,$$

so that

$$u_P = \tfrac{1}{2}(u_Q + u_R) + \frac{1}{2} \int_{QR} h(x) \, dx$$

or

$$u(x, y) = \tfrac{1}{2}\{f(x-y) + f(x+y)\} + \frac{1}{2} \int_{x-y}^{x+y} h(\tau) \, d\tau.$$

When $P$ is in $D_4$, take $C$ to be $PQRS$, as in Fig. 5; the inclined lines are characteristics. Then

$$0 = \int_{RS} u_y \, dx - \int_{SP} du + \int_{PQ} du - \int_{QR} du$$

$$= \int_{RS} h(x) \, dx + u_S - 2u_P - u_R,$$

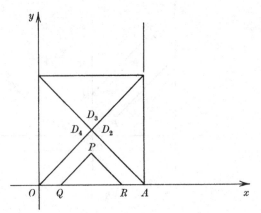

Fig. 4.

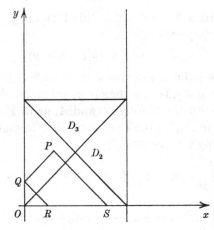

Fig. 5.

so that
$$u_P = \tfrac{1}{2}(u_S - u_R) + \frac{1}{2}\int_{RS} h(x)\,dx$$

or
$$u(x,y) = \tfrac{1}{2}\{f(y+x) - f(y-x)\} + \frac{1}{2}\int_{y-x}^{y+x} h(\tau)\,d\tau.$$

Similarly if $P(x,y)$ is in $D_2$,
$$u(x,y) = \tfrac{1}{2}\{f(x-y) - \tfrac{1}{2}f(2l-x-y)\} + \frac{1}{2}\int_{x-y}^{2l-x-y} h(\tau)\,d\tau.$$

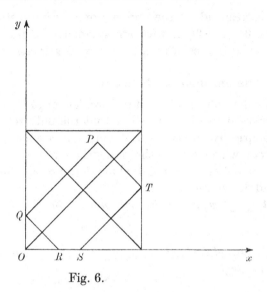

Fig. 6.

Lastly if $P(x,y)$ is in $D_3$, take $C$ to be $PQRST$ of Fig. 6, where the inclined lines are characteristics. Then

$$0 = \int_{RS} u_y\,dx + \int_{ST} du - \int_{TP} du + \int_{PQ} du - \int_{QR} du$$

$$= \int_{RS} h(x)\,dx - u_S - 2u_P - u_R,$$

so that

$$u_P = -\tfrac{1}{2}(u_R + u_S) + \frac{1}{2}\int_{RS} h(x)\,dx$$

or

$$u(x,y) = -\tfrac{1}{2}\{f(y-x) + f(2l-x-y)\} + \frac{1}{2}\int_{y-x}^{2l-x-y} h(\tau)\,d\tau.$$

All these formulae can be put into one rule by extending the range of definition of $f$ and $h$ by the following rules:

(i) $f(x)$ and $g(x)$ are periodic of period $2l$,

(ii) $f(-x) = -f(x), h(-x) = -h(x)$,

(iii) $f(2l-x) = -f(x), h(2l-x) = -h(x)$.

The solution then is, for all $x$ and $y$,

$$u(x,y) = \tfrac{1}{2}\{f(x-y) + f(x+y)\} + \frac{1}{2}\int_{x-y}^{x+y} h(\tau)\,d\tau.$$

This corresponds to the problem of an infinite string stopped at the points $0, \pm l, \pm 2l, \ldots$ with antisymmetry of the initial state about each node; the resulting motion is antisymmetrical about each node.

## 4.8   The uniqueness theorem

It is evident physically that the problem of §4.7 has a unique solution. To prove it, we have to show that the only solution when $f$ and $h$ are identically zero is $u \equiv 0$. We prove here a more general uniqueness theorem which covers the case when the end conditions may be either $u = 0$ or $u_x = 0$ and which does not assume continuity of the second derivatives.

*Let $u(x, y)$, $u_x$ and $u_y$ be continuous in the closed rectangle*

$$\bar{\Delta} : 0 \leqslant x \leqslant l, 0 \leqslant y \leqslant Y.$$

*Let the second derivatives $u_{xx}, u_{xy}, u_{yx}, u_{yy}$ exist and be bounded in the open rectangle*

$$\Delta : 0 < x < l, 0 < y < Y,$$

*and satisfy there*     $u_{xx} = y_{yy}, \quad u_{xy} = u_{yx}.$

*Let $u(x, 0) = u_y(x, 0) = 0$ on $0 \leqslant x \leqslant l$.*
*Let $u_x(x, y) u_y(x, y) = 0$ for $x = 0$ and $x = l, 0 \leqslant y \leqslant Y$.*
*Then $u$ is identically zero in $\bar{\Delta}$.*

We need two lemmas:

(i) *If $f(x)$ has a bounded derivative $f'(x)$ at each point of a closed interval $a \leqslant x \leqslant b$, then $f'(x)$ is integrable in Lebesgue's sense and*

$$\int_a^b f'(x)\, dx = f(b) - f(a).$$

(ii) *If $f(x, t)$ is integrable in Lebesgue's sense over $a \leqslant x \leqslant b$ for each fixed $t$ in $\alpha \leqslant t \leqslant \beta$, and if $f_t(x, t)$ exists and is bounded in $a \leqslant x \leqslant b$, $\alpha \leqslant t \leqslant \beta$, then*

$$\frac{d}{dt} \int_a^b f(x, t)\, dx = \int_a^b f_t(x, t)\, dx.$$

Consider     $I(a, b, y) = \displaystyle\int_a^b (u_x^2 + u_y^2)\, dx,$

where $a$ and $b$ are constants such that $0 < a < b < l$. This integral is a continuous function of $a, b, y$ and tends to

$$J(y) = \int_0^l (u_x^2 + u_y^2)\, dx.$$

when $a \to +0, b \to l - 0$. $J(y)$ is a continuous function of $y$ in $0 \leqslant y \leqslant Y$, and vanishes when $y = 0$.

By (ii),

$$\frac{d}{dy} I(a, b, y) = \int_a^b \frac{\partial}{\partial y} (u_x^2 + u_y^2) \, dx,$$

since

$$\frac{\partial}{\partial y} (u_x^2 + u_y^2) = 2 u_x u_{xy} + 2 u_y u_{yy}$$

is bounded in

$$0 < \alpha \leqslant y \leqslant \beta < Y, \quad 0 < a \leqslant x \leqslant b < l.$$

But since $u_{xx} = u_{yy}$, $u_{xy} = u_{yx}$

$$\frac{\partial}{\partial y} (u_x^2 + u_y^2) = 2 u_x u_{yx} + 2 u_{xx} u_y = 2 \frac{\partial}{\partial x} (u_x u_y).$$

Hence, by (i)

$$\frac{d}{dy} I(a, b, y) = \int_a^b \frac{\partial}{\partial y} (u_x^2 + u_y^2) \, dx = 2 \int_a^b \frac{\partial}{\partial x} (u_x u_y) \, dx$$

$$= 2 u_x(b, y) \, u_y(b, y) - 2 u_x(a, y) \, u_y(a, y).$$

But $u_x(x, y) \, u_y(x, y)$ is uniformly continuous in $\bar{\Delta}$ and so tends to zero uniformly with respect to $y$ as $x \to +0$ or $x \to l - 0$. Hence

$$\frac{d}{dy} I(a, b, y) \to 0$$

as $a \to +0, b \to l - 0$, uniformly with respect to $y$ in $0 < \alpha \leqslant y \leqslant \beta < Y$. It follows that

$$\frac{d}{dy} J(y) = 0$$

for $0 < y < Y$, and so $J(y)$ is a constant. But $J(y)$ is continuous on $0 \leqslant y \leqslant Y$ and vanishes when $y = 0$. Hence

$$\int_0^l (u_x^2 + u_y^2) \, dx = 0$$

for $0 \leqslant y \leqslant Y$. This implies that $u_x$ and $u_y$ vanish on $\bar{\Delta}$, and so $u$ is constant on $\bar{\Delta}$. But, by hypothesis, $u = 0$ for $0 \leqslant x \leqslant l, y = 0$. Hence $u$ is identically zero on $\bar{\Delta}$.

## 4.9 The use of Fourier series

The violin string problem of §4.7 can also be solved by the method of separation of variables, the solution being in the form of Fourier series.

If we put $u = XY$, where $X$ is a function of $x$ alone, $Y$ a function of $y$ alone, in $u_{xx} - u_{yy} = 0$, we get

$$\frac{X''}{X} = \frac{Y''}{Y},$$

dashes denoting differentiation with respect to the relevant variable. Hence

$$X'' + k^2 X = 0, \quad Y'' + k^2 Y = 0,$$

when $k^2$ is the separation constant. This gives a particular solution

$$u = (A \cos kx + B \sin kx)(a \cos ky + b \sin ky)$$

if $k \neq 0$. If $u$ vanishes when $x = 0$ and when $x = l$ for all values of $y$, $A$ is zero and $k = n\pi/l$ where $n$ is a positive integer. If $k = 0$ the solution is

$$u = (A + Bx)(a + by)$$

and the end conditions give $A = B = 0$, so that $u \equiv 0$.

As the equation is linear, we get a formal solution

$$u = \sum_1^\infty \left( a_n \cos \frac{n\pi y}{l} + b_n \sin \frac{n\pi y}{l} \right) \sin \frac{n\pi x}{l}. \tag{1}$$

If the initial conditions are $u = f(x), u_y = g(x) \ (0 \leqslant x \leqslant l)$, we should expect that

$$f(x) = \sum_1^\infty a_n \sin \frac{n\pi x}{l}, \quad h(x) = \sum_1^\infty \frac{n\pi}{l} b_n \sin \frac{n\pi x}{l}, \tag{2}$$

where $a_n$ and $b_n$ are given by Fourier's rule

$$a_n = \frac{2}{l} \int_0^l f(t) \sin \frac{n\pi t}{l} \, dt, \quad b_n = \frac{2}{n\pi} \int_0^l h(t) \sin \frac{n\pi t}{l} \, dt.$$

There are no simple conditions to ensure that the series (1) and (2) are convergent. This is purely formal; but with special initial data, it may happen that the series (1) is convergent and gives the desired solution.

## 4.10   The equation of telegraphy

The equation satisfied by the potential $V$ in the propagation of signals along a uniform telegraph cable with self-inductance $L$, resistance $R$ and capacity $C$ per unit length is

$$LC \frac{\partial^2 V}{\partial t^2} + RC \frac{\partial V}{\partial t} = \frac{\partial^2 V}{\partial x^2}.$$

We may take $LC = 1$, $RC = 2$ by changing the units of $x$ and $t$. So we consider

$$\frac{\partial^2 V}{\partial x^2} - \frac{\partial^2 V}{\partial y^2} - 2\frac{\partial V}{\partial y} = 0.$$

If we put $V = ue^{-y}$, we get

$$\frac{\partial^2 u}{\partial x^2} - \frac{\partial^2 u}{\partial y^2} + u = 0.$$

We treat this equation in an unrigorous manner as an example of the Fourier integral method of solving the problem of Cauchy. We want the solution which satisfies for all $x$ the conditions $u = f(x), u_y = h(x)$ when $y = 0$.

If we represent $u$ by a complex Fourier integral

$$u(x,y) = \frac{1}{\sqrt{(2\pi)}}\int_{-\infty}^{\infty} U(\xi,y)\,e^{ix\xi}\,d\xi,$$

the differential equation becomes

$$\int_{-\infty}^{\infty}\left\{\frac{\partial^2 U}{\partial y^2} + (\xi^2 - 1)\,U\right\}e^{ix\xi}\,d\xi = 0,$$

which can be satisfied by

$$U = F(\xi)\cos\{y\sqrt{(\xi^2-1)}\} + H(\xi)\frac{\sin\{y\sqrt{(\xi^2-1)}\}}{\sqrt{(\xi^2-1)}},$$

where $F(\xi)$ and $H(\xi)$ are arbitrary functions.

This gives the formal solution

$$u(x,y) = \frac{1}{\sqrt{(2\pi)}}\int_{-\infty}^{\infty}\left[F(\xi)\cos\{y\sqrt{(\xi^2-1)}\}\right.$$
$$\left. + \frac{H(\xi)}{\sqrt{(\xi^2-1)}}\sin\{y\sqrt{(\xi^2-1)}\}\right]e^{ix\xi}\,d\xi,$$

where $F$ and $G$ have to be determined from the initial conditions, which give

$$f(x) = \frac{1}{\sqrt{(2\pi)}}\int_{-\infty}^{\infty} F(\xi)\,e^{ix\xi}\,d\xi,$$

$$h(x) = \frac{1}{\sqrt{(2\pi)}}\int_{-\infty}^{\infty} H(\xi)\,e^{ix\xi}\,d\xi.$$

Using the inversion formula, we get

$$F(\xi) = \frac{1}{\sqrt{(2\pi)}}\int_{-\infty}^{\infty} e^{-i\xi\tau}f(\tau)\,d\tau,$$

$$H(\xi) = \frac{1}{\sqrt{(2\pi)}}\int_{-\infty}^{\infty} e^{-i\xi\tau}h(\tau)\,d\tau.$$

Hence

$$u(x, y) = \frac{1}{2\pi} \iint_{-\infty}^{\infty} e^{i\xi(x-\tau)} \cos\{y\sqrt{(\xi^2-1)}\} f(\tau) \, d\xi \, d\tau$$

$$+ \frac{1}{2\pi} \iint_{-\infty}^{\infty} e^{i\xi(x-\tau)} \sin\frac{\{y\sqrt{(\xi^2-1)}\}}{\sqrt{(\xi^2-1)}} h(\tau) \, d\xi \, d\tau$$

$$= \frac{1}{2\pi} \int_{-\infty}^{\infty} Q(x-\tau, y) h(\tau) \, d\tau + \frac{1}{2\pi} \frac{\partial}{\partial y} \int_{-\infty}^{\infty} Q(x-\tau, y) f(\tau) \, d\tau,$$

where

$$Q(x, y) = \int_{-\infty}^{\infty} e^{i x \xi} \frac{\sin\{y\sqrt{(\xi^2-1)}\}}{\sqrt{(\xi^2-1)}} \, d\xi.$$

It can be shown that, if $y > 0$, $Q(x, y) = \pi I_0\{\sqrt{(y^2-x^2)}\}$ when $x$ lies between $\pm y$, and is zero elsewhere. $I_0$ denotes the Bessel function of imaginary argument and order zero. In this way, we get formally the solution

$$u(x, y) = \frac{1}{2} \int_{x-y}^{x+y} I_0[\sqrt{(y^2-(x-\tau)^2)}] h(\tau) \, d\tau$$

$$+ \frac{1}{2} \frac{\partial}{\partial y} \int_{x-y}^{x+y} I_0[\sqrt{(y^2-(x-\tau)^2)}] f(\tau) \, d\tau.$$

We obtain this solution by other methods in the next chapter.

### Exercises

**1.** If $u(x, y)$ satisfies $u_{xy} = 0$, prove that

$$\int_C u_\xi(\xi, \eta) \, d\xi$$

taken round any simple closed curve is zero. Deduce that, if $u = f(x)$, $\partial u/\partial x = g(x)$ when $x+y = 0$, then

$$u(x, y) = f(-y) + \int_{-y}^{x} g(\xi) \, d\xi.$$

**2.** If $u(x, t)$ satisfies the wave equation

$$\frac{\partial^2 u}{\partial x^2} - \frac{1}{c^2} \frac{\partial^2 u}{\partial t^2} = 0,$$

where $c$ is a constant, show that

$$\int_C \{u_\tau(\xi, \tau) \, d\xi + c^2 u_\xi(\xi, \tau) \, d\tau\}$$

round any simple closed curve is zero. Deduce that the solution which satisfies the conditions $u = f(x)$, $u_t = g(x)$ when $t = 0$ is

$$u(x, t) = \tfrac{1}{2}f(x+ct) + \tfrac{1}{2}f(x-ct) + \frac{1}{2c}\int_{x-ct}^{x+ct} g(\xi)\, d\xi.$$

Find the particular solution when $f(x) = \alpha \sin x$, $g(x) = \beta \sin x$, where $\alpha$ and $\beta$ are constants.

3. Find the solution of $\quad \dfrac{\partial^2 u}{\partial x^2} - \dfrac{\partial^2 u}{\partial y^2} = 0$

given that $u = 3x^2$, $u_y = 2x$ when $y = 0$.

4. Find the solution of $\quad \dfrac{\partial^2 u}{\partial x^2} - \dfrac{\partial^2 u}{\partial y^2} = 0$

in $x \geqslant 0, y \geqslant 0$, given that $u = 3x^2, \partial u/\partial y = 2x$ when $y = 0, x \geqslant 0$, and $u = 0$ when $x = 0, y \geqslant 0$. Examine the continuity of $u$ and its derivatives.

5. Find the solution of $\quad \dfrac{\partial^2 u}{\partial x^2} - \dfrac{\partial^2 u}{\partial y^2} = 0$

in the half-plane $y > 0$ given that, on $y = 0$, $u = \sin \pi x$ when $0 \leqslant x \leqslant 1$, $u = 0$ when $x > 1$ and when $x < 0$, and $u_y = 0$ for all $x$. Solve the same equation given that, on $y = 0$, $u = 0$ for all $x$, $u_y = \sin \pi x$ when $0 \leqslant x \leqslant 1$, $u_y = 0$ when $x > 1$ and when $x \leqslant 0$.

6. Find the solution in $x > 0, y > 0$ of

$$\frac{\partial^2 u}{\partial x^2} - \frac{\partial^2 u}{\partial y^2} = 0$$

given that $u = 0$ on the axes and that, on $y = 0$, $u_y = x$ when $0 \leqslant x \leqslant 1$ and $u_y = 1$ when $x > 1$. Examine the continuity of $u$ and its derivatives.

7. Find by using Fourier series the solution in $-l \leqslant x \leqslant l, t \geqslant 0$ of

$$\frac{\partial^2 u}{\partial x^2} - \frac{1}{c^2}\frac{\partial^2 u}{\partial t^2} = 0$$

given that, when $t = 0$, $\quad u = \dfrac{h}{l}(l - |x|)$, $\quad u_t = 0$,

and that $u = 0$ for all $t$ when $x = \pm l$.

8. Show that the solution of the equation of telegraphy

$$\frac{\partial^2 u}{\partial x^2} - \frac{\partial^2 u}{\partial t^2} + u = 0$$

is $\quad u = \displaystyle\int_{-\infty}^{\infty} e^{iqx}\left[\theta(q)\cos\{t\sqrt{(q^2-1)}\} + \theta_1(q)\frac{\sin\{t\sqrt{(q^2-1)}\}}{\sqrt{(q^2-1)}}\right] dq$

if $$u = \int_{-\infty}^{\infty} \theta(q)\,e^{iqx}\,dq, \quad u_t = \int_{-\infty}^{\infty} \theta_1(q)\,e^{iqx}\,dq$$

when $t = 0$.

**9.** $u(x, y)$ is the solution of $u_{xx} - u_{yy} = 0$ with continuous first and second derivatives in $x \geqslant 0, y \geqslant 0$. It satisfies the Cauchy conditions $u = f(x)$, $u_y = g(x)$ on $y = 0, x \geqslant 0$ and $u = F(y)$, $u_x = H(y)$ on $x = 0, y \geqslant 0$, where $f(0) = F(0), g(0) = F'(0), f'(0) = H(0)$. Since the boundary is not duly inclined, the data are not independent. Prove that

$$f'(x) + g(x) = F'(x) + H(x),$$

when $x \geqslant 0$.

# 5

# RIEMANN'S METHOD

## 5.1 Adjoint linear operators

To a linear operator $L$, defined by

$$L(u) = au_{xx} + 2hu_{xy} + bu_{yy} + 2gu_x + 2fu_y + cu, \tag{1}$$

where the coefficients are continuously differentiable functions of $x$ and $y$ alone, there corresponds a unique linear operator $L^*$, called the *adjoint* of $L$, such that $vL(u) - uL^*(v)$ is a divergence

$$\partial H/\partial x + \partial K/\partial y.$$

If $D$ is a domain whose boundary is a regular closed curve $C$, it follows by Green's theorem that

$$\iint_D \{vL(u) - uL^*(v)\}\,dx\,dy = \int_C (lH + mK)\,ds,$$

where $(l, m)$ are the direction cosines of the outward normal to $C$.

Since $L^*$ does not depend on the particular domain $D$, we may find it by considering the particular case when $D$ is a rectangle with sides $x = x_0, x = x_1, y = y_0, y = y_1$. By integration by parts,

$$\iint_D vau_{xx}\,dx\,dy = \int_{y_0}^{y_1} [avu_x]_{x_0}^{x_1}\,dy - \iint_D (av)_x u_x\,dx\,dy$$

$$= \int_{y_0}^{y_1} [avu_x - (av)_x u]_{x_0}^{x_1}\,dy + \iint_D u(av)_{xx}\,dx\,dy$$

so that

$$\iint_D [vau_{xx} - u(av)_{xx}]\,dx\,dy = \int_{y_0}^{y_1} [avu_x - (av)_x u]_{x_0}^{x_1}\,dy.$$

Similarly

$$\iint_D [vhu_{xy} - u(hv)_{xy}]\,dx\,dy = \int_{y_0}^{y_1} [huv_y]_{x_0}^{x_1}\,dy - \int_{x_0}^{x_1} [(hv)_x u]_{y_0}^{y_1}\,dx$$

$$= \int_{x_0}^{x_1} [hvu_x]_{y_0}^{y_1}\,dx - \int_{y_0}^{y_1} [(hv)_y u]_{x_0}^{x_1}\,dy,$$

$$\iint_D [vbu_{yz} - u(bv)_{yy}]\,dx\,dy = \int_{x_0}^{x_1} [bvu_y - (bv)_y u]_{y_0}^{y_1}\,dx,$$

$$\iint_D [gvu_x + u(gv)_x]\,dx\,dy = \int_{y_0}^{y_1} [gvu]_{x_0}^{x_1}\,dy,$$

$$\iint_D [fvu_y + u(fv)_y]\,dx\,dy = \int_{x_0}^{x_1} [fvu]_{y_0}^{y_1}\,dx.$$

It follows that

$$L^*(v) = (av)_{xx} + 2(hv)_{xy} + (bv)_{yy} - 2(gv)_x - 2(fv)_y + cv, \qquad (2)$$

and that $\qquad H = avu_x - u(av)_x + hvu_y - u(hv)_y + 2guv, \qquad (3)$

$$K = hvu_x - (hv)_x + bvu_y - u(bv)_y + 2fuv. \qquad (4)$$

$H$ and $K$ are not unique; we can add to them $\partial\theta/\partial y$ and $-\partial\theta/\partial x$ respectively; but $L^*$ is unique. And the adjoint of $L^*$ is $L$.

$L$ is said to be *self-adjoint* if $L^* = L$. The condition for this is

$$a_x + h_y = 2g, \quad h_x + b_y = 2f$$

so that a self-adjoint operator can be written in the form

$$\frac{\partial}{\partial x}(au_x) + \frac{\partial}{\partial x}(hu_y) + \frac{\partial}{\partial y}(hu_x) + \frac{\partial}{\partial y}(bu_y) + cu.$$

## 5.2 Riemann's method

Riemann's method of solving the problem of Cauchy for a linear equation $L(u) = F(x,y)$ of hyperbolic type first appeared in his memoir† on the propagation of sound waves of finite amplitude. It depends on finding a particular solution, known as the Riemann–Green function, of the characteristic boundary value problem.

Suppose that we use characteristic variables. Then we have to solve $\qquad L(u) \equiv 2u_{xy} + 2gu_x + 2fu_y + cu = F(x,y)$

given $u, u_x, u_y$ on some duly-inclined regular arc $C$. The data satisfy the strip condition $du = u_x\,dx + u_y\,dy$ on $C$. The functions $g, f, c, F$ are assumed to be continuously differentiable functions of $x$ and $y$ alone.

Let the characteristics $y = y_0$ and $x = x_0$ through $P(x_0, y_0)$ cut $C$ in $Q$ and $R$ respectively. Let $D$ be the domain bounded by $PQ$, $PR$ and $C$; let $\Gamma$ be its boundary. Then, if $L(u) = F$, $L^*(v) = 0$, we have

$$\iint_D vF\,dx\,dy = \iint_D \{vL(u) - uL^*(v)\}\,dx\,dy = \iint_D \left(\frac{\partial H}{\partial x} + \frac{\partial K}{\partial y}\right)dx\,dy$$

$$= \int_\Gamma (-K\,dx + H\,dy)$$

† Gött. Abh. **8** (1860). p, 43. This paper will be found in Riemann's *Gesammelte Mathematische Werke* (Dover Press reprint, New York, 1953), pp. 156–78.

by Green's theorem, assuming that $u$ and $v$ have continuous second derivatives. Using equations (3) and (4) of §5.1, the integral round $\Gamma$ is equal to

$$-\int_{PQ}\{(vu)_x - 2u(v_x - fv)\}\,dx$$

$$+\int_{QR}(-K\,du + H\,dy) + \int_{RP}\{(vu)_y - 2u(v_y - gv)\}\,dy$$

$$= 2(uv)_P - (uv)_Q - (uv)_R + \int_{QR}(-K\,dx + H\,dy)$$

provided that $\quad v_x - fv = 0 \quad$ when $\quad y = y_0$,

$$v_y - gv = 0 \quad \text{when} \quad x = x_0,$$

that is, provided that

$$v(x, y_0) = \exp\int_{x_0}^{x} f(\xi, y_0)\,d\xi,$$

$$v(x_0, y) = \exp\int_{y_0}^{y} g(x_0, \eta)\,d\eta,$$

where we have taken $v(x_0, y_0)$ to be unity, as we may without loss of generality.

Hence $v$ is given by a characteristic boundary value problem for $L^*(v) = 0$. This problem has a solution, the Riemann–Green function, which we denote by $v(x, y; x_0, y_0)$. If $Q$ is $(x_1, y_0)$, and $R$ is $(x_0, y_1)$, it follows that

$$u(x_0, y_0) = \tfrac{1}{2}u(x_1, y_0)\,v(x_1, y_0; x_0, y_0) + \tfrac{1}{2}u(x_0, y_1)\,v(x_0, y_1; x_0, y_0)$$

$$+\frac{1}{2}\int_{QR}(K\,dx - H\,dy) + \frac{1}{2}\iint_{D} v(x, y; x_0, y_0)\,F(x, y)\,dx\,dy.$$

This is the required solution, since

$$H = vu_y - uv_y + 2guv, \quad K = uv_x - uv_x + 2fuv$$

are given on $QR$ by the Cauchy data.

If $v^*(x, y; x_1, y_1)$ is the Riemann–Green function for the adjoint equation $L^*(v) = 0$,

$$v^*(x_0, y_0; x_1, y_1) = v(x_1, y_1; x_0, y_0).$$

In particular if $L$ is self-adjoint.

$$v(x_0, y_0; x_1, y_1) = v(x, y_{11}; x_0, y_0).$$

To prove this symmetry property, let $\Gamma$ be the rectangle $P_0QP_1R$ with opposite corners $P_0(x_0, y_0)$, $P_1(x_1, y_1)$. Then

$$\int_\Gamma (vv_y^* - v^*v_y + 2gvv^*)\,dy - (vv_x^* - v^*v_x + 2fvv^*)\,dx = 0.$$

On $y = y_0, v_x = fv$; on $x = x_0, v_y = gv$. On $y = y_1, v_x^* = -fv^*$; on $x = x_1, v_y^* = -gv^*$. The contribution of the line from $P_0$ to $Q$ $(x_1, y_0)$ is

$$-\int_{x_0}^{x_1} (vv_x^* + v^*v_x)\,dx = -[vv^*]_{P_0}^Q = v_{P_0}^* - (vv^*)_Q.$$

Similarly the contributions of $QP_1, P_1R, RP_0$ are respectively

$$(vv^*)_Q - v_{P_1}, \quad (vv^*)_R - v_{P_1}, \quad v_{P_0}^* - (vv^*)_R.$$

Hence the value of the integral round $\Gamma$ is $2v_{P_0}^* - 2v_{P_1}$, and so

$$v_{P_0}^* = v_{P_1},$$

which was to be proved.

## 5.1    Another form of Riemann's method

Riemann's method also applies when the equation is given in the form

$$L(u) \equiv u_{xx} - u_{yy} + 2gu_x + 2fu_y + cu = F(x, y)$$

with characteristics $x \pm y = $ constant. The adjoint operator is

$$L^*(v) \equiv v_{xx} - v_{yy} - 2gv_x - 2fv_y + (c - g_x - f_y)\,v.$$

Then
$$vL(u) - uL^*(v) = \frac{\partial H}{\partial x} + \frac{\partial K}{\partial y},$$

where
$$H = vu_x - uv_x + 2guv, \quad K = -vu_y + uv_y + 2fuv.$$

If $D$ is a domain bounded by a regular closed curve $\Gamma$,

$$\iint_D vF\,dx\,dy = \int_\Gamma (-K\,dx + H\,dy).$$

Suppose that the Cauchy data are carried by a duly-inclined regular arc $C$. Let the characteristic $y - x = y_0 - x_0$ through $P_0(x_0, y_0)$ cut $C$ in $Q$; let $y + x = y_0 + x_0$ cut $C$ in $R$. Take $\Gamma$ to be the segment $P_0Q$, the arc $QR$ and the segment $RP_0$.

Since $dy = dx$ on $P_0Q$

$$\int_{P_0Q} (-K\,dx + H\,dy) = \int_{P_0Q} H\,dx - K\,dy$$

$$= \int_{P_0Q} v\,du - u\,dv + 2guv\,dx - 2fuv\,dy$$

$$= \int_{P_0Q} v\,du + u\,dv - 2\int_{P_0Q} u(v_x + y_y + fv - gv)\,dx$$

$$= (uv)_Q - (uv)_{P_0},$$

provided that $\qquad\qquad v_x + v_y = (g-f)\,v$

on $y = y_0$.

Again, on $RP_0$, $dx = -dy$. Hence

$$\int_{RP_0} -K\,dx + H\,dy = -\int_{RP_0} H\,dx - K\,dy$$

$$= -\int_{RP_0} v\,du - u\,dv + 2guv\,dx - 2fuv\,dy$$

$$= -\int_{RP_0} v\,du + u\,dv + 2\int_{RP_0} u(v_x - v_y - fv - gv)\,dx$$

$$= (uv)_R - (uv)_{P_0},$$

provided that $\qquad\qquad v_x - v_y = (g+f)\,v$

on $x + y = x_0 + y_0$.

We again have a characteristic boundary value problem, which determines a Riemann–Green function $v(x, y; x_0, y_0)$, the constant of integration being chosen so that $v(x_0, y_0; x_0, y_0) = 1$. The resulting solution is

$$u(x_0, y_0) = \tfrac{1}{2}(uv)_Q + \tfrac{1}{2}(uv)_R + \frac{1}{2}\int_{QR} (-K\,dx + H\,dy)$$

$$- \frac{1}{2}\iint_D v(x, y; x_0, y_0)\,F(x, y)\,dx\,dy.$$

## 5.4   Determination of the Riemann–Green function

The difficulty in Riemann's solution is the determination of the Riemann–Green function. The method replaces a Cauchy problem by a characteristic boundary value problem.

It suffices to consider the case when the independent variables are characteristic variables, so that the linear operator is

$$L(u) = 2u_{xy} + 2gu_x + 2fu_y + c.$$

If we make the change of dependent variable $u = \phi U$, and divide through by $\phi$, we get a linear operator

$$M(U) = 2U_{xy} + 2GU_x + 2FU_y + CU,$$

where

$$G = g + \frac{\phi_y}{\phi}, \quad F = f + \frac{\phi_x}{x}, \quad C = c + 2g\frac{\phi_x}{\phi} + 2f\frac{\phi_y}{\phi} + \frac{\phi_{xy}}{\phi}.$$

The Riemann–Green function for $L(u)$, $v(x, y; x_0, y_0)$, satisfies the adjoint equation $L^*(u) = 0$ and the conditions

$$v_x = fv \quad \text{on} \quad y = y_0, \quad v_x = gv \quad \text{on} \quad x = x_0,$$

and

$$v(x_0, y_0; x_0, y_0) = 1.$$

It easily follows that the Riemann–Green function for $M(U) = 0$ is

$$V(x, y; x_0, y_0) = \frac{\phi(x, y)}{\phi(x_0, y_0)} \, v(x, y; x_0, y_0).$$

For example, if we put $u = (x + y) U$ in $u_{xy} = 0$, for which the Riemann–Green function is constant, we get

$$U_{xy} + \frac{U_x + U_y}{x + y} = 0$$

for which the Riemann–Green function is therefore

$$V(x, y; x_0, y_0) = \frac{x + y}{x_0 + y_0}.$$

The change of dependent variable is useful if we can choose $\phi$ so that $M(U) = 0$ is self-adjoint. This occurs if

$$f = -\frac{\partial}{\partial x} \log \phi, \quad g = -\frac{\partial}{\partial y} \log \phi;$$

such a transformation is thus possible when $g_x = f_y$.

The Riemann–Green function is the solution of a characteristic boundary value problem, and does not depend in any way on the arc carrying the Cauchy data. If it is possible to solve by some other method the problem of Cauchy with a simple curve carrying the data, a comparison of this solution and the Riemann solution should give the Riemann–Green function. In the case of the two equations discussed by Riemann, it was possible to do this; he solved the problem of Cauchy with data on a straight line by using Fourier cosine transforms.

The method given below is suggested by Hadamard's observation in his *Lectures on Cauchy's Problem* that the Riemann–Green function is the coefficient of the logarithmic term in his elementary solution.

## 5.5   A series formula for the Riemann–Green function

The Riemann–Green function for

$$L(u) = u_{xy} + gu_x + fu_y + cu = 0$$

satisfies      $L^*(v) = v_{xy} - gv_x - fv_y + (c - g_x - f_y)\,v = 0$

under the conditions

$$v_x = fv \quad \text{on} \quad y = y_0, \quad v_y = gv \quad \text{on} \quad x = x_0,$$

and              $v(x_0, y_0; x_0, y_0) = 1.$

Following Hadamard, we try

$$v = \sum_0^\infty \frac{v_j \Gamma^j}{j!\,j!}, \tag{1}$$

where $v_j$ are functions of $x$ and $y$ to be determined and

$$\Gamma = (x - x_0)\,(y - y_0).$$

Assuming that it is legitimate to differentiate term-by-term, we have

$$L^*(v) = \sum_0^\infty \frac{\Gamma^j}{j!\,j!} L^*(v_j) + \sum_1^\infty \frac{\Gamma^{j-1}}{j!\,(j-1)!} \left\{ (x - x_0) \frac{\partial v_j}{\partial x} + (y - y_0) \frac{\partial v_j}{\partial y} \right\}$$

$$- \sum_1^\infty \frac{\Gamma^{j-1}}{j!\,(j-1)!} \{(x - x_0)f + (y - y_0)\,g\}\,v_j + \sum_1^\infty \frac{\Gamma^{j-1}}{(i-1)!\,(j-1)!}\,v_j.$$

Since $L^*(v) = 0$, equating to zero the coefficients of powers of $\Gamma$, we obtain

$$jL^*(v_{j-1}) + jv_j + (x - x_0) \frac{\partial v_j}{\partial x} + (y - y_0) \frac{\partial v_j}{\partial y} - f(x - x_0)\,v_j - g(y - y_0)\,v_j = 0 \tag{2}$$

for $j = 1, 2, \ldots$. The coefficient $v_0$ is at our disposal; once it is fixed, equation (2) gives successively the coefficients $v_1, v_2, \ldots$.

On the characteristic $y = y_0$, $v = v_0$ and $\partial v/\partial x = \partial v_0/\partial x$. Hence

$$\frac{\partial}{\partial x}\,v_0(x, y_0) = f(x, y_0)\,v_0(x, y_0).$$

If we take $v_0 = 1$ at $(x_0, y_0)$, we get

$$\log v_0(x, y_0) = \int_{x_0}^x f(\xi, y_0)\,d\xi. \tag{3}$$

Similarly
$$\log v_0(x_0, y) = \int_{y_0}^{y} g(x_0, \eta)\, d\eta. \tag{4}$$

Thus we know $v_0(x, y)$ on the two characteristics $x = x_0, y = y_0$ and have to assign $v_0$ elsewhere. There are many ways of doing this; each gives rise to a different formula for the Riemann–Green function.

We follow Hadamard and define
$$\log v_0(x, y) = \int_{(x_0, y_0)}^{(x, y)} f(\xi, \eta)\, d\xi + g(\xi, \eta)\, d\eta, \tag{5}$$

where integration is along the straight line from $(x_0, y_0)$ to $(x, y)$. This function reduces to that given in (3) on the characteristic $y = y_0$ and to that in (4) on $x = x_0$. The integral in (5) would be changed if we altered the path, since $f(\xi, \eta)\, d\xi + g(\xi, \eta)\, d\eta$ is not necessarily an exact differential.

If $x = x_0 + r\cos\theta, y = y_0 + r\sin\theta$, we have
$$\xi = x_0 + s\cos\theta, \quad \eta = y_0 + s\sin\theta$$

where $\theta$ is constant and $s$ varies from 0 to $r$. Equation (5) then becomes
$$\log v_0(x, y) = \int_0^r \{(\xi - x_0) f(\xi, \eta) + (\eta - y_0) g(\xi, \eta)\} \frac{ds}{s}. \tag{6}$$

If we denote the value of $v_0$ given by (6) by $\Omega$, we get
$$\frac{\partial \Omega}{\partial r} = \frac{\Omega}{r} \{(x - x_0) f(x, y) + (y - y_0) g(x, y)\}.$$

Using polar coordinates, (2) becomes
$$r \frac{\partial v_j}{\partial r} - \frac{r}{\Omega} \frac{\partial \Omega}{\partial r} v_j + j v_j = -j L^*(v_{j-1}),$$

or
$$\frac{\partial}{\partial r}\left(r^j \frac{v_j}{\Omega}\right) = -\frac{j r^{j-1}}{\Omega} L^*(v_{j-1}).$$

Hence since $v_j$ is finite when $r = 0$,
$$v_j = -\frac{j\Omega}{r^j} \int_0^r \frac{s^{j-1}}{\Omega} L^*(v_{j-1})\, ds, \tag{7}$$

where, in $L^*(v_{j-1}), \xi = x_0 + s\cos\theta, \eta = y_0 + s\sin\theta$.

Equation (7) determines successively all the coefficients $v_j$. The convergence of the resulting series can be proved by the method of dominant functions, as Hadamard does for his elementary solution. But it may not be possible to carry out the integrations. A knowledge of the first few terms sometimes suggests the form of the Riemann–Green function which can then be found by elementary methods.

## 5.6 The equation of telegraphy

The equation of telegraphy

$$L(u) \equiv \frac{\partial^2 u}{\partial x^2} - \frac{\partial^2 u}{\partial t^2} + u = 0 \tag{1}$$

is self-adjoint; and

$$vL(u) - uL(v) = \frac{\partial}{\partial x}(vu_x - uv_x) - \frac{\partial}{\partial t}(vu_t - uv_t).$$

Hence, if $u$ and $v$ are two solutions with continuous second derivatives within and on a regular closed curve $\Gamma$

$$\int_\Gamma (vu_t - uv_t)\,dx + (vu_x - uv_x)\,dt = 0. \tag{2}$$

To solve the initial value problem, to find $u$ in $t > 0$ given that

$$u = f(x), \quad u_t = h(x),$$

when $t = 0$, we take $\Gamma$ to be the triangle with vertices $P_0(x_0, t_0)$, $Q(x_0 - t_0, 0)$, $R(x_0 + t_0, 0)$. The inclined sides of the triangle are characteristics. On $P_0 Q$, $dx = dt$; on $P_0 R$, $dx = -dt$. Hence equation (2) becomes

$$\int_{QR} (vu_t - uv_t)\,dx - \int_{RP_0} (v\,du - u\,dv) + \int_{P_0 Q} (v\,du - u\,dv) = 0,$$

whence

$$\int_{QR} (vu_t - uv_t)\,dx - [uv]_R^{P_0} + 2\int_{RP_0} u\,dv + [uv]_{P_0}^Q - 2\int_{P_0 Q} u\,dv = 0.$$

If $v$ is constant and equal to unity on $P_0 Q$ and $P_0 R$, this gives

$$u_{P_0} = \tfrac{1}{2}u_Q + \tfrac{1}{2}u_R + \frac{1}{2}\int_{QR} (vu_t - uv_t)\,dx. \tag{3}$$

The function $v$ is the Riemann–Green function which we have to determine in order to complete the solution.

We try to find the Riemann–Green function as a series of the form

$$v = v_0 + \sum_1^\infty \frac{v_j}{j!j!}\,\Gamma^j,$$

where $$\Gamma = (x - x_0)^2 - (t - t_0)^2.$$

The conditions on the characteristics $P_0 Q$ and $P_0 R$ are satisfied by taking $v_0 \equiv 1$. The differential equation $L(v) = 0$ becomes

$$\sum_0^\infty \frac{\Gamma^j}{j!j!} L(v_j) + 4 \sum_1^\infty \frac{\Gamma^{j-1}}{(j-1)!j!} \left\{ (x-x_0)\frac{\partial v_j}{\partial x} + (t-t_0)\frac{\partial v_j}{\partial t} \right\}$$

$$+ 4\sum_1^\infty \frac{\Gamma^{j-1}}{(j-1)!(j-1)!} v_j = 0.$$

Hence $\quad L(v_j) + \dfrac{4}{j+1}\left[ (x-x_0)\dfrac{\partial v_{j+1}}{\partial x} + (t-t_0)\dfrac{\partial v_{j+1}}{\partial t} \right] + 4v_{j+1} = 0.$

This is satisfied when $v_j$ is constant and

$$j_{j+1} = -\tfrac{1}{4}L(v_j) = -\tfrac{1}{4}v_j.$$

Hence $\qquad\qquad v_j = (-\tfrac{1}{4})^j v_0 = (-\tfrac{1}{4})^j.$

Thus the Riemann–Green function is

$$v(x,t;x_0,t_0) = \sum_0^\infty \frac{(-1)^j}{j!j!} \frac{\Gamma^j}{4^j} = J_0(\sqrt{\Gamma}),$$

where $J_0$ is the Bessel function of order zero. In our initial value problem, $\Gamma$ is negative, and so

$$v = I_0[\sqrt{\{(t-t_0)^2 - (x-x_0)^2\}}],$$

where $I_0$ is the Bessel function 'of imaginary argument'. This result can be easily verified. The solution (3) then becomes

$$u(x_0,t_0) = \tfrac{1}{2}f(x_0-t_0) + \tfrac{1}{2}f(x_0+t_0) + \frac{1}{2}\int_{x_0-t_0}^{x_0+t_0} h(x)\, I_0[\sqrt{\{t_0^2 - (x-x_0)^2\}}]\, dx$$

$$+ \frac{1}{2}\int_{x_0-t_0}^{x_0+t_0} f(x)\frac{\partial}{\partial t_0} I_0[\sqrt{\{t_0^2 - (x-x_0)^2\}}]\, dx.$$

## 5.7 More examples of the Riemann–Green function

The Riemann–Green function $v(x,y;x_0,y_0)$ for the self-adjoint equation

$$\frac{\partial^2 u}{\partial x\, \partial y} + \frac{\nu(1-\nu)}{(x+y)^2} u = 0, \tag{1}$$

where $\nu$ is a constant, is constant on the characteristics $x = x_0, y = y_0$, and so is equal to unity when

$$\Gamma = (x-x_0)(y-y_0)$$

vanishes. Hence we try

$$v(x,y;x_0,y_0) = \sum_0^\infty \frac{v_j \Gamma^j}{j!j!}.$$

with $v_0 = 1$.

The coefficients $v_j$ satisfy the recurrence relation

$$(x-x_0)\frac{\partial v_j}{\partial x}+(y-y_0)\frac{\partial v_j}{\partial y}+jv_j+j\frac{\partial^2 v_{j-1}}{\partial x\,\partial y}+j\frac{\nu(1-\nu)}{(x+y)^2}\,v_{j-1}=0.$$

Hence, if $v_j = K_j(x+y)^j$ where $K_j$ is a constant,

$$K_j = -K_{j-1}(\nu+j-1)(1-\nu+j-1)/(x_0+y_0).$$

Since $K_0 = 1$,     $K_j = \dfrac{(-1)^j}{(x_0+y_0)^j}\dfrac{\Gamma(\nu+j)\,\Gamma(1-\nu+j)}{\Gamma(\nu)\,\Gamma(1-\nu)}.$

Therefore   $v(x,y;x_0,y_0) = F\left(\nu,1-\nu;1;-\dfrac{(x-x_0)(y-y_0)}{(x+y)(x_0+y_0)}\right),$

which can also be written as

$$v(x,y;x_0,y_0) = P_{-\nu}\left(1+2\frac{(x-x_0)(y-y_0)}{(x+y)(x_0+y_0)}\right).$$

If, in the equation

$$\frac{\partial^2 u}{\partial x\,\partial y}+\frac{\nu}{x+y}\frac{\partial u}{\partial x}+\frac{\nu}{x+y}\frac{\partial u}{\partial y}=0, \tag{2}$$

where $\nu$ is a constant, we put $u = U(x+y)^{-\nu}$, we find that $U$ satisfies equation (1) and hence the Riemann–Green function for (2) is

$$v(x,y;x_0,y_0) = \frac{(x+y)^\nu}{(x_0+y_0)^\nu}\,F\left(\nu,1-\nu;1;-\frac{(x-x_0)(y-y_0)}{(x+y)(x_0+y_0)}\right). \tag{3}$$

With a different choice of $v_0$, we can obtain an alternative form

$$v(x,y;x_0,y_0) = \frac{(x+y)^{2\nu}}{(x+y_0)^\nu(x_0+y)^\nu}\,F\left(\nu,\nu;1;\frac{(x-x_0)(y-y_0)}{(x+y_0)(x_0+y)}\right). \tag{4}$$

If the constants $\mu$ and $\nu$ in

$$\frac{\partial^2 u}{\partial x\,\partial y}+\frac{\mu}{x+y}\frac{\partial u}{\partial y}+\frac{\nu}{(x+y)}\frac{\partial u}{\partial y}=0$$

are distinct, we cannot reduce it to self-adjoint form. If we try to find a Riemann–Green function of the form

$$v = \sum_0^\infty \frac{v_j}{j!j!}\,\Gamma^j,$$

a possible form for $v_0$ is

$$v_0 = \exp\left\{\int_{y_0}^y \frac{\mu}{x+\eta}\,d\eta+\int_{x_0}^x \frac{\nu}{\xi+y}\,d\xi\right\} = \frac{(x+y)^{\mu+\nu}}{(x+y_0)^\mu(x_0+y)^\nu}.$$

Comparing this with the formula (4) for the case $\mu = \nu$, we substitute in the adjoint equation

$$v = \frac{(x+y)^{\mu+\nu}}{(x+y_0)^{\mu} (x_0+y)^{\nu}} F(t),$$

where

$$t = \frac{(x-x_0)(y-y_0)}{(x+y_0)(x_0+y)}.$$

It turns out that

$$t(1-t)\frac{d^2 F}{dt^2} + \{1-(\mu+\nu+1)t\}\frac{dF}{dt} - \mu\nu F = 0$$

of which the solution which takes the value 1 at $t = 0$ is $F(\mu, \nu; 1; t)$. Hence

$$v(x,y;x_0,y_0) = \frac{(x+y)^{\mu+\nu}}{(x+y_0)^{\mu} (x_0+y)^{\nu}} F\left(\mu, \nu; 1; \frac{(x-x_0)(y-y_0)}{(x+y_0)(x_0+y)}\right),$$

which reduces to (4) when $\mu = \nu$.

With a different choice of $v_0$, we could get the alternative form

$$v(x,y;x_0,y_0) = \frac{(x+y)^{\nu} (x_0+y)^{\mu-\nu}}{(x_0+y_0)^{\mu}} F\left(\mu, 1-\nu; 1; -\frac{(x-x_0)(y-y_0)}{(x+y)(x_0+y_0)}\right)$$

corresponding to (3).

## Exercises

**1.** Verify that the Riemann–Green function for

$$\frac{\partial^2 u}{\partial x \partial y} - \frac{2}{(x+y)^2} u = 0$$

is
$$v(x,y;x_0,y_0) = \frac{(x+y_0)(x_0+y)+(x-x_0)(y-y_0)}{(x+y)(x_0+y_0)}.$$

Use Riemann's method to show that the solution which satisfies the conditions $u = 0$, $u_x = x^2$ on $y = x$ is

$$u = \tfrac{1}{4}(x-y)(x+y)^2.$$

**2.** Prove that the Riemann–Green function for $u_{xy} + u = 0$ is $J_0(2\sqrt{\Gamma})$, where $\Gamma = (x-x_0)(y-y_0)$.

**3.** Show that the Riemann–Green function for

$$\frac{\partial^2 u}{\partial x \partial y} + \frac{2}{x+y}\left(\frac{\partial u}{\partial x} + \frac{\partial u}{\partial y}\right) = 0$$

is
$$u(x,y;x_0,y_0) = \frac{x+y}{(x_0+y_0)^3}\{(x+y)(x_0+y_0)+2(x-x_0)(y-y_0)\}.$$

Use Riemann's method to show that the solution which satisfies the conditions $u = 0$, $\partial u/\partial x = 3x^2$ on $y = x$ is

$$u = 2x^3 - 3x^2y + 3xy^2 - y^3.$$

**4.** Show that the Riemann–Green function for

$$\frac{\partial^2 u}{\partial x^2} - \frac{\partial^2 u}{\partial y^2} - \frac{2}{x}\frac{\partial u}{\partial x} = 0$$

is

$$v(x, y; x_0, y_0) = \frac{x^2 + x_0^2 - (y - y_0)^2}{2x^2}.$$

Hence find the solution in $x > y > 0$ which satisfies the conditions $u = f(x)$, $u_y = h(x)$ when $y = 0$.

**5.** Prove that the Riemann–Green function for

$$\frac{\partial^2 u}{\partial x^2} - \frac{\partial^2 u}{\partial y^2} + \frac{\nu(1-\nu)}{x^2} u = 0$$

is

$$v(x, y; x_0, y_0) = F(\nu, 1 - \nu; 1; \Gamma/(4xx_0)),$$

where

$$\Gamma = (y - y_0)^2 - (x - x_0)^2.$$

Hence find the solution which satisfies the conditions $u = f(x)$, $u_y = h(x)$ when $y = 0$.

**6.** Prove that the Riemann–Green function for

$$\frac{\partial^2 u}{\partial x \partial y} + \frac{a}{x}\frac{\partial u}{\partial x} + \frac{b}{x}\frac{\partial u}{\partial y} = 0,$$

where $a$ and $b$ are constants is

$$v(x, y; x_0, y_0) = \left(\frac{x}{x_0}\right)^b \exp\left\{\frac{a(y - y_0)}{x_0}\right\} {}_1F_1(1 - b; 1; -a\Gamma/(xx_0)),$$

where

$$\Gamma = (x - x_0)(y - y_0),$$

and

$${}_1F_1(\alpha; \beta; t) = \frac{\Gamma(\beta)}{\Gamma(\alpha)} \sum_0^\infty \frac{\Gamma(\alpha + n)}{\Gamma(\beta + n)} \frac{t^n}{n!}.$$

**7** $v(x, y; x_0, y_0)$ is the Riemann–Green function of

$$L(u) \equiv u_{xy} + gu_x + fu_y + cu = 0.$$

$u(x, y; x_1, y_1)$ is the solution of $L(u) = 0$ which satisfies the conditions $u_x = -fu$ on $y = y_1$, $u_y = -gy$ on $x = x_1$. By applying Green's transformation to

$$\iint_D (vL(u) - uL^*(v))\, dx\, dy,$$

where $D$ is the rectangle with opposite corners $(x_0, y_0)$ and $(x_1, y_0)$ and $(x_1, y_1)$, prove that

$$u(x_0, y_0; x_1, y_1) = v(x_1, y_1; x_0, y_0).$$

# 6

# THE EQUATION OF WAVE MOTIONS

## 6.1 Spherical waves

The equation of wave motions in space of $n$ dimensions is

$$\nabla^2 U - \frac{\partial^2 U}{\partial t^2} = 0$$

with an appropriate choice of units, $\nabla^2$ being Laplace's operator. In spherical polar coordinates, this is

$$\frac{\partial^2 U}{\partial r^2} + \frac{n-1}{r}\frac{\partial U}{\partial r} + \frac{1}{r^2}\Lambda(U) - \frac{\partial^2 U}{\partial t^2} = 0,$$

where $\Lambda$ is a linear operator containing only derivatives with respect to the angle variables. A solution $U$ with spherical symmetry about the origin does not involve the angle variables, and so satisfies

$$\frac{\partial^2 U}{\partial r^2} + \frac{n-1}{r}\frac{\partial U}{\partial r} - \frac{\partial^2 U}{\partial t^2} = 0. \tag{1}$$

This can be reduced to the self-adjoint form

$$\frac{\partial^2 u}{\partial r^2} - \frac{\partial^2 u}{\partial t^2} - \frac{(n-1)(n-3)}{4r^2}u = 0 \tag{2}$$

by putting $U = ur^{-(n-1)/2}$.

When $n = 3$, this is the one-dimensional wave equation. Hence the general solution of the wave equation with spherical symmetry in three dimensional space is

$$U = \frac{1}{r}\{\Phi(r-t) + \Psi(r+t)\}. \tag{3}$$

The term involving $\Phi$ represents expanding waves, that in $\Psi$ contracting waves.

If $U = f(r), U_t = h(r)$ when $t = 0, r \geqslant 0$, where $f''$ and $h'$ are continuous, $U$ is given for $t > 0$ by

$$U = \frac{1}{2r}\left\{(r-t)f(r-t) + (r+t)f(r+t) + \int_{r-t}^{r+t} sh(s)\,ds\right\}, \tag{4}$$

provided that $r \geqslant t$. But when $t > r$, this solution fails because $f(r)$

and $h(r)$ are not defined for $r < 0$. This is what we should expect; Cauchy data on $t = 0, r \geqslant 0$ determine $U$ only up to the characteristic $r = t$ in the $rt$-plane. To determine $U$ when $t > r$ another condition must be satisfied on $r = 0$.

If we require $U$ to be finite at $r = 0$ for all $t \geqslant 0$, $u$ must vanish for $r = 0$. Now

$$u = \tfrac{1}{2}(r-t)f(r-t) + \tfrac{1}{2}(r+t)f(r+t) + \frac{1}{2}\int_{r-t}^{r+t} sh(s)\,ds,$$

when $0 \leqslant t \leqslant r$, and $\quad u = \tfrac{1}{2}\phi(t-r) + \tfrac{1}{2}\psi(t+r)$ \hfill (5)

in $0 \leqslant r \leqslant t$, where $\quad \phi(t) + \psi(t) = 0,$

for $t \geqslant 0$. But since $u_r + u_t$ is constant on every characteristic $r+t = $ constant, we have

$$\psi'(t+r) = (r+t)f'(r+t) + f(r+t) + (r+t)h(r+t),$$

or $\quad\quad\quad\quad \psi'(t) = tf'(t) + f(t) + th(t).$

Therefore $\quad\quad\quad\quad \psi(t) = tf(t) + \int_0^t sh(s)\,ds + \psi(0)$

and $\quad\quad\quad\quad \phi(t) = -tf(t) - \int_0^t sh(s)\,ds - \psi(0).$

Inserting these values in (5), we find that, when $t \geqslant r$,

$$U = \frac{1}{2r}\left\{-(t-r)f(t-r) + (t+r)f(t+r) + \int_{t-r}^{t+r} sh(s)\,ds\right\}.$$

If $f(s)$ and $h(s)$ are zero for $s \geqslant a$, then, for a fixed $r \geqslant a$, $U$ remains zero until $t = r - a$ and again when $t > r + a$. Thd initial bounded disturbance is thus propagated with a sharp head and tail.

## 6.2   Cylindrical waves

The equation of wave motions in two-dimensional space is

$$\frac{\partial^2 U}{\partial r^2} + \frac{1}{r}\frac{\partial U}{\partial r} + \frac{1}{r^2}\frac{\partial^2 U}{\partial \theta^2} - \frac{\partial^2 U}{\partial t^2} = 0,$$

in polar coordinates. This equation can be solved by the method of separation of variables. In particular, if $U$ is independent of $\theta$, it is of the form $V(r)e^{it\tau}$ where $\tau$ is a constant and

$$\frac{d^2 V}{dr^2} + \frac{1}{r}\frac{dV}{dr} + \tau^2 V = 0,$$

which is Bessel's equation of order zero. There are two independent
solutions $J_0(\tau r)$ and $Y_0(\tau r)$; but it is preferable to use the Hankel
functions†

$$H_0^{(1)}(\tau r) = J_0(\tau r) + iY_0(\tau r), \quad H_0^{(2)}(\tau r) = J_0(\tau r) - iY_0(\tau r)$$

because of their simpler behaviour when $\tau r$ is large.

We have, then, two solutions symmetrical about the origin

$$U_1 = e^{i\tau t}H_0^{(1)}(\tau r),$$
$$U_2 = e^{i\tau t}H_0^{(2)}(\tau r).$$

When $\tau$ is real and $\tau r$ large,

$$U_1 \sim \left(\frac{2}{\pi\tau r}\right)^{\frac{1}{2}} \exp\left(i\tau(t+r) - \tfrac{1}{4}\pi i\right),$$

$$U_2 \sim \left(\frac{2}{\pi\tau r}\right)^{\frac{1}{2}} \exp\left(i\tau(t-r) + \tfrac{1}{4}\pi i\right),$$

so that the real parts of $U_1$ and $U_2$ represent contracting and expanding
cylindrical waves respectively which vary simply-harmonically with
the time.

Using the integral formulae for the Hankel functions, we find that

$$U_1 = \frac{2}{\pi i}\int_0^\infty \exp\left(it\tau + ir\tau \cosh\phi\right)d\phi,$$

$$U_2 = -\frac{2}{\pi i}\int_0^\infty \exp\left(it\tau - ir\tau \cosh\phi\right)d\phi.$$

It follows that a general solution of the two-dimensional wave
equation with symmetry about the origin is

$$U = \int_{-\infty}^\infty d\tau \int_0^\infty d\phi \exp\left(i\tau(t - r\cosh\phi)\right)f(\tau)$$

$$+ \int_{-\infty}^\infty d\tau \int_0^\infty d\phi \exp\left(i\tau(t + r\cosh\phi)\right)g(\tau),$$

or $\qquad U = \int_0^\infty F(t - r\cosh\phi)\,d\phi + \int_0^\infty G(t + r\cosh\phi)\,d\phi.$ \hfill (1)

This result, which corresponds to the solution (3) for spherical
waves, has been obtained by a purely formal argument. To verify
that, if $F''$ is continuous,

$$U = \int_0^\infty F(t - r\cosh\phi)\,d\phi \hfill (2)$$

† See, for example, Copson, *Functions of a Complex Variable* (Oxford, 1935), p. 323.

satisfies the equation

$$\frac{\partial^2 U}{dr^2} + \frac{1}{r}\frac{\partial U}{\partial r} - \frac{\partial^2 U}{\partial t^2} = 0$$

is quite straightforward, provided that we can justify differentiation under the sign of integration and provided

$$\sinh\phi\, F'(t - r\cosh\phi) \to 0$$

as $\phi \to +\infty$.

The function (2) can be regarded as the cylindrical wave function due to a source of strength $2\pi F(t)$ at $r = 0$. A peculiarity of the two-dimensional propagation of waves which it describes is the existence of a 'tail' to the disturbance. For if $F(t) = 0$ when $t > t_0$, we have

$$U = \int_s^\infty F(t - r\cosh\phi)\,d\phi$$

when $t > t_0 + r$, where $r\cosh s = t - t_0$; this expression does not generally vanish. In this respect, waves in two dimension differ essentially from those in spaces of one or three dimensions.

A different integral formula for a cylindrical wave function

$$U = \int_0^\pi f(t - r\cos\phi)\,d\phi$$

was given by Poisson; this is certainly valid if $f''$ is continuous. None of these formulae is of much use in solving the initial value problem for the equation of cylindrical waves.

### 6.3 Poisson's mean value solution

*If $\rho(x, y, z)$ has continuous partial derivatives of the first and second orders, the equation of wave motions*

$$\frac{\partial^2 u}{\partial x^2} + \frac{\partial^2 u}{\partial y^2} + \frac{\partial^2 u}{\partial z^2} = \frac{\partial^2 u}{\partial t^2} \tag{1}$$

*has the solution* $\qquad u = t\mathcal{M}(\rho; x, y, z; t),$ $\qquad(2)$

*when $t > 0$, where $\mathcal{M}$ denotes the mean value of $\rho$ over the sphere with centre $(x, y, z)$ and radius $t$; this solution satisfies the initial conditions $u = 0$, $u_t = \rho$ when $t = 0$ and has continuous derivatives of the first and second orders.*

The proof uses Boussinesq's spherical potential

$$V(x, y, z, t) = \iint_S \frac{\rho(\xi, \eta, \zeta)}{\sqrt{\{(\xi - x)^2 + (\eta - y)^2 + (\zeta - z)^2\}}}\,dS,$$

where $S$ is the sphere with centre $(x, y, z)$ and radius $t$. If $(l, m, n)$ are the direction cosines of the outward normal to $S$,

$$V(x, y, z, t) = t \iint_\Omega \rho(x + lt, y + mt, z + nt) \, d\Omega, \qquad (3)$$

where $\Omega$ is the unit sphere $l^2 + m^2 + n^2 = 1$. Evidently $V$ has continuous derivatives of the first and second orders, and vanishes when $t = 0$.

Now
$$\nabla^2 V(x, y, z, t) = t \iint_\Omega \nabla^2 \rho(x + lt, y + mt, z + nt) \, d\Omega$$

$$= \frac{1}{t} \iint_S \left\{ \frac{\partial^2 \rho}{\partial \xi^2} + \frac{\partial^2 \rho}{\partial \eta^2} + \frac{\partial^2 \rho}{\partial \zeta^2} \right\} dS.$$

Also
$$\frac{\partial V}{\partial t} = \frac{V}{t} + t \iint_\Omega \left( l \frac{\partial \rho}{\partial x} + m \frac{\partial \rho}{\partial y} + n \frac{\partial \rho}{\partial z} \right) d\Omega$$

$$= \frac{V}{t} + \frac{1}{t} \iint_S \left( l \frac{\partial \rho}{\partial \xi} + m \frac{\partial \rho}{\partial \eta} + n \frac{\partial \rho}{\partial \zeta} \right) dS$$

$$= \frac{V}{t} + \frac{1}{t} \iiint_D \left( \frac{\partial^2 \rho}{\partial \xi^2} + \frac{\partial^2 \rho}{\partial \eta^2} + \frac{\partial^2 \rho}{\partial \zeta^2} \right) d\xi \, d\eta \, d\zeta, \qquad (4)$$

where $D$ is the interior of $S$. This may be written as

$$\frac{\partial V}{dt} = \frac{V}{t} + \frac{1}{t} \int_0^t d\tau \iint_\Omega \tau^2 \nabla^2 \rho(x + l\tau, y + m\tau, z + n\tau) \, d\Omega.$$

Hence
$$\frac{\partial^2 V}{\partial t^2} = \frac{1}{t} \frac{\partial V}{\partial t} - \frac{V}{t^2} - \frac{1}{t^2} \iiint_D \nabla^2 \rho \, d\xi \, d\eta \, d\zeta$$

$$+ t \iint_\Omega \nabla^2 \rho(x + lt, y + mt, z + nt) \, d\Omega.$$

By (4)
$$\frac{\partial^2 V}{\partial t^2} = t \iint_\Omega \nabla^2 \rho(x + lt, y + mt, z + nt) \, d\Omega = \nabla^2 V, \qquad (5)$$

so that $V$ satisfies the equation of wave motions. Moreover

$$\frac{\partial V}{\partial t} = \iint_\Omega \rho(x + lt, y + mt, z + nt) \, d\Omega + t \iint_\Omega \left( l \frac{\partial \rho}{\partial x} + m \frac{\partial \rho}{\partial y} + n \frac{\partial \rho}{\partial z} \right) d\Omega,$$

so that
$$V_t(x, y, z, 0) = 4\pi \rho(x, y, z). \qquad (6)$$

Since $V = 4\pi t \mathscr{M}$, it follows that

$$u = t \mathscr{M}(\rho; x, y, z; t)$$

is a solution of the equation of wave motions which has the required differentiability properties and satisfies the initial conditions $u = 0$

$u_t = \rho$. We see later that it is unique. Moreover, if $|\rho| < \epsilon$, we have $|u| < t\epsilon \leqslant T\epsilon$ in $0 \leqslant t \leqslant T$, so that the problem is well posed.

If $u = V_t(x, y, z; t)/4\pi$ it follows from (6) and (5) that

$$u(x, y, z; 0) = \rho(x, y, z), \quad u_t(x, y, z, 0) = 0.$$

Hence
$$u = \frac{\partial}{\partial t}\{t\mathcal{M}(\rho; x, y, z; t)\}$$

satisfies the equation of wave motions under the initial conditions $u = \rho$, $u_t = 0$. In order that this solution should have continuous derivatives of the second order, it suffices that $\rho$ has continuous derivatives of the third order. Again, the solution is unique. And if $|\rho| < \epsilon$, $|\mathrm{grad}\,\rho| < \epsilon$, we have $|u| < \epsilon + t\epsilon \leqslant (1+T)\epsilon$ in $0 \leqslant t \leqslant T$, so that the problem is well posed.

To sum up, if $f(x, y, z)$ has continuous derivatives of the third order and $h(x, y, z)$ of the second order, a solution of the wave equation in $t > 0$ satisfying the initial conditions $u = f$, $u_t = h$ is

$$u = t\mathcal{M}(h; x, y, z; t) + \frac{\partial}{\partial t}\{t\mathcal{M}(f; x, y, z; t)\}. \tag{7}$$

We show later that this solution is unique.

The expression on the right of (7) depends only on the initial data on the sphere with centre $(x, y, z)$ and radius $t$. If $f$ and $h$ are zero except on some bounded closed set $F$, and if the greatest and least distances from $F$ of some point $(x, y, z)$ not in $F$ are $R_1$ and $R_2$, then $u = 0$ when $t > R_1$ and when $0 < t < R_2$, since in these cases the sphere over which the mean is taken does not then cut $F$. The solution represents a wave motion with a sharply defined front and rear; there is no residual after-effect.

## 6.4 The method of descent

If we use the method of §6.3 with the initial conditions

$$u(x, y, z, 0) = f(x, y), \quad u_t(x, y, z, 0) = h(x, y), \tag{1}$$

the two mean values in (7) do not involve $z$, and so the solution satisfies

$$\frac{\partial^2 u}{\partial x^2} + \frac{\partial^2 u}{\partial y^2} = \frac{\partial^2 u}{\partial t^2}. \tag{2}$$

This device was called by Hadamard the method of descent.

With this sort of data,

$$t\mathcal{M}(f;x,y;t) = \frac{1}{4\pi t}\iint_D f(x+\xi,y+\eta)\,dS,$$

where integration is over the sphere $\xi^2+\eta^2+\zeta^2 = t^2$. Hence

$$t\mathcal{M}(f;x,y;t) = \frac{1}{2\pi}\iint_{\Sigma_0} f(x+\xi,y+\eta)\,\frac{d\xi\,d\eta}{\sqrt{(t^2-\xi^2-\eta^2)}},$$

where $\Sigma_0$ is the disc $\xi^2+\eta^2 \leqslant t^2$, since $dS = t\,d\xi\,d\eta/|\zeta|$; a factor 2 arises since there are two hemispheres in $\zeta \geqslant 0$ and $\zeta \leqslant 0$ whose projections on $\zeta = 0$ are both the disc $\Sigma_0$. The solution in $t > 0$ of the initial value problem (1) for the equation of wave motions in the plane is thus

$$\begin{aligned}
u = \frac{1}{2\pi}\iint_\Sigma h(\xi,\eta)\,&\frac{d\xi\,d\eta}{\sqrt{\{t^2-(x-\xi)^2-(y-\eta)^2\}}}\\
&+\frac{1}{2\pi}\frac{\partial}{\partial t}\iint_\Sigma f(\xi,\eta)\,\frac{d\xi\,d\eta}{\sqrt{\{t^2-(x-\xi)^2-(y-\eta)^2\}}},
\end{aligned}\qquad(3)$$

where integration is over the disc $\Sigma$ on which $(x-\xi)^2+(y-\eta)^2 \leqslant t^2$.

Suppose that $f$ and $h$ are zero except on a bounded closed set $F$. If $(x,y)$ is a point outside $F$ at a distance $R$ from $F$, both integrals in (3) vanish when $0 \leqslant t < R$, but not when $t > R$. As $t \to +\infty$, the integrals tend to integrals over the whole plane, which are usually not zero. The solution (3) represents a wave motion which has a sharply defined start but leaves a residual after-effect.

We obtain formula (3) again by a different method in §7.4 where we show that, if $h$ has continuous second order derivatives and $f$ has continuous third order derivatives, then (3) gives the solution of (2) which satisfies the initial conditions and has continuous second order derivatives.

If $f$ and $h$ are defined only on some domain $D$ of the initial plane $t = 0$, formula (3) determines $u(x,y;t)$ when $t$ is positive only if the disc $\Sigma$ lies in $D$. This determines the domain of influence of $D$. For example, if $D$ is the disc $\xi^2+\eta^2 \leqslant a^2$ on the plane $t = 0$, the domain of influence in $t \geqslant 0$ is $x^2+y^2 \leqslant (a-t)^2$, $0 \leqslant t \leqslant a$.

## 6.5  The uniqueness theorem

To prove that the solution of the equation

$$\frac{\partial^2 u}{\partial x^2}+\frac{\partial^2 u}{\partial y^2} = \frac{\partial^2 u}{\partial t^2}$$

found in §6.4 is unique, we have to show that, if $u = 0$, $u_t = 0$ when $t = 0$, then $u = 0$ for all $t > 0$.

Regard $(x, y, t)$ as Cartesian coordinates in a three-dimensional Euclidean space. Let $D$ be the volume defined by

$$(x - x_0)^2 + (y - y_0)^2 \leqslant (t - t_0)^2, \quad 0 \leqslant t \leqslant t_0.$$

Then if $u$ has continuous second order derivatives, we have, by Green's theorem,

$$\iiint_D \frac{\partial u}{\partial t} \left( \frac{\partial^2 u}{\partial t^2} - \frac{\partial^2 u}{\partial x^2} - \frac{\partial^2 u}{\partial y^2} \right) dx \, dy \, dt$$

$$= \frac{1}{2} \iiint_D \left\{ \frac{\partial}{\partial t} (u_x^2 + u_y^2 + u_t^2) - 2 \frac{\partial}{\partial x} (u_x u_t) - 2 \frac{\partial}{\partial y} (u_y u_t) \right\} dx \, dy \, dt$$

$$= \frac{1}{2} \iint_S \{ n(u_x^2 + u_y^2 + u_t^2) - 2l u_x u_t - 2m u_y u_t \} dS,$$

where $(l, m, n)$ are the direction cosines of the outward normal to the boundary $S$ of $D$. If $u$ satisfies the wave equation, this surface integral vanishes.

$S$ consists of two parts, a disc in the plane $t = 0$ and a portion $S_0$ of the cone

$$(x - x_0)^2 + (y - y_0)^2 = (t - t_0)^2$$

cut off by the plane $t = 0$. On $S_0$, $n = 1/\sqrt{2}$ and so $l^2 + m^2 = n^2$. If we multiply through by $n$, we get

$$\iint_{S_0} \{ (l^2 + m^2) u_t^2 + n^2 u_x^2 + n^2 y_y^2 - 2l n u_x u_t - 2m n u_y u_t \} dS = 0,$$

or

$$\iint_{S_0} \{ (l u_t - n u_x)^2 + (m u_t - n u_y)^2 \} dS = 0.$$

The integrand must be zero, and so

$$\frac{u_x}{l} = \frac{u_y}{m} = \frac{u_t}{n}$$

on $S_0$. Denote each fraction by $v$. Then, on a generator of $S_0$,

$$du = u_x dx + u_y dy + u_z dz = v(l \, dx + m \, dy + n \, dz) = 0$$

since the normal with direction cosines $(l, m, n)$ is perpendicular to the generator. Therefore $u$ is constant on every generator, and so is zero since $u = 0$ when $t = 0$. In particular, $u(x_0, y_0, t_0)$ is zero, which was to be proved.

There is a variant of this proof in which we take $D$ to be the truncated cone

$$(x - x_0)^2 + (y - y_0)^2 \leqslant (t - t_0)^2, \quad 0 \leqslant t \leqslant t_1,$$

where $t_1 < t_0$. As before, we get

$$\iint_S \{n(u_x^2 + u_y^2 + u_t^2) - 2lu_x u_t - 2mu_y u_t\} \, dS = 0,$$

where $S$ is now the boundary of the truncated cone. On the base, the integrand vanishes. It follows that

$$\frac{1}{n} \iint_{S_1} \{(lu_t - nu_x)^2 + (mu_t - nu_y)^2\} \, dS + \iint_{\Sigma_1} (u_x^2 + u_y^2 + u_t^2) \, dx \, dy = 0,$$

where $n = 1/\sqrt{2}$ on the curved boundary $S_1$ of $D$, and $\Sigma_1$ is the disc in which $t = t_1$ cuts $D$. Therefore

$$\iint_{\Sigma_1} (u_x^2 + u_y^2 + u_t^2) \, dx \, dy = 0,$$

so that $u_x, u_y, u_t$ all vanish on $\Sigma_1$. As $t_1$ is arbitrary, $u_x, u_y, u_t$ vanish everywhere in $D$, and so $u$ is a constant. As $u$ vanishes when $t = 0$, $u$ is zero everywhere in $D$.

These proofs also apply to the equation of wave motions with any number of space variables, though the details are more complicated.

## 6.6  The Euler–Poisson–Darboux equation

In §6.1 we saw that spherically symmetrical solutions of the equation of wave motions in space of $m+1$ dimensions satisfy the equation

$$\frac{\partial^2 u}{\partial r^2} + \frac{m}{r} \frac{\partial u}{\partial r} - \frac{\partial^2 u}{\partial t^2} = 0. \tag{1}$$

If we introduce the characteristic variables $x = r+t$, $y = r-t$, we obtain the equation of Euler, Poisson and Darboux

$$\frac{\partial^2 u}{\partial x \, \partial y} + \frac{N}{x+y} \left( \frac{\partial u}{\partial x} + \frac{\partial u}{\partial y} \right) = 0, \tag{2}$$

where $N = \frac{1}{2}m$. We do not restrict our discussion to integral values of $2N$; the equation arises in some physical problems when $2N$ is not an integer.

Nor shall we restrict our attention to the case when $x$ and $y$ are real. Laplace's equation

$$\sum_1^{m+1} \frac{\partial^2 u}{\partial x_k^2} + \frac{\partial^2 u}{\partial y^2} = 0$$

has axially symmetrical solutions, which are functions of $y$ and

$r = \sqrt{\sum_{1}^{m+1} x_k^2}$. These solutions satisfy

$$\frac{\partial^2 u}{\partial r^2} + \frac{m}{r} \frac{\partial u}{\partial r} + \frac{\partial^2 u}{\partial y^2} = 0.$$

The characteristic variables are now conjugate complex numbers

$$z = r + iy, \quad \bar{z} = r - iy.$$

In terms of these variables, the equation becomes

$$\frac{\partial^2 u}{\partial z \partial \bar{z}} + \frac{N}{z+\bar{z}} \left( \frac{\partial u}{\partial z} + \frac{\partial u}{\partial \bar{z}} \right) = 0.$$

The self-adjoint form of (2),

$$\frac{\partial^2 v}{\partial x \partial y} + \frac{N(1-N)}{(x+y)^2} v = 0 \tag{3}$$

is obtained by putting $u = v(x+y)^{-N}$. Equation (3) is unaltered if we replace $N$ by $1 - N$. It follows that, if $u$ is a solution of (2),

$$U = u(x+y)^{2N-1}$$

is a solution of $\qquad \dfrac{\partial^2 U}{\partial x \partial y} + \dfrac{1-N}{x+y} \left( \dfrac{\partial U}{\partial x} + \dfrac{\partial U}{\partial y} \right) = 0.$

Again, if we put $u_x + u_y = (x+y)\,v$ in (2), we have

$$u_{xy} + Nv = 0.$$

Hence $\qquad (u_x + u_y)_{xy} + N(v_x + v_y) = 0,$

or $\qquad \{(x+y)\,v\}_{xy} + N(v_x + v_y) = 0.$

Therefore $v$ satisfies

$$\frac{\partial^2 v}{\partial x \partial y} + \frac{N+1}{x+y} \left( \frac{\partial v}{\partial x} + \frac{\partial v}{\partial y} \right) = 0.$$

These recurrence formulae enable us to deduce from a solution of (2) a solution of the same equation with $N$ replaced by $1 + N$ or $1 - N$. When $N$ is not an integer, we need only consider the case when $N$ lies between 0 and 1. Considerable use has been made of these formulae by Weinstein, to whom they are due.

The general solution of (2) when $N = 0$ is

$$u_0 = \Phi(x) + \Psi(y).$$

By the first relation,

$$u_1 = \frac{1}{x+y} \{ \Phi(x) + \Psi(y) \}$$

is the general solution when $N = 1$. In particular

$$u_1 = \frac{\Phi(x)}{x+y}$$

is a solution when $N = 1$. Therefore

$$u_2 = \frac{1}{x+y}\left(\frac{\partial u_1}{\partial x} + \frac{\partial u_1}{\partial y}\right) = \frac{\partial}{\partial x}\frac{\Phi(x)}{(x+y)^2}$$

is a solution when $N = 2$; so also is

$$\frac{\partial}{\partial y}\frac{\Psi(x)}{(x+y)^2}.$$

Similarly      $u_3 = \frac{1}{x+y}\left(\frac{\partial u_2}{\partial x} + \frac{\partial u_2}{\partial y}\right) = \frac{\partial^2}{\partial x^2}\frac{\Phi(x)}{(x+y)^3}$

is a solution when $N = 3$. And so on. For any integer $N$, the general solution of (2) is

$$u = \frac{\partial^{N-1}}{\partial x^{N-1}}\frac{\Phi(x)}{(x+y)^N} + \frac{\partial^{N-1}}{\partial y^{N-1}}\frac{\Psi(y)}{(x+y)^N}, \tag{4}$$

where $\Phi$ and $\Psi$ are arbitrary functions, and $N$ is a positive integer.

The first term in (4) is

$$\frac{(N-1)!}{2\pi i}\int_C\frac{\Phi(\zeta)\,d\zeta}{(\zeta-x)^N(\zeta+y)^N}$$

if $C$ is a simple closed contour surrounding $x$ but not $-y$, provided that $\Phi(\zeta)$ is an analytic function of the complex variable $\zeta$, regular in and on $C$. By the first recurrence formula

$$(x+y)^{1-2N}\int_C\frac{\phi(\zeta)}{(\zeta-x)^{1-N}(\zeta+y)^{1-N}}$$

is also a solution when $N$ is zero or a negative integer. This suggests that we should consider complex integrals of the form

$$\int_C\frac{\Phi(\zeta)}{(\zeta-x)^N(\zeta+y)^N}\,d\zeta, \tag{5}$$

for general values of the constant $N$.

Suppose that $\Phi(\zeta)$ is an analytic function regular in a domain which contains the disc $|\zeta| \leqslant R$. Let $x$ and $y$ be two points of the disc. The integrand is not one-valued in the disc if $N$ is not an integer; it has branch points at $x$ and $-y$. If we choose a particular branch at, say, $R$ and take $C$ to be a figure of eight starting at $R$, enclosing $x$ in the positive sense, $-y$ in the negative sense, and returning to $R$, the integrand returns to its original value. We can deform $C$, so long as it remains a figure of eight enclosing $x$ in the positive sense, $-y$ in the negative sense, without altering the value of the integral. Moreover, the integral is an analytic function of $x$ and of $y$; its partial derivatives can be calculated by differentiation under the sign of integration. Since $(\zeta - x)^{-N}(\zeta + y)^{-N}$ satisfies the Euler–Poisson–Darboux equation, so also does the integral.

Alternatively, if $\Phi(\zeta)$ is an integral function, we can choose a particular branch and take $C$ to be a path which starts at infinity, passes round $x$ in the positive sense and returns to infinity.

## 6.7  Poisson's solutions

Suppose that $x$ and $y$ are real, and that $C$ is the figure of eight contour of §6.6. Choose the branch of $(\zeta - x)^N (\zeta + y)^N$ which is real and positive when $\zeta$ is real and greater than $\max(|x|, |y|)$.

Consider the case when $x > -y$. We can deform $C$ into a small circle of radius $\epsilon$ and centre $x$, described in the positive sense, the real axis from $x - \epsilon$ to $-y + \delta$, a small circle of radius $\delta$ and centre $-y$, described in the negative sense, the real axis from $-y + \delta$ to $x - \epsilon$. In

$$u_1 = \int_C \frac{\Phi(\zeta)}{(\zeta - x)^N (\zeta + y)^N} \, d\xi,$$

the integrals round the small circles tend to zero as $\epsilon$ and $\delta$ tend to zero if $N < 1$. Then

$$u_1 = 2i \sin N\pi \int_{-y}^{x} (x - \xi)^{-N} (y + \xi)^{-N} \Phi(\xi) \, d\xi.$$

$u_1$ vanishes if $N$ is zero or a negative integer. If we put

$$\zeta = \tfrac{1}{2}\{(x - y) + (x + y) \cos \psi\}$$

and drop a factor which is unimportant if $0 < N < 1$, we obtain

$$u_1 = \left(\frac{x+y}{2}\right)^{1-2N} \int_0^{\pi} \Phi\{\tfrac{1}{2}(x - y) + \tfrac{1}{2}(x + y) \cos \psi\} \sin^{1-2N}\psi \, d\psi.$$

Restoring the original variables

$$r = 4(x + y), \quad t = \tfrac{1}{2}(x - y),$$

we find that a solution of

$$\frac{\partial^2 u}{\partial r^2} + \frac{m}{r} \frac{\partial u}{\partial r} - \frac{\partial^2 u}{\partial t^2} = 0$$

is

$$u_1 = r^{1-m} \int_0^{\pi} \Phi(t + r \cos \psi) \sin^{1-m} \psi \, d\psi,$$

when $0 < m < 2$. We assumed that $x+y = 2r$ is positive; but the argument can be modified when $x+y$ is negative, with the same result.

Similarly, from

$$u_2 = (x+y)^{1-2N} \int_C \frac{\Psi(\zeta)}{(\zeta-x)^{1-N}(\zeta+y)^{1-N}} d\zeta$$

we find the solution

$$u_2 = \int_0^\pi \Psi(t+r\cos\psi)\sin^{m-1}\psi \, d\psi,$$

when $0 < m < 2$. The two solutions are identical when $m = 1$.

The proof assumed that $\Phi$ and $\Psi$ are analytic, but the formulae give solutions which have continuous derivatives of the first and second orders if $\Phi''$ is continuous.

When $m = 1$,

$$u_1 = \int_0^\pi \Phi(t+r\cos\psi)\,d\psi.$$

To get a second solution, consider the limit as $m \to 1$ of

$$\frac{1}{m-1}\int_0^\pi \Psi(t+r\cos\psi)(\sin^{m-1}\psi - r^{1-m}\sin^{1-m}\psi)\,d\psi.$$

This gives the other solution

$$u_3 = \int_0^\pi \Psi(t+r\cos\psi)\log(r\sin^2\psi)\,d\psi$$

when $m = 1$.

All these formulae are due to Poisson, who did not derive them in this way.

## 6.8  The formulae of Volterra and Hobson

In the solution

$$u = \int_C \frac{\Phi(\zeta)\,d\zeta}{(\zeta-x)^N(\zeta+y)^N},$$

where $x > -y$, take $C$ to be the path starting at $+\infty$ on the real axis encircling $x$ in the positive sense and returning to $+\infty$. Start with $(\zeta-x)^N(\zeta+y)^N$ real and positive when $\zeta$ is real and larger than $|x|$ and $|y|$. If $\Phi(\zeta)$ is analytic, we may suppose $N$ is not zero or a negative integer, since the integral then vanishes.

Deform $C$ into the real axis from $+\infty$ to $x+\epsilon$, followed by $|\zeta-x| = \epsilon$ described in the positive sense, and then the real axis from $x+\epsilon$ to $+\infty$. The integral round the small circle tends to zero if $N < 1$, which we assume to be the case. Then

$$u = (\exp(-2N\pi i)-1)\int_x^\infty \Phi(\xi)(\xi-x)^{-N}(\xi+y)^{-N}d\xi.$$

If we put $\qquad \xi = \frac{1}{2}(x-y) + \frac{1}{2}(x+y)\cosh\psi$

and drop the non-zero numerical factor, we get

$$u = \{\tfrac{1}{2}(x+y)\}^{1-2N} \int_0^\infty \Phi\{\tfrac{1}{2}(x-y) + \tfrac{1}{2}(x+y)\cosh\psi\} \sinh^{1-2N}\psi\, d\psi.$$

Restoring the original variables, we obtain Hobson's solution of

$$\frac{\partial^2 u}{\partial r^2} + \frac{m}{r}\frac{\partial u}{\partial r} - \frac{\partial^2 u}{\partial t^2} = 0,$$

namely $\qquad u_1 = r^{1-m} \int_0^\infty \Phi(t + r\cosh\psi)\sinh^{1-m}\psi\, d\psi.$

This solution in the special case $m = 1$ was found by Volterra.† We do not need to assume that $\Phi$ is analytic. The result is true if $m < 2$, $\Phi''$ is continuous and $\Phi$ behaves suitably at infinity. Since the equation is unaltered if we replace $r$ by $-r$, a second solution is

$$u_2 = r^{1-m}\int_0^\infty \Phi(t - r\cosh\psi)\sinh^{1-m}\psi\, d\psi.$$

The first solution $u_1$ represents converging waves, the second $u_2$ expanding waves.

Using the first recurrence formula, we get solutions

$$u_3 = \int_0^\infty \Phi(t + r\cosh\psi)\sinh^{m-1}\psi\, d\psi$$

and $\qquad u_4 = \int_0^\infty \Phi(t - r\cosh\psi)\sinh^{m-1}\psi\, d\psi$

valid when $m > 0$.

When $m = 1$, $u_2$ and $u_4$ are the same. A second solution is then

$$\lim_{m\to 1}\frac{1}{m-1}\int_0^\infty \Phi(t - r\cosh\psi)(\sinh^{m-1}\psi - r^{1-m}\sinh^{1-m}\psi)\, d\psi$$

$$= \int_0^\infty \Phi(t - r\cosh\psi)\log(r\sinh^2\psi)\, d\psi.$$

Similarly, since $u_1$ and $u_3$ are the same when $m = 1$, the second solution is then

$$\int_0^\infty \Phi(t + r\cosh\psi)\log(r\sinh^2\psi)\, d\psi.$$

† Volterra, *Acta math.* **18** (1894), 161.
   Hobson, *Proc. London Math. Soc.* (1) **22** (1891), 431–449.

### Exercises

**1.** Use Poisson's formula to find the solution of the equation

$$u_{tt} - u_{xx} - u_{yy} - u_{zz} = 0,$$

given that      $u = 0$,   $u_t = x^2 + xy + z^2$   when   $t = 0$.

**2.** Use Poisson's formula to solve the wave equation, given that, when $t = 0$, $u = 0$, and $u_t = 1$ for $x^2 + y^2 + z^2 \leqslant a^2$, $u_t = 0$ for $x^2 + y^2 + z^2 \geqslant a^2$. Examine the discontinuities of the solution.

**3.** $u$ is a solution of $u_{tt} - u_{xx} - u_{yy} - u_{zz} = 0$ of the form $v(r, t) \cos \theta$ in spherical polar coordinates. Prove that

$$\frac{\partial^2 v}{\partial r^2} + \frac{2}{r} \frac{\partial v}{\partial r} - \frac{2}{r^2} v = \frac{\partial^2 v}{\partial t^2}.$$

Hence show that      $u = \cos \theta \dfrac{\partial}{\partial r} \dfrac{\phi(r-t) + \psi(r+t)}{r}$,

where $\phi$ and $\psi$ are arbitrary functions. Obtain the same result from the fact that, if $u$ is a wave function, so also is $u_x$.

**4.** Show that $(1 - xw)^{-N}(1 + yw)^{-N}$ satisfies the equation

$$\frac{\partial^2 u}{\partial x \partial y} + \frac{N}{(x+y)} \left( \frac{\partial u}{\partial x} + \frac{\partial u}{\partial y} \right) = 0.$$

By considering its expansion in ascending powers of $w$, prove that the equation has homogeneous polynomial solutions

$$F_n(x, y) = \sum_{p+q=n} \frac{(N)_p (N)_q}{p! \, q!} x^p (-y)^q,$$

where $(N)_p = N(N+1) \dots (N+p-1)$. Prove that, if $K = |N|$ and

$$R = \max(|x|, |y|)$$

then      $|F_n(x, y)| \leqslant \dfrac{(2K)_n}{n!} R^n.$

**5.** Prove that Tricomi's equation of mixed type $y u_{xx} + u_{yy} = 0$ has imaginary characteristics $x \pm \frac{2}{3} i y^{\frac{3}{2}} = $ constant in the elliptic half-plane $y > 0$ and real characteristics $x \pm \frac{2}{3}(-y)^{\frac{3}{2}} = $ constant in the hyperbolic half-plane $y < 0$. Prove that, if $y < 0$, the equation can be written as

$$\frac{\partial^2 u}{\partial r^2} + \frac{1}{3r} \frac{\partial u}{\partial r} - \frac{\partial^2 u}{\partial t^2} = 0,$$

where $r = \frac{2}{3}(-y)^{\frac{3}{2}}$, $t = x$.

**6.** If $u$ satisfies
$$\frac{\partial^2 u}{\partial x \partial y} + \frac{M}{x+y}\frac{\partial u}{\partial x} + \frac{N}{x+y}\frac{\partial u}{\partial y} = 0,$$

where $M$ and $N$ are real constants, prove that $v = (x+y)^{M+N-1}u$ satisfies the same equation with $M$ and $N$ replaced by $1-N$ and $1-M$ respectively

**7.** Show that $u = (\zeta-x)^{-N}(\zeta+y)^{-M}$ satisfies the equation of Ex. 6. By considering
$$u = \int_C \frac{\Phi(\zeta)\,d\zeta}{(\zeta-x)^N(\zeta+y)^M},$$

where $C$ is the double circuit contour of fig. 7, deduce the solution

$$u_1 = \int_{-y}^{x} \Phi(\xi)\,(x-\xi)^{-N}(\xi+y)^{-M}\,d\xi,$$

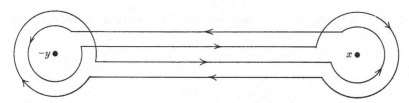

Fig. 7 (after Whittaker & Watson, *A Course of Modern Analysis*).

when $0 < M < 1, 0 < N < 1$. Obtain also the solution

$$u_2 = (x+y)^{1-M-N}\int_{-y}^{x} \Phi(\xi)\,(x-\xi)^{M-1}(y+\xi)^{N-1}\,d\xi.$$

These solutions are the same when $M+N = 1$. Obtain the second solution when $M+N = 1$.

**8.** If $u$ satisfies the equation of Ex. 6, show that
$$v = \frac{Mu_x + Nu_y}{x+y},$$

satisfies
$$\frac{\partial^2 v}{\partial x \partial y} + \frac{M+1}{x+y}\frac{\partial v}{\partial x} + \frac{N+1}{x+y}\frac{\partial v}{\partial y} = 0.$$

**9.** $v(x,y,z;t;\tau)$ is the solution of the equation of wave motions
$$L(v) \equiv \frac{\partial^2 v}{\partial t^2} - \frac{\partial^2 v}{\partial x^2} - \frac{\partial^2 v}{\partial y^2} - \frac{\partial^2 v}{\partial z^2} = 0$$

which satisfies the initial conditions
$$v(x,y,z;\tau;\tau) = 0, \quad v_t(x,y,z;\tau;\tau) = h(x,y,z;\tau)$$

for all $\tau$. If
$$u(x,y,z;t) = \int_0^t v(x,y,z;t;\tau)\,d\tau$$

WAVE MOTIONS

prove that $u$ satisfies the equation $L(u) = h(x, y, z; t)$ under the initial conditions $u(x, y, z; 0) = 0$, $u_t(x, y, z; 0) = 0$.

By using Poisson's mean value solution, show that

$$u(x, y, z; t) = \frac{1}{4\pi} \iiint_{R \leqslant t} \frac{h(\xi, \eta, \zeta; t - R)}{R} \, d\xi \, d\eta \, d\zeta,$$

where $$R^2 = (x - \xi)^2 + (y - \eta)^2 + (z - \zeta)^2.$$

**10.** $u(x, t)$ is the solution of $u_{tt} - u_{xx} = h(x, t)$ which satisfies the initial conditions $u = 0$, $u_t = 0$ when $t = 0$. By considering the integral of $u_{tt} - u_{xx}$ over a suitable triangle, prove that

$$u(x, t) = \frac{1}{2} \int_0^t d\tau \int_{x - \tau}^{x + \tau} d\xi \, h(\xi, t - \tau).$$

Obtain this result also by a modification of the method of the previous example.

7

# MARCEL RIESZ'S METHOD

## 7.1 A comparison with potential theory

Let $u$ be a solution of Laplace's equation $\nabla^2 u = 0$ which has continuous derivatives of the second order inside and on a simple regular closed surface $\Sigma$, and let $P_0(x_0, y_0, z_0)$ be any point inside $\Sigma$. The elementary solution which plays an important part in potential theory is

$$v = \frac{1}{\sqrt{\{(x-x_0)^2 + (y-y_0)^2 + (z-z_0)^2\}}}.$$

If we apply Green's transformation to

$$\iiint_V (u\nabla^2 v - v\nabla^2 u)\,dx\,dy\,dz = 0,$$

where $V$ is bounded externally by $\Sigma$ and internally by a small sphere $S$ with centre $(x_0, y_0, z_0)$, $v$ being this elementary solution, we obtain

$$\iint_\Sigma \left(u\frac{\partial v}{\partial N} - v\frac{\partial u}{\partial N}\right) dS + \iint_S \left(u\frac{\partial v}{\partial N} - v\frac{\partial u}{\partial N}\right) dS = 0,$$

where $\partial/\partial N$ denotes differentiation along the normal drawn out of $V$. If we now make the radius of $S$ tend to zero, we find that

$$u(x_0, y_0, z_0) = \frac{1}{4\pi}\iint_\Sigma \left(v\frac{\partial u}{\partial N} - u\frac{\partial v}{\partial N}\right) dS,$$

which expresses $u$ at a point $P_0$ inside $\Sigma$ in terms of the values taken by $u$ and $\partial u/\partial N$ on $\Sigma$. It does not solve the problem of Cauchy, since we cannot assign $u$ and $\partial u/\partial N$ arbitrarily on $\Sigma$; given $u$ on $\Sigma$, $\partial u/\partial N$ is determined there.

If we replace $x$ and $y$ by $ix$ and $iy$ respectively and $z$ by $t$, Laplace's equation becomes the two-dimensional equation of wave motions

$$L(u) \equiv \frac{\partial^2 u}{\partial t^2} - \frac{\partial^2 u}{\partial x^2} - \frac{\xi^2 u}{\partial y^2} = 0.$$

We might expect that an application of Green's transformation to

$$\iiint_V \{uL(v) - vL(u)\}\,dx\,dy\,dt = 0$$

with

$$v = \frac{1}{\sqrt{\{(t-t_0)^2 - (x-x_0)^2 - (y-y_0)^2\}}},$$

[ 107 ]

would enable us to find $u$ in terms of the boundary values of $u$ and $\partial u/\partial N$ on a surface $\Sigma$ in the space with $(x, y, t)$ as Cartesian coordinates. But we immediately meet difficulties which do not appear in potential theory.

In the first place, $v$ is real only when

$$(x - x_0)^2 + (y - y_0)^2 \leqslant (t - t_0)^2,$$

so we take $V$ to be bounded in part by the characteristic cone

$$(x - x_0)^2 + (y - y_0)^2 = (t - t_0)^2.$$

Next, the surface $S$ on which we have the Cauchy data must be duly inclined. The characteristic cone then cuts $S$ in a simple closed curve, bounding an area $\Sigma$ on $S$. The boundary of $V$ is $\Sigma$ and the characteristic cone. For example, if $S$ is the plane $t = 0$ so that we are dealing with an initial value problem, $V$ is given by

$$(x - x_0)^2 + (y - y_0)^2 \leqslant (t - t_0)^2, \quad 0 \leqslant t \leqslant t_0.$$

The next difficulty is more serious. On the characteristic cone, $v$ and its derivatives are infinite. There are two classical methods of getting over this difficulty. Volterra used, instead of $v$, its integral

$$\cosh^{-1} \frac{|t - t_0|}{\sqrt{\{(x - x_0)^2 + (y - y_0)^2\}}},$$

which behaves satisfactorily on the characteristic cone but has a line of singularities on the axis of the cone. By cutting out of $V$ the singularities on the axis by means of a small coaxial cylinder and then applying Green's transformation, Volterra obtained a formula for

$$\int^{t_0} u(x_0, y_0, t)\, dt$$

and hence a formula for $u$. The values of $u$ and $\partial u/\partial N$ on the characteristic cone do not appear in the solution because of the properties of Volterra's function. Hadamard, in his *Lectures on Cauchy's Problem*, on the other hand, did not try to avoid the occurrence of divergent integrals, but developed a method of picking out the 'finite part' of a divergent integral.

Marcel Riesz has shown how the difficulties of Hadamard's method disappear if we introduce a complex parameter $\alpha$ whose real part can be chosen so large that all the integrals converge. The solution is then obtained by the analytic continuation of a function of a complex variable. Moreover Riesz's method is applicable to the wave equation in a Euclidean space of any number of dimensions and more generally to the wave equation in a Riemannian space with positive definite metric. What Riesz does is to introduce a generalization of the Riemann–Liouville integral of fractional order of a function of one variable.†

## 7.2  The Riesz integral of functional order
The problem of Cauchy for the equation

$$L(u) \equiv \frac{\partial^2 u}{\partial t^2} - \sum_{k=1}^{n} \frac{\partial^2 u}{\partial x_k^2} = 0 \tag{1}$$

is to find $u$, given the values taken by $u$ and its first derivatives on some duly-inclined $n$-dimensional manifold $S$ in the $(n+1)$-dimensional Euclidean space with rectangular Cartesian coordinates

$$(x_1, x_2, ..., x_n, t).$$

In the case of the initial value problem, $S$ is $t = 0$. As $t$ plays a special part in equation (1), we treat it separately and denote a point in the space–time manifold by $(\mathbf{x}; t)$ where $\mathbf{x}$ is the vector $(x_1, x_2, ..., x_n)$.

The equation of the characteristic cone with vertex $P(\mathbf{x}; t)$ is

$$\Gamma \equiv (t-\tau)^2 - |\mathbf{x} - \boldsymbol{\xi}|^2 = 0$$

where $(\boldsymbol{\xi}; \tau)$ is a variable point in the space–time manifold. This cone divides space–time into three domains, viz. $\Gamma > 0, \tau > t$; $\Gamma > 0, \tau < t$; $\Gamma < 0$. The region in which $\Gamma < 0$ lies outside the cone. $\Gamma = 0$ is a double cone; the part for which $\tau \leqslant t$ is called the retrograde cone. Since, by hypothesis, $S$ is duly inclined, the retrograde cone cuts $S$ is a simple closed curve, if $t$ is large enough. We denote by $V(P, S)$ the domain bounded by $S$ and the retrograde cone. In the particular case of the initial value problem, $t$ is positive and $V(P, S)$ is the set of points $(\boldsymbol{\xi}; \tau)$ for which $\Gamma \geqslant 0, 0 \leqslant \tau \leqslant t$. The Riesz integral of fractional order $\alpha$ of a continuous function $f$ is defined to be

$$I^\alpha f(\mathbf{x}; t) = \frac{1}{H(\alpha, n)} \int_{V(P, S)} f(\boldsymbol{\xi}; \tau) \, \Gamma^{(\alpha-n-1)/2} d\xi_1 d\xi_2 ... d\xi_n d\tau,$$

where $t$ is chosen so that $V(P, S)$ is not an empty set.

The constant $H(\alpha, n)$ is chosen so that $I^\alpha e^t = e^t$ when $S$ is $\tau = -\infty$. Then

$$H(\alpha, n) e^t = \int_V e^\tau \, \Gamma^{(\alpha-n-1)/2} d\boldsymbol{\xi} \, d\tau,$$

† *Acta math.* **81** (1949), pp. 1–223.

where $V$ is the region $\Gamma \geqslant 0$, $\tau \leqslant t$. By a change of origin, we may assume that $x = 0$; and if we put $\tau = t - \tau'$, we get

$$H(\alpha, n) = \int_W e^{-\tau'} \Gamma^{(\alpha - n - 1)/2} d\xi \, d\tau',$$

where $W$ is defined by

$$\Gamma = \tau'^2 - r^2 \geqslant 0, \quad \tau' \geqslant 0,$$

and

$$r^2 = \xi_1^2 + \xi_2^2 + \ldots + \xi_n^2.$$

If we replace $(\xi_1, \xi_2, \ldots, \xi_n)$ by spherical polar coordinates and drop the dashes, we have

$$H(\alpha, n) = \int_0^\infty d\tau \int_0^\tau dr \int_{\Omega_n} d\Omega_n \, e^{-\tau} (\tau^2 - r^2)^{(\alpha - n - 1)/2} r^{n-1},$$

where $d\Omega_n$ is the element of solid angle in $n$-dimensional Euclidean space, and so

$$H(\alpha, n) = \Omega_n \int_0^\infty d\tau \int_0^\tau dr \, e^{-\tau} r^{n-1} (\tau^2 - r^2)^{(\alpha - n - 1)/2}.$$

Since the total solid angle $\Omega_n$ is $2\pi^{n/2}/\Gamma(\tfrac{1}{2}n)$, this gives

$$H(\alpha, n) = 2^{\alpha-1} \pi^{(n-1)/2} \Gamma(\tfrac{1}{2}\alpha) \, \Gamma(\tfrac{1}{2}\alpha - \tfrac{1}{2}n + \tfrac{1}{2}).$$

With this choice of $H(\alpha, n)$, $I^\alpha f(x; t)$ is an analytic function of the complex variable $\alpha$, regular when $\mathrm{re}\,\alpha > n - 1$.

The index notation is used because

$$I^\alpha I^\beta f(x; t) = I^{\alpha+\beta} f(x; t).$$

We prove this first when the real parts of $\alpha$ and $\beta$ exceed $n + 1$, and then use analytic continuation. When we do this, all the integrands are continuous.

If the real parts of $\alpha$ and $\beta$ exceed $n + 1$,

$$H(\alpha, n) \, H(\beta, n) \, I^\alpha I^\beta f(x; t)$$

$$= \int_{V(P, S)} d\xi \, d\tau \int_{V(Q, S)} d\xi' \, d\tau' f(\xi'; \tau') \, \Gamma^{(\alpha - n - 1)/2} \Gamma_1^{(\beta - n - 1)/2},$$

where $Q$ is the point $(\xi; \tau)$ and $\Gamma_1 = (\tau - \tau')^2 - |\xi - \xi'|^2$. When $f$ is continuous we can invert the order of integration to get

$$H(\alpha, n) \, H(\beta, n) \, I^\alpha I^\beta f(x; t) = \int_{V(P, S)} f(\xi'; \tau') \, g(x, t; \xi', \tau') \, d\xi' \, d\tau',$$

where

$$g(x, t; \xi', \tau') = \int_W \Gamma^{(\alpha - n - 1)/2} \Gamma_1^{(\beta - n - 1)/2} d\xi \, d\tau.$$

The domain $W$ is defined by $\Gamma > 0, \Gamma_1 > 0, \tau' < \tau < t$. But

$$g(x, t; \xi', \tau') = g(x - \xi', t - \tau'; 0, 0),$$

and
$$g(x, t; 0, 0) = \int_{W_0} \Gamma^{(\alpha-n-1)/2} \Gamma_0^{(\beta-n-1)/2} d\xi \, d\tau,$$

where
$$\Gamma_0 = \tau^2 - \xi^2,$$

and $W_0$ is defined by $\Gamma > 0, \Gamma_0 > 0, 0 < \tau < t$.

By a notation of axes, we can replace $x$ in this integral by

$$(r, 0, 0, \ldots, 0),$$

where $r = |x|$. Then, by a Lorentz transformation

$$x_1 = x_1' \cosh \gamma + t' \sinh \gamma,$$

$$t = x_1' \sinh \gamma + t' \cosh \gamma$$

we can replace $r$ by zero, and $t$ by $s = \sqrt{(t^2 - r^2)}$. Hence

$$g(x, t; 0, 0) = g(0, s; 0, 0).$$

But $\quad g(0, s; 0, 0) = \int_{W_1} \{(s-\tau)^2 - \xi^2\}^{(\alpha-n-1)/2} (\tau^2 - \xi)^{2(\beta-n-1)/2} d\xi \, d\tau,$

where $\quad \tau^2 - \xi^2 \geqslant 0, \quad (s-\tau)^2 - \xi^2 \geqslant 0, \quad 0 \leqslant \tau \leqslant s$

on $W_1$. Therefore $\quad g(0, s; 0, 0) = K(\alpha, \beta) s^{\alpha+\beta-n-1},$

and so $\quad g(x, t; \xi', \tau') = K(\alpha, \beta) \Gamma_2^{(\alpha+\beta-n-1)/2},$

where $\quad \Gamma_2 = (t - \tau')^2 - |x - \xi'|^2.$
Hence

$$H(\alpha, n) H(\beta, n) I^\alpha I^\beta f(x; t) = K(\alpha, \beta) \int_{V(P, S)} f(\xi', \tau') \Gamma_2^{(\alpha+\beta-n-1)/2} d\xi' \, d\tau'$$

$$= K(\alpha, \beta) H(\alpha + \beta; n) I^{\alpha+\beta} f(x; t).$$

The coefficients $H(\alpha, n), H(\beta, n), K(\alpha, \beta), H(\alpha + \beta, n)$ do not depend on the function $f(x; t)$ nor on the choice of the hypersurface $S$. Considering the case when $f(x; t) = e^t$ and $S$ is $t = -\infty$, we see that

$$H(\alpha, n) H(\beta, n) = K(\alpha, \beta) H(\alpha + \beta, n),$$

and so $I^\alpha I^\beta f(x; t) = I^{\alpha+\beta} f(x; t)$ when the real parts of $\alpha$ and $\beta$ exceed $n + 1$. By analytical continuation, the result holds when the real parts of $\alpha$ and $\beta$ exceed $n - 1$.

Lastly, when the real part of $\alpha$ is sufficiently large, we can differentiate under the sign of integration in the equation

$$I^{\alpha+2}f(x;t) = \frac{1}{H(\alpha+2,n)} \int_{V(P,S)} f(\xi,\tau)\,\Gamma^{(\alpha-n+1)/2}d\xi\,d\tau$$

despite the fact that the variables $x_1, x_2, \ldots, x_n, t$ occur not only in the integrand but also in the equations defining the boundary of $V(P,S)$. For, if $h > 0$, we have

$$\frac{H(\alpha+2,n)}{h} \{I^{\alpha+2}f(x;t+h) - I^{\alpha+2}f(x;t)\}$$

$$= \int_{V(P,S)} f(\xi;\tau)\{\Gamma_+^{(\alpha-n+1)/2} - \Gamma^{(\alpha-n+1)/2}\}\frac{d\xi\,d\tau}{h}$$

$$+ \int_{V_+} f(\xi;\tau)\,\Gamma_+^{(\alpha-n+1)/2}\frac{d\xi\,d\tau}{h},$$

where                     $\Gamma_+ = (t+h-\tau)^2 - |x-\xi|^2,$

and $V_+$ is the conical shell bounded by $\Gamma = 0$, $\Gamma_+ = 0$ and $S$. The volume of the shell is $O(h)$ and $\Gamma_+$ vanishes on the outer boundary of the shell. Hence the second term tends to zero as $h \to 0$ if re $\alpha > n-1$. A similar argument applies when $h$ is negative. Hence, when re $\alpha > n+1$,

$$H(\alpha+2,n)\frac{\partial}{\partial t}I^{\alpha+2}f(x;t)$$

$$= (\alpha-n+1)\int_{V(P,S)} f(\xi;\tau)\,\Gamma^{(\alpha-n+1)/2}(t-\tau)\,d\xi\,d\tau.$$

Similarly, when re $\alpha > n+3$,

$$H(\alpha+2,n)\frac{\partial^2}{\partial t^2}I^{\alpha+2}f(x;t)$$

$$= (\alpha-n+1)(\alpha-n-1)\int_{V(P,S)} f(\xi;\tau)\,\Gamma^{(\alpha-n-3)/2}(t-\tau)^2\,d\xi\,d\tau$$

$$+ (\alpha-n+1)\int_{V(P,S)} f(\xi,\tau)\,\Gamma^{(\alpha-n-1)/2}d\xi\,d\tau,$$

and

$$H(\alpha+2,n)\frac{\partial^2}{\partial x_1^2}I^{\alpha+2}f(x;t)$$

$$= (\alpha-n+1)(\alpha-n-1)\int_{V(P,S)} f(\xi;\tau)\,\Gamma^{(\alpha-n-3)/2}(x_1-\xi_1)^2\,d\xi\,d\tau$$

$$- (\alpha-n-1)\int_{V(P,S)} f(\xi;\tau)\,\Gamma^{(\alpha-n-1)/2}d\xi\,d\tau.$$

Therefore, when $\operatorname{re} \alpha > n+3$,

$$H(\alpha+2, n) LI^{\alpha+2}f(x; t) = \alpha(\alpha-n+1)\int_{V(P, S)} f(\xi; \tau)\, \Gamma^{(\alpha-n-1)/2} d\xi\, d\tau$$

$$= \alpha(\alpha-n+1) H(\alpha, n)\, I^\alpha f(x; t)$$

or $\qquad\qquad\qquad L\Gamma^{\alpha+2}f(x; t) = I^\alpha f(x; t).$

This result holds when $\operatorname{re}\alpha > n-1$, since the expressions on both sides of the equation are analytic functions, regular when $\operatorname{re}\alpha > n-1$.

The operators $L$ and $I$ do not commute. For

$$LI^{\alpha+2}f(x; t) - I^{\alpha+2}Lf(x; t) = \frac{1}{H(\alpha+2, n)}\int_{V(P, S)} (f\Lambda v - v\Lambda f)\, d\xi\, d\tau,$$

where $\qquad\qquad\qquad v = \Gamma^{(\alpha-n+1)/2},$

and $\qquad\qquad\qquad \Lambda f(\xi; \tau) = \dfrac{\partial^2 f}{\partial \tau^2} - \sum_1^n \dfrac{\partial^2 f}{\partial \xi_k^2}.$

The integral can be transformed into a integral over the boundary of $V(P, S)$. The integral over the part of the retrograde cone vanishes when $\operatorname{re}\alpha > n+1$, but the integral over the relevant part of $S$ does not in general vanish.

In what follows we write $H(\alpha)$ for $H(\alpha, n)$, the dimension $n$ being evident from the context.

## 7.3   The analytical continuation of Riesz's integral

In the initial value problem for the equation of wave motions, the Cauchy conditions are satisfied on $t = 0$ and the solution is required for $t > 0$. In this case,

$$I^\alpha f(x; t) = \frac{1}{H(\alpha)}\int_V f(\xi, \tau)\, \Gamma^{(\alpha-n-1)/2} d\tau\, d\xi, \qquad (1)$$

where $V$ is the region on which $\Gamma \geqslant 0$, $0 \leqslant \tau \leqslant t$. If we write $\xi_0 = \sqrt{\Gamma}$, this becomes

$$I^\alpha f(x; t) = \frac{1}{H(\alpha)}\int_W f(\xi; t-r)\, \frac{\xi_0^{\alpha-n}}{r}\, d\xi_0\, d\xi, \qquad (2)$$

where $\qquad\qquad\qquad r^2 = \xi_0^2 + \sum_1^n (\xi_k - x_k)^2,$

and $W$ is the hemisphere $0 \leqslant r \leqslant t$, $\xi_0 \geqslant 0$ in the $(n+1)$-dimensional space with Cartesian coordinates $(\xi_0, \xi_1, \xi_2, ..., \xi_n)$. Sometimes it is more convenient to use polar coordinates

$$\xi_0 = r\cos\theta, \quad \xi_k = x_k + l_k r\sin\theta,$$

where $\sum_1^n l_k^2 = 1$. In polar coordinates, (2) becomes

$$I^\alpha f(x;t) = \frac{1}{H(\alpha)} \int_0^t dr \int_0^{\frac{1}{2}\pi} d\theta \int_{\Omega_n} d\Omega_n f(x + lr \sin\theta; t - r)$$
$$\times r^{\alpha-1} \sin^{n-1}\theta \cos^{\alpha-n}\theta, \quad (3)$$

where $l$ is the vector $(l_1, l_2, ..., l_n)$ and $d\Omega_n$ is the element of solid angle in $n$-dimensional space. From this it follows at once that $I^\alpha f(x;t)$ is an analytic function of $\alpha$, regular in $\operatorname{re}\alpha > \max(0, n-1)$, on the assumption that $f(x;t)$ is continuous. To extend the region of regularity, we assume that $f$ is continuously differentiable to a suitable order.

Since $(\alpha - n + 1) H(\alpha) = H(\alpha+2)/\alpha$, we may write (2) in the form

$$I^\alpha f(x;t) = \frac{\alpha}{H(\alpha+2)} \int_W f(\xi; t-r) \frac{1}{r} \frac{d\xi_0^{\alpha-n+1}}{d\xi_0} d\xi_0 d\xi.$$

Integrating by parts, we obtain, when $\operatorname{re}\alpha > n - 1$,

$$I^\alpha f(x;t) = \frac{\alpha}{H(\alpha+2)} \int_W [f(\xi; t-r) + r f_t(\xi; t-r)] \frac{\xi_0^{\alpha-n+2}}{r^3} d\xi_0 d\xi$$
$$+ \frac{\alpha}{H(\alpha+2)} \int_\Sigma f(\xi; 0) \frac{\xi_0^{\alpha-n+1}}{t} d\xi, \quad (4)$$

where $\Sigma$ is $\qquad \xi_0^2 + \sum_1^n (\xi_k - x_k)^2 = t^2, \quad 0 \leqslant \xi_0 \leqslant t.$

On $\Sigma$, $\qquad \xi = x + l\rho$, where $\rho = t \sin\theta$.

Then, on $\Sigma$, $\quad d\xi = \rho^{n-1} d\rho d\Omega_n = t^n \sin^{n-1}\theta \cos\theta d\theta d\Omega.$

The last term in (4) is

$$J_3 = \frac{\alpha t^\alpha}{H(\alpha+2)} \int_0^{\frac{1}{2}\pi} d\theta \int_{\Omega_n} d\Omega_n f(x + lt \sin\theta; 0) \sin^{n-1}\theta \cos^{\alpha-n+2}\theta.$$

When $t > 0$, this is an analytic function regular when $\operatorname{re}\alpha > n - 3$.

The term involving $f_t$, which by hypothesis is continuous, is

$$J_2 = \frac{\alpha}{H(\alpha+2)} \int_0^t dr \int_0^{\frac{1}{2}\pi} d\theta$$
$$\times \int_{\Omega_n} d\Omega_n f_t(x + lr \sin\theta; t - r) r^\alpha \sin^{n-1}\theta \cos^{\alpha-n+2}\theta$$

which is an analytic function of $\alpha$, regular when $\operatorname{re}\alpha > \max(-1, n-3)$. The remaining term is

$$J_1 = \frac{\alpha}{H(\alpha+2)} \int_0^t dr \int_0^{\frac{1}{2}\pi} d\theta \int_{\Omega_n} d\Omega_n f(x + lr \sin\theta; t - r)$$
$$\times r^{\alpha-1} \sin^{n-1}\theta \cos^{\alpha-n+2}\theta$$

which is regular when $\operatorname{re}\alpha > \max(0, n-3)$.

(i) When $n = 2$, $2\pi J_3$ is equal to

$$\frac{\alpha t^\alpha}{\Gamma(\alpha+1)} \int_0^{\frac{1}{2}\pi} d\theta \sin\theta \cos^\alpha\theta \int_0^{2\pi} d\phi\, f(\mathbf{x}+\mathbf{1}t\sin\theta; 0),$$

where $\mathbf{x} = (x,y)$, $\mathbf{1} = (\cos\phi, \sin\phi)$ so that $d\Omega_2 = d\phi$. This is an analytic function of $\alpha$, when $t > 0$, regular when re $\alpha > -1$. Hence $J_3$ is zero when $\alpha = 0$.

Next, $2\pi J_2$ is equal to

$$\frac{\alpha}{\Gamma(\alpha+1)} \int_0^t dr\, r^\alpha \int_0^{\frac{1}{2}\pi} d\theta \sin\theta \cos^\alpha\theta \int_0^{2\pi} d\phi\, f_t(\mathbf{x}+\mathbf{1}r\sin\theta; t-r).$$

This again is regular when re $\alpha > -1$; and vanishes when $\alpha = 0$,
Lastly, when re $\alpha > 0$, $2\pi J_1$ is equal to

$$\frac{\alpha}{\Gamma(\alpha+1)} \int_0^t dr\, r^{\alpha-1} \int_0^{\frac{1}{2}\pi} d\theta \sin\theta \cos^\alpha\theta \int_0^{2\pi} d\phi\, f(\mathbf{x}+\mathbf{1}r\sin\theta; t-r).$$

Integration by parts is valid since $f$ is continuously differentiable. Hence $2\pi J_1$ is equal to

$$\frac{t^\alpha}{\Gamma(\alpha+1)} \int_0^{\frac{1}{2}\pi} d\theta \sin\theta \cos^\alpha\theta \int_0^{2\pi} d\phi\, f(\mathbf{x}+\mathbf{1}t\sin\theta; 0)$$

$$-\frac{1}{\Gamma(\alpha+1)} \int_0^t dr\, r^\alpha \int_0^{\frac{1}{2}\pi} d\theta \sin\theta \cos^\alpha\theta \int_0^{2\pi} d\phi\, \frac{\partial}{\partial r} f(\mathbf{x}+\mathbf{1}r\sin\theta; t-r)$$

which is regular when re $\alpha > -1$. When $\alpha$ is zero, $2\pi J_1$ becomes

$$\int_0^{\frac{1}{2}\pi} d\theta \sin\theta \int_0^{2\pi} d\phi\, f(\mathbf{x}+\mathbf{1}t\sin\theta\sin\theta; 0)$$

$$-\int_0^{\frac{1}{2}\pi} d\theta \sin\theta \int_0^{2\pi} d\phi \int_0^t dr\, \frac{\partial}{\partial r} f(\mathbf{x}+\mathbf{1}r\sin\theta; t-r)$$

$$=\int_0^{\frac{1}{2}\pi} d\theta \sin\theta \int_0^{2\pi} d\phi\, \mathbf{f}(\mathbf{x}, t) = 2\pi \mathbf{f}(\mathbf{x}; t).$$

Therefore, when $n = 2$ and $f(x, y; t)$ is continuously differentiable, Riesz's integral $I^\alpha f(x, y; t)$ is regular in re $\alpha > -1$, and

$$I^0 f(x, y; t) = f(x, y; t).$$

(ii) When $n = 3$ and $\mathrm{re}\,\alpha > 0$,

$$J_3 = \frac{t^\alpha}{2^\alpha \pi \Gamma(\tfrac{1}{2}\alpha)\,\Gamma(\tfrac{1}{2}\alpha)} \int_0^{\frac{1}{2}\pi} d\theta \int_{\Omega_3} d\Omega_3 \sin^2\theta \cos^{\alpha-1}\theta f(x + lt\sin\theta; 0),$$

where $l$ is the vector $(\sin\phi\cos\psi, \sin\phi\sin\psi, \cos\phi)$ and $d\Omega_3 = \sin\phi\,d\phi\,d\psi$. If $f$ is continuously differentiable, we may integrate by parts with respect to $\theta$ and obtain

$$J_3 = \frac{t^\alpha}{2^{\alpha+1}\pi\Gamma(1+\tfrac{1}{2}\alpha)\,\Gamma(\tfrac{1}{2}\alpha)} \int_0^{\frac{1}{2}\pi} d\theta \int_{\Omega_3} d\Omega_3 \cos^\alpha\theta$$

$$\times \frac{\partial}{\partial\theta}\{\sin\theta f(x + lt\sin\theta; 0)\}.$$

Hence $J_3$ is regular when $\mathrm{re}\,\alpha > -1$ and vanishes when $\alpha = 0$.

To deal with $J_1$ and $J_2$ we make the additional assumption that $f$ has continuous derivatives of the second order. When $\mathrm{re}\,\alpha > 0$, we integrate by parts in

$$J_2 = \frac{1}{2^\alpha\pi\Gamma(\tfrac{1}{2}\alpha)\,\Gamma(\tfrac{1}{2}\alpha)} \int_0^t dr \int_0^{\frac{1}{2}\pi} d\theta \int_{\Omega_3} d\Omega_3$$

$$\times f_t(x + lr\sin\theta; t - r)\, r^\alpha \sin^2\theta \cos^{\alpha-1}\theta$$

and obtain

$$J_2 = \frac{1}{2^{\alpha+1}\pi\Gamma(1+\tfrac{1}{2}\alpha)\,\Gamma(\tfrac{1}{2}\alpha)} \int_0^t dr \int_0^{\frac{1}{2}\pi} d\theta \int_{\Omega_3} d\Omega_3$$

$$\times r^\alpha \cos^\alpha\theta \frac{\partial}{\partial\theta} f_t(x + lr\sin\theta; t - r).$$

Hence $J_2$ is also regular when $\mathrm{re}\,\alpha > -1$ and vanishes when $\alpha = 0$.

Lastly, when $\mathrm{re}\,\alpha > 0$, we may again integrate by parts with respect to $\theta$ in

$$J_1 = \frac{1}{2^\alpha\pi\Gamma(\tfrac{1}{2}\alpha)\,\Gamma(\tfrac{1}{2}\alpha)} \int_0^t dr \int_0^{\frac{1}{2}\pi} d\theta \int_{\Omega_3} d\Omega_3$$

$$\times f(x + lr\sin\theta; t - r)\, r^{\alpha-1}\sin^2\theta \cos^{\alpha-1}\theta$$

and obtain

$$J_1 = \frac{1}{2^{\alpha+1}\pi\Gamma(1+\tfrac{1}{2}\alpha)\,\Gamma(\tfrac{1}{2}\alpha)} \int_0^t dr \int_0^{\frac{1}{2}\pi} d\theta \int_{\Omega_3} d\Omega_3\, r^{\alpha-1}\cos^\alpha\theta F$$

where

$$F = \frac{\partial}{\partial\theta}(f(x + lr\sin\theta; t - r)\sin\theta),$$

since, by hypothesis, $F$ is continuous. In fact, $F$ is continuously differentiable, so we may integrate by parts with respect to $r$, to get

$$J_1 = \frac{1}{2^{\alpha+2}\pi\Gamma(1+\tfrac{1}{2}\alpha)\,\Gamma(1+\tfrac{1}{2}\alpha)} \left\{ \int_0^{\frac{1}{2}\pi} d\theta \int_{\Omega_3} d\Omega_3 \cos^\alpha\theta\, t^\alpha[F]_{r=t} \right.$$

$$\left. - \int_0^{\frac{1}{2}\pi} d\theta \int_{\Omega_3} d\Omega_3 \int_0^t dr\, r^\alpha \cos^\alpha\theta \frac{\partial F}{\partial r} \right\}.$$

Thus $J_1$ is regular when $\operatorname{re}\alpha > -1$. When $\alpha = 0$,

$$J_1 = \frac{1}{4\pi}\int_0^{\frac{1}{2}\pi} d\theta \int_{\Omega_3} d\Omega_3 [F]_{r=t} - \frac{1}{4\pi}\int_0^{\frac{1}{2}\pi} d\theta \int_{\Omega_3} d\Omega_3 \int_0^t r \frac{\partial F}{\partial r}$$

$$= \frac{1}{4\pi}\int_0^{\frac{1}{2}\pi} d\theta \int_{\Omega_3} d\Omega_3 [F]_{r=0}$$

$$= \frac{1}{4\pi}\int_{\Omega_3} d\Omega_3 \int_0^{\frac{1}{2}\pi} d\theta \frac{\partial}{\partial\theta}\{f(\boldsymbol{x};t)\sin\theta\}$$

$$= f(\boldsymbol{x};t).$$

Thus when $f(\boldsymbol{x};t)$ has continuous derivatives of the second order, $I^{\alpha}f(\boldsymbol{x};t)$ is an analytic function of $\alpha$, regular in $\operatorname{re}\alpha > -1$, and

$$I^0 f(\boldsymbol{x};t) = f(\boldsymbol{x};t).$$

(iii) The argument can be repeated for larger values of $n$. $I^{\alpha}f(\boldsymbol{x};t)$ is regular in $\operatorname{re}\alpha > -1$ provided that we assume that $f(\boldsymbol{x};t)$ has continuous partial derivatives of a sufficiently high order. The cases $n = 2$ and $n = 3$ suffice for our applications. The simple case $n = 1$ occurs in Ex. 2 at the end of the chapter.

## 7.4   Cauchy's problem for the non-homogeneous wave equation in two dimensions

If
$$Lu = \frac{\partial^2 u}{\partial t^2} - \frac{\partial^2 u}{\partial x^2} - \frac{\partial^2 u}{\partial y^2},$$

the problem is to solve      $Lu = F(x,y;t),$          (1)
when $t > 0$, given that

$$u = f(x,y), \quad u_t = h(x,y),$$

when $t = 0$. Here $f, h$ and $F$ are continuous functions, with continuous partial derivatives of orders which emerge in the sequel. The solution is to have continuous second order derivatives.

If $v$ is a function of $\xi, \eta, \tau$, it is convenient to write

$$\Lambda v = \frac{\partial^2 v}{\partial \tau^2} - \frac{\partial^2 v}{\partial \xi^2} - \frac{\partial^2 v}{\partial \eta^2}.$$

Let $P$ be any fixed point $(x,y;t)$ in space–time with $t > 0$, and let $V$ be the region $\Gamma \geqslant 0, 0 \leqslant \tau \leqslant t$ where

$$\Gamma = (t-\tau)^2 - (x-\xi)^2 - (y-\eta)^2.$$

By Green's transformation

$$\int_V \{u(\xi,\eta;\tau)\,\Lambda v(\xi,\eta;\tau) - v(\xi,\eta;\tau)\,\Lambda u(\xi,\eta;\tau)\}\,d\xi\,d\eta\,d\tau$$

$$= \int_S \left\{ \nu\left( u\frac{\partial v}{\partial \tau} - v\frac{\partial u}{\partial \tau}\right) - \lambda\left( u\frac{\partial v}{\partial \xi} - v\frac{\partial u}{\partial \xi}\right) - \mu\left( u\frac{\partial v}{\partial \eta} - v\frac{\partial u}{\partial \eta}\right) \right\}dS,$$

where $S$ is the boundary of $V$ and $(\lambda,\mu,\nu)$ are the direction cosines of the normal to $S$ drawn out of $V$. The transformation is valid when $u$ and $v$ have continuous derivatives of the second order.

Let $u(x,y;t)$ be the desired solution of (1), and let $v = \Gamma^{(\alpha-1)/2}$. Then $\Lambda u = F(\xi,\eta;\tau)$, and $\Lambda v = \alpha(\alpha-1)\,\Gamma^{(\alpha-3)/2}$. Then, when re $\alpha > 3$,

$$\int_V \{\alpha(\alpha-1)\,u\Gamma^{(\alpha-3)/2} - F\Gamma^{(\alpha-1)/2}\}\,d\xi\,d\eta\,d\tau$$

$$= \int_S [(\alpha-1)\,\Gamma^{(\alpha-3)/2}\,u\{\lambda(\xi-x) + \mu(\eta-y) + \nu(\tau-t)\}$$
$$+ \Gamma^{(\alpha-1)/2}\{\lambda u_\xi + \mu u_\eta - \nu u_\tau\}]\,dS.$$

Since re $\alpha > 3$, the integral over the retrograde cone part of $S$ vanishes; the rest of $S$ is $\Sigma$, specified by

$$\tau = 0, \quad \Gamma_0 \equiv t^2 - (x-\xi)^2 - (y-\eta)^2 \geqslant 0.$$

On $\Sigma$, $\lambda = \mu = 0, \nu = -1$. Hence

$$\int_V \{\alpha(\alpha-1)\,u\Gamma^{(\alpha-3)/2} - F\Gamma^{(\alpha-1)/2}\}\,d\xi\,d\eta\,d\tau$$

$$= \int_\Sigma \{\Gamma_0^{(\alpha-1)/2}u_\tau + (\alpha-1)\,\Gamma_0^{(\alpha-3)/2}\,tu\}\,d\xi\,d\eta.$$

Using the initial conditions, this is, in Riesz's notation,

$$I^\alpha u - I^{\alpha+2}F = \frac{1}{2\pi\Gamma(\alpha+1)}\int_\Sigma \Gamma_0^{(\alpha-1)/2}h(\xi,\eta)\,d\xi\,d\eta$$

$$+ \frac{1}{2\pi\Gamma(\alpha+1)}\frac{\partial}{\partial t}\int_\Sigma \Gamma_0^{(\alpha-1)/2}f(\xi,\eta)\,d\xi\,d\eta.$$

Now the expressions on each side of this equation are analytic functions of $\alpha$, regular when re $\alpha > -1$; although we have proved the result only for re $\alpha > 3$, it is true in this larger domain. In particular, taking $\alpha = 0$, we have

$$u(x,y;t) = I^2F(x,y;t) + \frac{1}{2\pi}\int_\Sigma \Gamma_0^{-\frac{1}{2}}h(\xi,\eta)\,d\xi\,d\eta$$

$$+ \frac{1}{2\pi}\frac{\partial}{\partial t}\int_\Sigma \Gamma_0^{-\frac{1}{2}}f(\xi,\eta)\,d\xi\,d\eta,$$

or $\quad u(x,y;t) = \dfrac{1}{2\pi}\displaystyle\int_V F(\xi,\eta;\tau)\{(t-\tau)^2-(x-\xi)^2-(y-\eta)^2\}^{-\frac{1}{2}}d\xi\,d\eta\,d\tau$

$\qquad\qquad +\dfrac{1}{2\pi}\displaystyle\int_\Sigma h(\xi,\eta)\{t^2-(x-\xi)^2-(y-\eta)^2\}^{-\frac{1}{2}}d\xi\,d\eta$

$\qquad\qquad +\dfrac{1}{2\pi}\dfrac{\partial}{\partial t}\displaystyle\int_\Sigma f(\xi,\eta)\{t^2-(x-\xi)^2-(y-\eta)^2\}^{-\frac{1}{2}}d\xi\,d\eta. \quad (2)$

When $F$ is identically zero, this reduces to the solution found in §6.4.

If we put $\qquad \xi = x+\rho\cos\phi, \quad \eta = y+\rho\sin\phi,$

the first term in (2) becomes

$\Phi_1(x,y;t) = \dfrac{1}{2\pi}\displaystyle\int_0^t d\tau\int_0^{t-r}\rho\,d\rho\int_0^{2\pi}d\phi$

$\qquad\qquad\qquad \times F(x+\rho\cos\phi,y+\rho\sin\phi;\tau)\{(t-\tau)^2-\rho^2\}^{-\frac{1}{2}}$

$\qquad\quad = \dfrac{1}{2\pi}\displaystyle\int_0^t d\tau\int_0^\tau\rho\,d\rho\int_0^{2\pi}d\phi$

$\qquad\qquad\qquad \times F(x+\rho\cos\phi,y+\rho\sin\phi;t-\tau)\{\tau^2-\rho^2\}^{-\frac{1}{2}}.$

If we then put $\rho = t\rho', \tau = t\tau'$, we get

$\Phi_1(x,y;t) = \dfrac{t^2}{2\pi}\displaystyle\int_0^1 d\tau\int_0^\tau\rho\,d\rho\int_0^{2\pi}d\phi$

$\qquad\qquad\qquad \times F(x+t\rho\cos\phi,y+t\rho\sin\phi;t-t\tau)(\tau^2-\rho^2)^{-\frac{1}{2}} \quad (3)$

dropping the dashes. This representation as a repeated integral is valid since $F$ is, by hypothesis, continuous. The expression on the right of (3) is continuous in $t \geqslant 0$. If $F(x,y;t)$ has continuous derivatives of the second order, so also has $\Phi_1(x,y;t)$; and $\Phi_1$ and $\partial\Phi_1/\partial t$ vanish when $t = 0$.

The second term in (2) is

$\Phi_2(x,y;t) = \dfrac{t}{2\pi}\displaystyle\int_0^1\rho\,d\rho\int_0^{2\pi}d\phi\,h(x+t\rho\cos\phi,y+t\rho\sin\phi)(1-\rho^2)^{-\frac{1}{2}},$

which vanishes when $t = 0$. Since $h$ is continuous and

$\dfrac{1}{t}\{\Phi_2(x,y;t)-\Phi_2(x,y;0)\}$

$\qquad = \dfrac{1}{2\pi}\displaystyle\int_0^1\rho\,d\rho\int_0^{2\pi}d\phi\,h(x+t\rho\cos\phi,y+t\rho\sin\phi)(1-\rho^2)^{-\frac{1}{2}}$

$\partial \Phi_2/\partial t$ exists and is equal to

$$\frac{1}{2\pi} h(x,y) \int_0^1 \frac{\rho \, d\rho}{\sqrt{(1-\rho^2)}} \int_0^{2\pi} d\phi = h(x,y).$$

If $h$ has continuous derivatives of the second order, so also has $\Phi_2(x,y;t)$.

The third term in (2) is

$$\Phi_3(x,y;t) = \frac{1}{2\pi} \frac{\partial}{\partial t} t \int_0^1 \rho \, d\rho \int_0^{2\pi} d\phi f(x+t\rho \cos\phi, y+t\rho \sin\phi)(1-\rho^2)^{-\frac{1}{2}}$$

$$= \frac{1}{2\pi} \int_0^1 \rho \, d\rho \int_0^{2\pi} d\phi f(x+t\rho \cos\phi, y+t\rho \sin\phi)(1-\rho^2)^{-\frac{1}{2}}$$

$$+ \frac{1}{2\pi} t \int_0^1 \rho \, d\rho \int_0^{2\pi} d\phi \frac{\partial f(x+t\rho \cos\phi, y+t\rho \sin\phi)}{\partial t}(1-\rho^2)^{-\frac{1}{2}},$$

differentiation under the sign of integration being certainly valid when the partial derivatives $f_x$ and $f_y$ are continuous; and then $\Phi_3(x,y;0) = f(x,y)$. In order that $\Phi_3(x,y;t)$ may have continuous partial derivatives of the second order, it suffices that $f$ should have continuous derivatives of the third order.

We have thus proved that equation (2) solves the Cauchy problem for

$$\frac{\partial^2 u}{\partial t^2} - \frac{\partial^2 u}{\partial x^2} - \frac{\partial^2 u}{\partial y^2} = F(x,y;t);$$

when $F(x,y;t)$ and $h(x,y)$ have continuous partial derivatives of the second order, $f(x,y)$ of the third order, the solution (2) has continuous second order derivatives.

## 7.5   The equation of wave motions in three dimensions

The Riesz solution for

$$Lu \equiv \frac{\partial^2 u}{\partial t^2} - \frac{\partial^2 u}{\partial x^2} - \frac{\partial^2 u}{\partial y^2} - \frac{\xi^2 u}{\partial z^2} = F(x,y;z,t) \tag{1}$$

in three spatial dimensions starts like the corresponding solution in two dimensions. The initial conditions are

$$u = f(x,y,z), \quad u_t = h(x,y,z), \tag{2}$$

when $t = 0$. Here $f$, $h$ and $F$ are continuous functions, differentiable to orders which suffice to ensure the existence of a solution with continuous second derivatives.

With any fixed point $(x, y, z; t)$ in space–time, let $V$ be the region in $\tau \geqslant 0$ bounded by $\Gamma = 0$ and $\tau = 0$, where

$$\Gamma = (t - \tau)^2 - (x - \xi)^2 - (y - \eta)^2 - (z - \zeta)^2.$$

If $u$ and $v$ have continuous second derivatives, Green's transformation gives

$$\int_V \{u(\xi, \eta, \zeta; \tau)\, \Lambda v(\xi, \eta, \zeta; \tau) - v(\xi, \eta, \zeta; \tau)\, \Lambda u(\xi, \eta, \zeta; \tau)\}\, d\xi\, d\eta\, d\zeta\, d\tau$$

$$= \int_S \left\{ \varpi \left( u \frac{\partial v}{\partial \tau} - v \frac{\partial u}{\partial \tau} \right) - \lambda \left( u \frac{\partial v}{\partial \xi} - v \frac{\partial u}{\partial \xi} \right) \right.$$

$$\left. - \mu \left( u \frac{\partial v}{\partial \eta} - v \frac{\partial u}{\partial \eta} \right) - \nu \left( u \frac{\partial v}{\partial \zeta} - v \frac{\partial u}{\partial \zeta} \right) \right\} dS,$$

where $S$ is the boundary of $V$ and $(\lambda, \mu, \nu, \varpi)$ are the direction cosines of the outward normal. As before, $\Lambda u = u_{\tau\tau} - u_{\xi\xi} - u_{\eta\eta} - u_{\zeta\zeta}$.

Let $u$ be the desired solution of our problem, and let

$$v = \Gamma^{(\alpha-2)/2}.$$

Then $\qquad\qquad\qquad \Lambda v = \alpha(\alpha - 2)\, \Gamma^{(\alpha-4)/2}.$

We assume in the first instance that $\mathrm{re}\,\alpha > 4$, and then use analytical continuation. We have

$$\int_V \{\alpha(\alpha - 2)\, u(\xi, \eta, \zeta; \tau)\, \Gamma^{(\alpha-4)/2} - F(\xi, \eta, \zeta; \tau)\, \Gamma^{(\alpha-2)/2}\}\, d\xi\, d\eta\, d\zeta\, d\tau$$

$$= \int_S [(\alpha - 2)\, u\Gamma^{(\alpha-4)/2}\{\lambda(\xi - x) + \mu(\eta - y) + \nu(\zeta - z) + \varpi(\tau - t)\}$$

$$+ \Gamma^{(\alpha-2)/2}\left\{ \lambda \frac{\partial u}{\partial \xi} + \mu \frac{\partial u}{\partial \eta} + \nu \frac{\partial u}{\partial \zeta} - \varpi \frac{\partial u}{\partial \tau} \right\} ]\, dS.$$

If we divide through by

$$H(\alpha + 2) = 2^{\alpha+1} \pi \Gamma(\tfrac{1}{2}\alpha)\, \Gamma(\tfrac{1}{2}\alpha + 1),$$

we obtain

$$I^\alpha u(x, y, z; t) - I^{\alpha+2} F(x, y, z; t)$$

$$= \frac{1}{H(\alpha + 2)} \int_\Sigma \left\{ \Gamma_0^{(\alpha-2)/2} \frac{\partial u(\xi, \eta, \zeta; 0)}{\partial \tau} \right.$$

$$\left. + (\alpha - 2)\, \Gamma_0^{(\alpha-4)/2}\, tu(\xi, \eta, \zeta; 0) \right\}\, d\xi\, d\eta\, d\zeta,$$

where $\qquad\qquad \Gamma_0 = t^2 - (x - \xi)^2 - (y - \eta)^2 - (z - \zeta)^2$

and $\Sigma$ is the region of the plane $\tau = 0$ on which $0 \leqslant \Gamma_0 \leqslant t^2$.

The reason for this is that when $\mathrm{re}\,\alpha > 4$, the integrand vanishes on the part of $S$ belonging to the retrograde cone. On $\Sigma$, the rest of $S$, we have $\tau = 0$, $\lambda = \mu = \nu = 0$ and $\varpi = -1$. Using the initial conditions,

$$I^\alpha u(x, y, z; t) - I^{\alpha+2} F(x, y, z; t)$$

$$= \frac{1}{H(\alpha+2)} \int_\Sigma \Gamma_0^{(\alpha-2)/2} h(\xi, \eta, \zeta) \, d\xi \, d\eta \, d\zeta$$

$$+ \frac{1}{H(\alpha+2)} \frac{\partial}{\partial t} \int_\Sigma \Gamma_0^{(\alpha-2)/2} f(\xi, \eta, \zeta) \, d\xi \, d\eta \, d\zeta. \qquad (3)$$

The expression on the left is an analytic function of $\alpha$, regular when $\operatorname{re} \alpha > -1$ when $u$ has continuous derivatives of the second order; its value when $\alpha = 0$ is

$$u(x, y, z; t) - I^2 F(x, y, z; t).$$

If we introduce spherical polar coordinates

$$\xi = x + lr, \quad \eta = y + mr, \quad \zeta = z + nr,$$

where $\qquad l = \sin \theta \cos \phi, \quad m = \sin \theta \sin \phi, \quad n = \cos \theta,$

the first term on the right of (3) is

$$J_1 = \frac{1}{H(\alpha+2)} \int_0^t r^2 \, dr \int_{\Omega_3} d\Omega_3 (t^2 - r^2)^{(\alpha-2)/2} h(x + lr, y + mr, z + nr),$$

where $\qquad\qquad\qquad d\Omega_3 = \sin \theta \, d\theta \, d\phi.$

and $\qquad\qquad\qquad 0 \leqslant \theta \leqslant \pi, \quad 0 \leqslant \phi \leqslant 2\pi.$

Integration by parts gives

$$J_1 = -\frac{1}{\alpha H(\alpha+2)} \int_{\Omega_3} d\Omega_3 \int_0^t dr \, rh(x + lr, y + mr, z + nr) \frac{d}{dr} (t^2 - r^2)^{\alpha/2}$$

$$= \frac{1}{\alpha H(\alpha+2)} \int_{\Omega_3} d\Omega_3 \int_0^t dr \, (t^2 - r^2)^{\alpha/2} \Big\{ h(x + lr, y + mr, z + nr)$$

$$+ r \frac{\partial}{\partial r} h(x + lr, y + mr, z + nr) \Big\},$$

which is valid if $h$ is continuously differentiable. Now

$$\lim_{\alpha \to 0} \alpha H(\alpha + 2) = \lim_{\alpha \to 0} 2^{\alpha+2} \pi \Gamma(\tfrac{1}{2}\alpha + 1) \Gamma(\tfrac{1}{2}\alpha + 1) = 4\pi.$$

Hence

$$\lim_{\alpha \to +0} J_1 = \frac{1}{4\pi} \int_{\Omega_3} d\Omega_3 \int_0^t dr \frac{\partial}{\partial r} \{ rh(x + lr, y + mr, z + nr) \}$$

$$= \frac{1}{4\pi} \int_{\Omega_3} th(x + lt, y + mt, z + nt) \, d\Omega_3$$

$$= t \mathcal{M}(h; x, y, z; t),$$

where $\mathcal{M}(h; x, y, z; t)$ is the mean value of $h(\xi, \eta, \zeta)$ over the sphere

$$(\xi - x)^2 + (\eta - y)^2 + (\zeta - z)^2 = t^2.$$

A similar expression follows for the second term on the right of (3).

Lastly, by §7.3 (2),

$$I^{\alpha+2} F(x, y, z; t) = \frac{\alpha}{2^{\alpha+2}\pi\{\Gamma(1 + \frac{1}{2}\alpha)\}^2} \int_W F(\xi, \eta, \zeta; t - R) \frac{\tau^{\alpha-1}}{R} d\xi\, d\eta\, d\zeta\, d\tau,$$

where $\qquad R^2 = (x - \xi)^2 + (y - \eta)^2 + (z - \zeta)^2 + \tau^2 = r^2 + \tau^2$

with a slight change of notation. Here $W$ is $0 \leqslant R \leqslant t, \tau \geqslant 0$. When we integrate by parts with respect to $\tau$, $\tau$ varies from 0 to $\sqrt{(t^2 - r^2)}$. We then have

$$\Gamma^{\alpha+2} F(x, y; z, t) = \frac{1}{2^{\alpha+2}\pi\{\Gamma(1 + \frac{1}{2}\alpha)\}^2} \int_V \left\{ \left[ \tau^{\alpha} \frac{F(\xi, \eta, \zeta; t - R)}{R} \right]_0^{\sqrt{(t^2 - r^2)}} \right.$$
$$\left. - \int_0^{\sqrt{(t^2 - r^2)}} \tau^{\alpha} \frac{\partial}{\partial \tau} \frac{F(\xi, \eta, \zeta; t - R)}{R} d\tau \right\} d\xi\, d\eta\, d\zeta,$$

where $V$ is $\qquad (x - \xi)^2 + (y - \eta)^2 + (z - \zeta)^2 \leqslant t^2.$

We have transformed $I^{\alpha+2} F(x, y, z; t)$ into a function regular when re $\alpha > -1$. Hence

$$I^2 F(x, y, z; t) = \frac{1}{4\pi} \int_V \left\{ \left[ \frac{F(\xi, \eta, \zeta; t - R)}{R} \right]_{\tau = \sqrt{(t^2 - r^2)}} \right.$$
$$\left. - \int_0^{\sqrt{(t^2 - r^2)}} \frac{\partial}{\partial \tau} \frac{F(\xi, \eta, \zeta; t - R)}{R} d\tau \right\} d\xi\, d\eta\, d\zeta$$
$$= \frac{1}{4\pi} \int_V \frac{F(\xi, \eta, \zeta; t - r)}{r} d\xi\, d\eta\, d\zeta.$$

It follows that

$$u(x, y, z; t) = t \mathcal{M}(h; x, y, z; t) + \frac{\partial}{\partial t} t \mathcal{M}(f; x, y, z; t)$$

$$+ \frac{1}{4\pi} \int_V \frac{F(\xi, \eta, \zeta; t - r)}{r} d\xi\, d\eta\, d\zeta \qquad (4)$$

is the required solution. It has continuous derivatives of the second order if $h(x, y, z)$ and $F(x, y, z; t)$ have continuous derivatives of the second order, and $f(x, y, z)$ has continuous derivatives of the third order.

Where $F$ is identically zero, this reduces to the formula (7) of §6.3, which we obtained by using Boussinesq's spherical potential.

The solution of (1) which satisfies the conditions $u = 0, u_t = 0$ on $t = -a$ is

$$u(x, y, z; t) = \frac{1}{4\pi} \int_{V_a} \frac{F(\xi, \eta, \zeta; t-r)}{r} \, d\xi \, d\eta \, d\zeta,$$

where $V_a$ is now

$$(\xi - x)^2 + (\eta - y)^2 + (\zeta - z)^2 \leqslant (t+a)^2.$$

If we make $a \to +\infty$, we get the well-known retarded potential

$$u(x, y, z; t) = \frac{1}{4\pi} \int_V \frac{F(\xi, \eta, \zeta; t-r)}{r} \, d\xi \, d\eta \, d\zeta,$$

where integration is through all space.

## 7.6   Babha's equation

In his theory of the meson, Bhabha found it necessary to solve the differential equation

$$\frac{\partial^2 u}{\partial t^2} - \frac{\partial^2 u}{\partial x^2} - \frac{\partial^2 u}{\partial y^2} - \frac{\partial^2 u}{\partial z^2} + k^2 u = F(x, y, z; t). \tag{1}$$

The theory of this equation is well known, but rather difficult. We show how the use of Riesz's operator leads to the solution very simply. For brevity, we omit the rigorous details.

The equation (1) is, in the usual notation,

$$Lu + k^2 u = F.$$

If $u$ and $u_t$ vanish when $t = 0$, $I^2 Lu = LI^2 u = u$. Hence

$$u + k^2 I^2 u = I^2 F.$$

This implies that $\quad k^{2j} I^{2j} u + k^{2j+2} I^{2j+2} u = I^{2j+2} F.$

Omitting convergence considerations,

$$u = (I^2 - k^2 I^4 + k^4 I^6 - \dots) F.$$

But

$$I^{2r} F(x, y, z; t) = \frac{1}{2^{2r-1} \pi (r-1)! \, (r-2)!} \int_V F(\xi, \eta, \zeta; \tau) \, \Gamma^{r-2} d\xi \, d\eta \, d\zeta \, d\tau,$$

where $\quad \Gamma = (t - \tau)^2 - (x - \xi)^2 - (y - \eta)^2 - (z - \zeta)^2$

and $V$ is the region on which $\Gamma \geqslant 0, 0 \leqslant \tau \leqslant t$. This formula holds when $r \geqslant 2$. Hence

$$u(x, y, z; t) = I^2 F(x, y, z; t)$$

$$+ \sum_1^\infty \frac{(-1)^n k^{2n}}{2^{2n+1} \pi (n-1)! \, n!} \int_V F(\xi, \eta, \zeta; \tau) \, \Gamma^{n-1} d\xi \, d\eta \, d\zeta \, d\tau.$$

If we write $\qquad r^2 = (x-\xi)^2 + (y-\eta)^2 + (z-\zeta)^2,$

$$R^2 = \Gamma = (t-\tau)^2 - r^2,$$

this becomes

$$u(x,y,z;t) = \frac{1}{4\pi} \int_W \frac{F(\xi,\eta,\zeta;t-r)}{r} \, d\xi \, d\eta \, d\zeta$$

$$- \frac{1}{4\pi} \int_V \frac{F(\xi,\eta,\zeta,\tau)}{R} J_1(kR) \, d\xi \, d\eta \, d\zeta \, d\tau,$$

where $W$ is the sphere $0 \leqslant r \leqslant t$, and $J_1(kR)$ is the Bessel function of order 1.

The form of the solution used by Bhabha differs slightly from this; his result can be obtained by assuming that $u$ and $u_t$ vanish on $t = -a$ and making $a \to +\infty$.

## 7.7 A mixed boundary and initial value problem

The problem is to find the solution of

$$Lu \equiv \frac{\partial^2 u}{\partial t^2} - \frac{\partial^2 u}{\partial x^2} - \frac{\partial^2 u}{\partial y^2} = F(x,y;t) \tag{1}$$

in $y > 0, t > 0$, given that $u = f(x,y), u_t = h(x,y)$ on $t = 0, y \geqslant 0$ and $u = \phi(x,t)$ on $y = 0, t \geqslant 0$. The data are assumed to be continuous, and, in particular, $f(x,0) = \phi(x,0), h(x,0) = \phi_t(x,0)$.

We may assume that $\phi(x,t) \equiv 0$. For, if not, the substitution $u = v + \phi(x,t)$ leads to the same problem with different $F$, $f$ and $h$ and with $v$ zero on $y = 0, t > 0$.

We wish to find $u(x,y;t)$ when $y > 0, t > 0$. The retrograde characteristic cone

$$\Gamma \equiv (t-\tau)^2 - (x-\xi)^2 - (y-\eta)^2 = 0$$

cuts the plane $\tau = 0$ in the circle

$$(\xi - x)^2 + (\eta - y)^2 = t^2.$$

If $t \leqslant y$, the disc with this circle as boundary lies in $\eta \geqslant 0$ on which $u$ and $u_t$ are given, so that the solution is that given in §7.4. But if $t > y$, the circle crosses $\eta = 0$ and bounds a disc on part of which $u$ and $u_t$ are unknown.

From now on, assume that $t > y > 0$. As in §7.4,

$$\int_V (\alpha(\alpha - 1)\, \Gamma^{(\alpha-3)/2} u - \Gamma^{(\alpha-1)/2} F) \, d\xi \, d\eta \, d\tau$$

$$= \int_S [(\alpha-1)\, \Gamma^{(\alpha-3)/2} u \{\lambda(\xi-x) + \mu(\eta-y) + \nu(\tau-t)\}$$

$$+ \Gamma^{(\alpha-1)/2} \{\lambda u_\xi + \mu u_\eta - \nu u_\tau\}] \, dS,$$

where $V$ is the region bounded by $S$ on which $\Gamma = 0$, $\eta = 0$ and $\tau = 0$, and $(\lambda, \mu, \nu)$ are the direction cosines of the normal to $S$ drawn out of $V$. Since $t > y > 0$, $S$ consists of three parts. The part of $S$ belonging to the retrograde cone contributes nothing when $\mathrm{re}\,\alpha > 3$. Next there is a segment $\Sigma_0$ of a disc in $\tau = 0$, defined by

$$(\xi - x)^2 + (\eta - y)^2 \leqslant t^2, \quad \eta \geqslant 0.$$

Lastly there is a region $\Sigma_1$ in $\eta = 0$ defined by

$$(t - \tau)^2 - (\xi - x)^2 \geqslant y^2, \quad 0 \leqslant \tau \leqslant t - y.$$

On $\Sigma_0$, $\lambda = \mu = 0, \nu = -1$; on $\Sigma_1$, $\lambda = \nu = 0, \mu = -1$. Using the conditions $u = f(\xi, \eta), u_t = h(\xi, \eta)$ on $\tau = 0, \eta \geqslant 0$, and $u = 0$ on $\eta = 0$, $\tau \geqslant 0$, we have

$$\frac{1}{2\pi\Gamma(\alpha - 1)} \int_V \Gamma^{(\alpha - 3)/2} u(\xi, \eta; \tau)\, d\xi\, d\eta\, d\tau$$

$$- \frac{1}{2\pi\Gamma(\alpha - 1)} \int_V \Gamma^{(\alpha - 1)/2} F(\xi, \eta; \tau)\, d\xi\, d\eta\, d\tau$$

$$= \frac{1}{2\pi\Gamma(\alpha + 1)} \left\{ \int_{\Sigma_0} \Gamma_0^{(\alpha - 1)/2} h(\xi, \eta)\, d\xi\, d\eta \right.$$

$$\left. + \frac{\partial}{\partial t} \int_{\Sigma_0} \Gamma_0^{(\alpha - 1)/2} f(\xi, \eta)\, d\xi\, d\eta - \int_{\Sigma_1} \Gamma_1^{(\alpha - 1)/2} \psi(\xi, \tau)\, d\xi\, d\tau \right\}, \quad (2)$$

where $$\Gamma_0 = (\Gamma)_{\tau = 0}, \quad \Gamma_1 = (\Gamma)_{\eta = 0}$$

and $\psi(x, t)$ is the unknown value of $u_y$ at $(x, 0; t)$.

We get over the difficulty that $u_y$ is unknown on $y = 0$ by a method of reflection. Let $$\bar{\Gamma} = (t - \tau)^2 - (x - \xi)^2 - (y + \eta)^2.$$

Then $\bar{\Gamma} = 0$ is the characteristic cone with vertex $(x, -y; t)$. The retrograde part of $\bar{\Gamma} = 0$ cuts off from $V$ a region $V_1$ whose boundary is part of the retrograde cone, the same area $\Sigma_1$ on $y = 0$, and a segment $\Sigma_2$ defined by

$$(\xi - x)^2 + (\eta + y)^2 \leqslant t^2, \quad y \geqslant 0.$$

On $\Sigma_2$, $u = f, u_t = h$. Repeating the argument, we find that, when $\mathrm{re}\,\alpha > 3$,

$$\frac{1}{2\pi\Gamma(\alpha - 1)} \int_{V_1} \bar{\Gamma}^{(\alpha - 3)/2} u(\xi, \eta; \tau)\, d\xi\, d\eta\, d\tau$$

$$- \frac{1}{2\pi\Gamma(\alpha + 1)} \int_{V_1} \bar{\Gamma}^{(\alpha - 1)/2} F(\xi, \eta; \tau)\, d\xi\, d\eta\, d\tau$$

is equal to

$$\frac{1}{2\pi\Gamma(\alpha+1)}\left\{\int_{\Sigma_0}\overline{\Gamma}_0^{(\alpha-1)/2}h(\xi,\eta)\,d\xi\,d\eta+\frac{\partial}{\partial t}\int_{\Sigma_2}\overline{\Gamma}_0^{(\alpha-1)/2}f(\xi,\eta)\,d\xi\,d\eta\right.$$

$$\left.-\int_{\Sigma_1}\Gamma_1^{(\alpha-1)/2}\psi(\xi,\tau)\,d\xi\,d\tau\right\},\quad(3)$$

where $\qquad \overline{\Gamma}_0=(\overline{\Gamma})_{\tau=0},\quad(\overline{\Gamma})_{\eta=0}=(\Gamma)_{\eta=0}=\Gamma_1.$

If we subtract (3) from (2) the terms involving $\psi$ cancel.

The expressions which appear in (2) and (3) are analytic functions of $\alpha$ regular when $\mathrm{re}\,\alpha>-1$.

Let $V_2$ and $V_3$ be the parts of $V$ on which $t-y\leqslant\tau\leqslant t$ and $0\leqslant\tau\leqslant t-y$. Then

$$\frac{1}{2\pi\Gamma(\alpha-1)}\int_{V_2}\Gamma^{(\alpha-3)/2}u(\xi,\eta;\tau)\,d\xi\,d\eta\,d\tau$$

is Riesz's integral of order $\alpha$ corresponding to the initial plane

$$\tau=t-y;$$

its limit as $\alpha\to0$ is $u(x,y;t)$. But the limit of

$$\frac{1}{2\pi\Gamma(\alpha-1)}\int_{V_3}\Gamma^{(\alpha-3)/2}u(\xi,\eta;\tau)\,d\xi\,d\eta\,d\tau$$

is zero since $(x,y;t)$ is outside $V_3$. Similarly the limit of

$$\frac{1}{2\pi\Gamma(\alpha-1)}\int_{V_1}\overline{\Gamma}^{(\alpha-3)/2}u(\xi,\eta;\tau)\,d\xi\,d\eta\,d\tau$$

is zero.

It follows from the limiting forms of (2) and (3) that

$$u(x,y;t)=\frac{1}{2\pi}\int_V\frac{F(\xi,\eta;\tau)}{\sqrt{\Gamma}}\,d\xi\,d\eta\,d\tau-\frac{1}{2\pi}\int_{V_1}\frac{F(\xi,\eta;\tau)}{\sqrt{\overline{\Gamma}}}\,d\xi\,d\eta\,d\tau$$

$$+\frac{1}{2\pi}\int_{\Sigma_0}\frac{h(\xi,\eta)}{\sqrt{\Gamma_0}}\,d\xi\,d\eta-\frac{1}{2\pi}\int_{\Sigma_2}\frac{h(\xi,\eta)}{\sqrt{\overline{\Gamma}_0}}\,d\xi\,d\eta$$

$$+\frac{1}{2\pi}\frac{\partial}{\partial t}\int_{\Sigma_0}\frac{f(\xi,\eta)}{\sqrt{\Gamma_0}}\,d\xi\,d\eta-\frac{1}{2\pi}\frac{\partial}{\partial t}\int_{\Sigma_2}\frac{f(\xi,\eta)}{\sqrt{\overline{\Gamma}_0}}\,d\xi\,d\eta.$$

This is precisely the result we should have got if we had extended the range of definition of $F,f$ and $h$ to $y<0$ by requiring them to be odd functions of $y$.

If we are given Cauchy data on $t=0,0\leqslant y\leqslant l$, and if $u$ is zero on $y=0$ and on $y=l$, we can solve the problem by requiring $F,f$ and $h$ to be odd functions of $y$, periodic in $u$ of period $2l$.

## 7.8 A note

The problems discussed in this chapter seem to have a rather restricted scope. The Riesz integral has proved of use in some branches of mathematical physics, such as the solution of Maxwell's equations, the theory of the meson field, quantum electrodynamics. Riesz has also developed integrals of fractional order associated with Laplace's operator and with the parabolic operator of the equation of heat.

For the connexion of Riesz's theory with semi-groups, see Einar Hille's American Mathematical Society's Colloquium Publication *Functional Analysis and Semi-Groups*, of which the first edition appeared in 1948.

### Exercises

1. If $f(x)$ is continuous, prove that

$$I^\alpha f(x) = \frac{1}{\Gamma(\alpha)} \int_0^x f(t)\,(x-t)^{\alpha-1} dt$$

is an analytic function of $\alpha$, regular in $\operatorname{re}\alpha > 0$. If $f(x)$ has a continuous derivative $f'(x)$, show that

$$I^\alpha f(x) = \frac{x^\alpha}{\Gamma(\alpha+1)} f(0) + \frac{1}{\Gamma(\alpha+1)} \int_0^x f'(t)\,(x-t)^\alpha dt$$

is regular in $\operatorname{re}\alpha > -1$, and hence that $I^0 f(x) = f(x)$.

Prove also that $I^\alpha I^\beta f(x) = I^{\alpha+\beta} f(x)$ and that

$$\frac{d}{dx} I^{\alpha+1} f(x) = I^\alpha f(x).$$

2. Marcel Riesz's operator associated with the differential operator $L$, where

$$Lu = \frac{\partial^2 u}{\partial t^2} - \frac{\partial^2 u}{\partial x^2},$$

and Cauchy data on $t = 0$ is

$$I^\alpha u(x;t) = \frac{1}{2^{\alpha-1}\{\Gamma(\tfrac{1}{2}\alpha)\}^2} \int_V u(\xi;\tau)\,\Gamma^{(\alpha-2)/2} d\xi\,d\tau,$$

where $\qquad \Gamma = (t-\tau)^2 - (x-\xi)^2$

and $V$ is defined by $\Gamma \geqslant 0$, $0 \leqslant \tau \leqslant t$. Prove the characteristic properties of $I^\alpha u(x;t)$.

$u(x;t)$ is the solution of $Lu - u = F(x;t)$ which satisfies the initial conditions $u = f(x)$, $u_t = h(x)$ when $t = 0$. Prove that

$$I^{2\alpha} u - I^{2\alpha+2} u = I^{2\alpha+2} F + \frac{1}{2^{2\alpha+1}\{\Gamma(\alpha+1)\}^2} \int_{x-t}^{x+t} \Gamma_0^\alpha h(\xi)\,d\xi$$

$$+ \frac{1}{2^{2\alpha+1}\{\Gamma(\alpha+1)\}^2} \frac{\partial}{\partial t} \int_{x-t}^{x+t} \Gamma_0^\alpha f(\xi)\,d\xi,$$

where $\Gamma_0$ is the value of $\Gamma$ when $\tau = 0$. Deduce that

$$u(x;t) = \frac{1}{2}\int_V F(\xi;\tau)\, I_0(\sqrt{\Gamma})\, d\xi\, d\tau + \frac{1}{2}\int_{x-t}^{x+t} g(\xi)\, I_0(\sqrt{\Gamma_0})\, d\xi$$

$$+ \frac{1}{2}\frac{\partial}{\partial t}\int_{x-t}^{x+t} f(\xi)\, I_0(\sqrt{\Gamma_0})\, d\xi.$$

**3.** $u(x,y;t)$ is the solution of

$$\frac{\partial^2 u}{\partial t^2} - \frac{\partial^2 u}{\partial x^2} - \frac{\partial^2 u}{\partial y^2} - u = F(x,y;t)$$

with continuous derivatives of the second order. When $t = 0$, $u = f(x,y)$, $u_t = h(x,y)$. The functions $F$ and $h$ have continuous derivatives of the second order, $f$ of the third order. Prove by Riesz's method that

$$u(x,y;t) = \frac{1}{2\pi}\int_V F(\xi,\eta;\tau)\frac{\cosh\sqrt{\Gamma}}{\sqrt{\Gamma}}\, d\xi\, d\eta\, d\tau + \frac{1}{2\pi}\int_\Sigma h(\xi,\eta)\frac{\cosh\sqrt{\Gamma_0}}{\sqrt{\Gamma_0}}\, d\xi\, d\eta$$

$$+ \frac{1}{2\pi}\frac{\partial}{\partial t}\int_\Sigma f(\xi,\eta)\frac{\cosh\sqrt{\Gamma_0}}{\sqrt{\Gamma_0}}\, d\xi\, d\eta,$$

where
$$\Gamma = (t-\tau)^2 - (x-\xi)^2 - (y-\eta)^2,$$
$$\Gamma_0 = t^2 - (x-\xi)^2 - (y-\eta)^2,$$

and where $V$ is $\Gamma \geqslant 0$, $0 \leqslant \tau \leqslant t$, and $\Sigma$ is $\Gamma_0 \geqslant 0$.

**4.** $u(x;t)$ is a function of the variables $x = (x_1, x_2, \ldots, x_{2m-1})$ and $t$ with continuous partial derivatives of order $m-1$. The function $U(x;t)$ is defined by

$$U(x;t) = \int_V u(\xi;\tau)\, d\xi\, d\tau,$$

where, on $V$,
$$\Gamma = (t-\tau)^2 - |x-\xi|^2 \geqslant 0, \quad 0 \leqslant \tau \leqslant t.$$

If
$$Lu = \frac{\partial^2 u}{\partial t^2} - \nabla^2 u,$$

where $\nabla^2$ is Laplace's operator in the $x$-space, prove that

$$L^m U(x;t) = 2^{2m-1}\pi^{m-1}(m-1)!\, u(x;t).$$

If $u$ is a solution of $Lu = 0$ which satisfies the initial conditions $u = f(x)$, $u_t = h(x)$ when $t = 0$, prove by applying Green's transformation to

$$\int_V \{u\Lambda\Gamma - \Gamma\Lambda u\}\, d\xi\, d\tau$$

that
$$U(x;t) = \frac{1}{4m}\int_W \Gamma_0 g(\xi)\, d\xi + \frac{1}{4m}\frac{\partial}{\partial t}\int_W \Gamma_0 f(\xi)\, d\xi,$$

where $\Gamma_0$ is the value of $\Gamma$ when $\tau = 0$ and on $W$, $\Gamma_0 = 0$. Deduce that

$$u(x;t) = \frac{1}{2^{2m+1}\pi^{m-1}m!} L^m \left\{ \int_W \Gamma_0 g(\xi)\, d\xi + \frac{\partial}{\partial t} \int_W \Gamma_0 f(\xi)\, d\xi \right\}$$

and hence that

$$u(x;t) = \frac{1}{2^{2m-1}\pi^{m-1}(m-1)!} L^{m-1} \left\{ \int_W g(\xi)\, d\xi + \frac{\partial}{\partial t} \int_W f(\xi)\, d\xi \right\}.$$

To what does this reduce when $m = 1$?

**5.** $D$ is the triangle bounded by the lines $\xi = x$, $\eta = y$, $\xi + \eta = 0$ where $x + y > 0$. An operator $J^\alpha$ is defined by

$$J^\alpha u(x,y) = \frac{1}{\{\Gamma(\alpha)\}^2} \int_D u(\xi,\eta)\,(x-\xi)^{\alpha-1}(y-\eta)^{\alpha-1}d\xi\,d\eta.$$

Prove that, if $u$ is continuous, $J^\alpha u$ is an analytic function of $\alpha$ regular in re $\alpha > 0$, and that, if $u$ has continuous first derivatives, $J^\alpha u$ tends to $u$ as $\alpha \to +0$. Show also that

$$J^\alpha J^\beta u(x,y) = J^{\alpha+\beta} u(x,y),$$

$$\frac{\partial^2}{\partial x\, \partial y} J^\alpha u(x,y) = J^{\alpha-1} u(x,y).$$

$u$ is the solution of $u_{xy} + u = F(x,y)$ which satisfies the conditions $u = 0$, $u_x = 0$ on $x + y = 0$. By applying Green's transformation to

$$\int_D (uv_{\xi\eta} - vu_{\xi\eta})\, d\xi\, d\eta,$$

where        $v = (x-\xi)^\alpha (y-\eta)^\alpha,$

show that, when re $\alpha > 0$,

$$J^\alpha u + J^{\alpha+1} u = J^{\alpha+1} F.$$

Deduce that

$$u(x,y) = \int_D F(\xi,\eta)\, J_0[\sqrt{\{2(x-\xi)(y-\eta)\}}]\, d\xi\, d\eta,$$

where $J_0$ is Bessel's function of order zero.

# 8

# POTENTIAL THEORY IN THE PLANE

## 8.1 Gravitation

The simplest equation of elliptic type is Laplace's equation $\nabla^2 u = 0$, where

$$\nabla^2 u = \frac{\partial^2 u}{\partial x_1^2} + \frac{\partial^2 u}{\partial x_2^2} + \dots + \frac{\partial^2 u}{\partial x_n^2}.$$

This equation with $n = 2$ or $3$ is of frequent occurrence in mathematical physics, notably in the theory of gravitation.

Newton's law of universal gravitation asserts that every particle of matter in the universe attracts every other particle with a force whose direction is that of the line joining them, and whose magnitude varies directly as the product of their masses and inversely as the square of their distance apart. With an appropriate choice of units, the attractive force is $mm'/r^2$ where $m$ and $m'$ are the masses of the particles and $r$ their distance apart.

If a particle of unit mass moves under the attraction of a particle of mass $m$ fixed at, say, the origin, the increase in the kinetic energy of the particle of unit mass as it moves from a position $P_0$ to a position $P$ is

$$\frac{m}{OP} - \frac{m}{OP_0}.$$

The expression

$$u = \frac{m}{OP}$$

is called the gravitational potential of $m$. If the coordinates of the variable point $P$ are $(x, y, z)$,

$$\frac{\partial^2 u}{\partial x^2} + \frac{\partial^2 u}{\partial y^2} + \frac{\partial^2 u}{\partial z^2} = 0;$$

and the force is $\operatorname{grad}(m/r)$.

By a generalisation of Newton's law, the attractive force due to a continuous distribution of matter of density $m(x, y, z)$ in a volume $V$ acting on a particle of unit mass at a point $(x, y, z)$ outside $V$ is equal to $\operatorname{grad} u$, where

$$u = \iiint_V m(\xi, \eta, \zeta) \frac{1}{R} \, d\xi \, d\eta \, d\zeta,$$

where $\qquad R^2 = (x-\xi)^2 + (y-\eta)^2 + (z-\zeta)^2.$

This potential satisfies Laplace's equation outside $V$, but in $V$ it satisfies Poisson's equation $\nabla^2 u = -4\pi m$.

Similar formulae hold for the gravitational potential of distributions of matter on curves or surfaces. For example, an infinitely long uniform straight thin wire of mass $m$ per unit length located on the $z$-axis attracts a particle of unit mass at $(x, y, z)$ with a force normal to the wire of magnitude

$$\int_{-\infty}^{\infty} m\,\frac{r}{R^3}\,d\zeta,$$

where $\qquad r^2 = x^2 + y^2, \quad R^2 = r^2 + (z-\zeta)^2.$

This is equal to $2m/r$ and is independent of $z$. If the unit particle moves from a position $P_0$ to a position $P$ under the attraction of the wire, its increase in kinetic energy is

$$2m \log \frac{r_0}{r},$$

where $r_0$ and $r$ are the distances of $P_0$ and $P$ from the wire. The expression

$$2m \log \frac{1}{r}$$

is the gravitational potential of the wire; its gradient is the attractive force.

More generally, if we have an infinitely long straight rod of cross section $D$ parallel to the $z$-axis, the density $\sigma(x, y)$ being independent of $z$, it attracts a unit particle at $(x, y, z)$ outside the rod with a force $2\,\mathrm{grad}\,u$, where

$$u = \iint_D \sigma(\xi, \eta) \log \frac{1}{R}\, d\xi\, d\eta,$$

where $\qquad R^2 = (x-\xi)^2 + (y-\eta)^2.$

Such a function $u$ is called a logarithmic potential. It is independent of $z$ and satisfies Laplace's equation outside the rod; but inside the rod, $\nabla^2 u = -2\pi\sigma(x, y)$.

There is one important difference between the Newtonian potential and the logarithmic potential. If $V$ is bounded,

$$u \equiv \iiint_V m(\xi, \eta, \zeta) \frac{d\xi\, d\eta\, d\zeta}{R} \sim \frac{1}{\sqrt{(x^2+y^2+z^2)}} \iiint_V m(\xi, \eta, \zeta)\, d\xi\, d\eta\, d\zeta$$

at great distances. But if $D$ is bounded,

$$u = \iint_D \sigma(\xi, \eta) \log \frac{1}{R}\, d\xi\, d\zeta \sim \log \frac{1}{\sqrt{(x^2+y^2)}} \iint_D \sigma(\xi, \eta)\, d\xi\, d\eta.$$

The Newtonian potential behaves like the potential of a particle, and vanishes at infinity. The logarithmic potential behaves like the potential of a thin wire, and is infinite at infinity.

In this chapter, we deal with the solution of Laplace's equation and Poisson's equation in the plane. Fundamental tools in the theory are the first and second identities of Green, which are consequences of Green's theorem enunciated in Note 5 of the Appendix.

Let $D$ be a domain bounded by a regular closed curve $C$. Let $\phi, \psi$ and their first partial derivatives be continuous in the closure of $D$. Then, if we put $u = \phi \psi_x, v = \phi \psi_y$ in Green's theorem, we have

$$\iint_D (\phi \nabla^2 \psi + \phi_x \psi_x + \phi_y \psi_y)\, dx\, dy = \int_C \phi \frac{\partial \psi}{\partial N}\, ds,$$

where $\partial \psi / \partial N$ is the derivative along the outward normal, provided that $\psi_{xx}$ and $\psi_{yy}$ exist and are bounded in $D$ and the double integrals of $\phi \psi_{xx}$ and $\phi \psi_{yy}$ over $D$ exist. This is Green's first identity. It holds when $\phi$ and $\psi$ are interchanged, provided that $\phi_{xx}$ and $\phi_{yy}$ exist and are bounded in $D$ and the double integrals of $\psi \phi_{xx}$ and $\psi \phi_{yy}$ over $D$ exist. By subtraction, we have Green's second identity

$$\iint_D (\phi \nabla^2 \psi - \psi \nabla^2 \phi)\, dx\, dy = \int_C \left( \phi \frac{\partial \psi}{\partial N} - \psi \frac{\partial \phi}{\partial N} \right) ds.$$

Simple conditions for the truth of these results are that $\phi$ and $\psi$ should have continuous second derivatives in $\bar{D}$, the closure of $D$.

$D$ may be a multiply-connected domain bounded by several non-intersecting regular closed curves. These results still hold, the integral over $C$ being replaced by the sum of the integrals over the boundary curves; $N$ is still the unit normal vector drawn out of the domain $D$.

## 8.2   Green's equivalent layer

It is convenient to use $(\xi, \eta)$ as the coordinates of the integration point and $(x, y)$ as the coordinates of a fixed point, usually the co-ordinates of the point at which the solution is desired; and we use $\Delta$ to denote Laplace's operator with $(\xi, \eta)$ as variables.

In Green's second identity, put $\psi = \log 1/R$ where

$$R^2 = (x - \xi)^2 + (y - \eta)^2.$$

Then if $C$ is a regular closed curve bounding a domain $D$, we have

$$\int_C \left\{ \phi(\xi, \eta) \frac{\partial}{\partial N} \log \frac{1}{R} - \frac{\partial \phi(\xi, \eta)}{\partial N} \log \frac{1}{R} \right\} ds = -\iint \log \frac{1}{R} \Delta \phi(\xi, \eta)\, d\xi\, d\eta.$$

This is certainly true if $\phi$ has continuous second derivatives in $\bar{D}$, provided that $(x, y)$ is not a point of $\bar{D}$.

If $(x, y)$ is a point of $D$, the region $R \leqslant \epsilon$ lies in $D$ for all sufficiently small values of $\epsilon$. Let $D_0$ be the domain $D$ when the region $R \leqslant \epsilon$ has been deleted; and let $C_0$ be $R = \epsilon$. Then

$$\int_C \left\{ \phi \frac{\partial}{\partial N} \log \frac{1}{R} - \frac{\partial \phi}{\partial N} \log \frac{1}{R} \right\} ds + \int_{C_0} \left\{ \phi \frac{\partial}{\partial N} \log \frac{1}{R} - \frac{\partial \phi}{\partial N} \log \frac{1}{R} \right\} ds$$

$$= -\iint_{D_0} \log \frac{1}{R} \, \Delta \phi \, d\xi \, d\eta.$$

On $C_0$, $\partial / \partial N = -\partial / \partial R$. The integral over $C_0$ is then

$$\int_{C_0} \left\{ \frac{\phi}{R} + \frac{\partial \phi}{\partial R} \log \frac{1}{R} \right\} ds$$

$$= \int_0^{2\pi} \left\{ \phi(x + \epsilon \cos \theta, y + \epsilon \sin \theta) + \epsilon \phi_\epsilon(x + \epsilon \cos \theta, y + \epsilon \sin \theta) \log \frac{1}{\epsilon} \right\} d\theta$$

$$\to 2\pi \phi(x, y)$$

as $\epsilon \to 0$, since $\phi$ is continuously differentiable. Therefore

$$\phi(x, y) = -\frac{1}{2\pi} \iint_D \log \frac{1}{R} . \Delta \phi . d\xi \, d\eta$$

$$- \frac{1}{2\pi} \int_C \left( \phi \frac{\partial}{\partial N} \log \frac{1}{R} - \log \frac{1}{R} \frac{\partial \phi}{\partial N} \right) ds, \quad (1)$$

when $(x, y)$ is a point of $D$. But when $(x, y)$ is outside the curve $C$, the value of the expression on the right of (1) is zero.

If $\phi$ is a solution of Laplace's equation, we have

$$\phi(x, y) = \frac{1}{2\pi} \int_C \log \frac{1}{R} \frac{\partial \phi}{\partial N} ds - \frac{1}{2\pi} \int_C \phi \frac{\partial}{\partial N} \log \frac{1}{R} ds. \quad (2)$$

The first term is the logarithmic potential of a distribution of matter of line density

$$\frac{1}{2\pi} \frac{\partial \phi}{\partial N}$$

on $C$. The second term is the logarithmic potential of a distribution of normally directed doublets of linear strength

$$\frac{1}{2\pi} \phi.$$

These two distributions on $C$ form *Green's equivalent layer*. It must be emphasised that the formula does not provide a solution of the problem of Cauchy; we cannot assign $\phi$ and $\partial\phi/\partial N$ independently on $C$.

If $\phi$ satisfies Poisson's equation $\nabla^2\phi = -2\pi\sigma$, there is an additional term

$$\iint_D \log\frac{1}{R}\,\sigma(\xi,\eta)\,d\xi\,d\eta$$

in (1). This is the potential of a distribution of matter of density $\sigma$ on $D$. If $\phi$ satisfies Poisson's equation everywhere, we may take $C$ to be a circle of large radius. If the integral over $C$ in (1) tends to zero as the radius tends to infinity, then

$$\phi(x,y) = \iint \log\frac{1}{R}\,\sigma(\xi,\eta)\,d\xi\,d\eta$$

holds everywhere, integration being over the whole plane.

### 8.3   Properties of the logarithmic potentials

*If $\sigma(x,y)$ is bounded and integrable over a domain $D$, the logarithmic potential*

$$u(x,y) = \iint_D \sigma(\xi,\eta) \log\frac{1}{R}\,d\xi\,d\eta,$$

*when $R^2 = (x-\xi)^2 + (y-\eta)^2$, is continuous everywhere. Its first derivatives are also continuous, and can be found by differentiating under the sign of integration. If $\sigma(x,y)$ has continuous first derivatives on $D$, the second derivatives of $u$ are continuous on $D$ and satisfy Poisson's equation $\nabla^2 u = -2\pi\sigma$.*

If we differentiate formally, we obtain the expressions

$$X = \iint_D \sigma(\xi,\eta)\frac{\xi-x}{R^2}\,d\xi\,d\eta, \quad Y = \iint_D \sigma(\xi,\eta)\frac{\eta-y}{R^2}\,d\xi\,d\eta.$$

We have to show that $u$, $X$, $Y$ are continuous and that grad $u = (X, Y)$.

If we introduce polar coordinates, $\xi = x + R\cos\theta, \eta = y + R\sin\theta$, we obtain

$$u(x,y) = \iint_D \sigma(x+R\cos\theta, y+R\sin\theta)\,R\log\frac{1}{R}\,dR\,d\theta,$$

with corresponding results for $X$ and $Y$. Since $\sigma(x,y)$ is bounded and integrable, the integrals defining $u$, $X$ and $Y$ converge everywhere. On the exterior of $D$, $u$ has continuous derivatives of all orders, obtained by differentiation under the sign of integration; it satisfies Laplace's equation there.

To deal with the case when $(x, y)$ is a point of $D$, introduce the function

$$f(R; \delta) = \tfrac{1}{2} + \log\frac{1}{\delta} - \frac{R^2}{2\delta^2} \quad (0 \leqslant R \leqslant \delta)$$

$$= \log\frac{1}{R} \quad (R \geqslant \delta),$$

where $\delta$ is so small that the disc $R < \delta$ lies in $D$; denote the disc by $D_0$. Then $f(R; \delta)$ is continuous and continuously differentiable everywhere.

If

$$u_\delta(x, y) = \iint_D \sigma(\xi, \eta) f(R; \delta)\, d\xi\, d\eta,$$

then

$$u - u_\delta = \iint_D \sigma(\xi, \eta) \left[\log\frac{1}{R} - \frac{1}{2} - \log\frac{1}{\delta} + \frac{R^2}{2\delta^2}\right] d\xi\, d\eta.$$

On $D_0$, the expression in square brackets is not negative; and there exists a constant $K$ such that $|\sigma| \leqslant K$ on $D$. Then

$$|u - u_\delta| \leqslant 2\pi K \int_0^\delta \left[-R\log R - \tfrac{1}{2}R + R\log\delta + \frac{R^3}{2\delta^2}\right] dR = \tfrac{1}{4}\pi K \delta^2.$$

Hence $u_\delta$ tends to $u$ uniformly as $\delta \to +0$. Since $u_\delta$ is continuous, so also is $u$.

Similarly,

$$X - \frac{\partial u_\delta}{\partial x} = \iint_{D_0} \sigma(\xi, \eta) \left[\frac{\xi - x}{R^2} - \frac{\xi - x}{\delta^2}\right] d\xi\, d\eta$$

$$= \iint_{D_0} \sigma(\xi, \eta) \frac{\delta^2 - R^2}{R^2\delta^2} (\xi - x)\, d\xi\, d\eta.$$

and so

$$\left|X - \frac{\partial u_\delta}{\partial x}\right| \leqslant 2\pi K \int_0^\delta \frac{\delta^2 - R^2}{\delta^2}\, dR = \frac{4\pi K \delta}{3}.$$

Hence, as $\delta \to +0$, $\partial u_\delta/\partial x$ converges uniformly to $X$. Since $\partial u_\delta/\partial x$ is continuous, $X$ is continuous everywhere, and $\partial u/\partial x = X$. In the same way, $Y$ is continuous, and $\partial u/\partial y = Y$.

On the exterior of $D$, that is, on the complement of the closure of $D$, $u$ has continuous derivatives of all orders and satisfies Laplace's equation. In order to deal with second derivatives on $D$, we assume that $\sigma$ has continuous first order derivatives. We could have assumed that $\sigma$ has piece-wise-continuous first derivatives, or, even that it satisfies a Hölder condition; these weaker conditions make the analysis more difficult.

Let $D_1$ be an open disc contained in $D$, and let $D_2$ be the rest of $D$. Write

$$u_1(x,y) = \iint_{D_1} \sigma(\xi,\eta) \log \frac{1}{R} d\xi d\eta, \quad u_2(x,y) = \iint_{D_2} \sigma(\xi,\eta) \log \frac{1}{R} d\xi d\eta.$$

The function $u_2$ has continuous derivatives of all orders on $D_1$ and satisfies Laplace's equation there.

If $(x,y)$ is any point of $D_1$,

$$\frac{\partial u_1}{\partial x} = - \iint_{D_1} \sigma(\xi,\eta) \frac{\partial}{\partial \xi} \log \frac{1}{R} d\xi d\eta.$$

Since $\sigma$ has continuous first derivatives, we have

$$\frac{\partial u_1}{\partial x} = - \iint_{D_1} \frac{\partial}{\partial \xi} \left( \sigma \log \frac{1}{R} \right) d\xi d\eta + \iint_{D_1} \frac{\partial \sigma}{\partial \xi} \log \frac{1}{R} d\xi d\eta$$

$$= - \int_{C_1} \lambda \sigma \log \frac{1}{R} ds + \iint_{D_1} \frac{\partial \sigma}{\partial \xi} \log \frac{1}{R} d\xi d\eta,$$

where $C_1$ is the boundary of the disc $D_1$ and $(\lambda, \mu)$ are the direction cosines of the outward normal. This expresses $\partial u_1/\partial x$ as the sum of the logarithmic potential of a continuous distribution of density $\partial \sigma/\partial x$ over $D_1$ and the logarithmic potential of a continuous linear distribution over $C_1$. Hence $\partial u_1/\partial x$ has at each point of $D_1$ a continuous derivative with respect to $x$ or $y$; and

$$\frac{\partial^2 u_1}{\partial x^2} = - \int_{C_1} \lambda \sigma \frac{\partial}{\partial x} \log \frac{1}{R} ds + \iint_{D_1} \frac{\partial \sigma}{\partial \xi} \frac{\partial}{\partial x} \log \frac{1}{R} d\xi d\eta$$

$$= \int_{C_1} \lambda \sigma \frac{\partial}{\partial \xi} \log \frac{1}{R} ds + \iint_{D_1} \frac{\partial \sigma}{\partial \xi} \frac{\xi - x}{R^2} d\xi d\eta.$$

Similarly for $\partial^2 u_1/\partial y^2$. Therefore $u$ itself has continuous second derivatives on $D$. Hence, on $D_1$,

$$\nabla^2 u = \nabla^2 u_1$$

$$= \int_{C_1} \sigma \frac{\partial}{\partial N} \log \frac{1}{R} ds + \iint_{D_1} \left( \frac{\partial \sigma}{\partial \xi} \frac{\xi - x}{R^2} + \frac{\partial \sigma}{\partial \eta} \frac{\eta - y}{R^2} \right) d\xi d\eta.$$

This holds when $(x,y)$ is any point of any disc $D_1$ contained in $D$. Now take $C_1$ to be the circle $R = \epsilon$ with centre $(x,y)$, $\epsilon$ being chosen so that $D_1$ is contained in $D$. Then

$$\int_{C_1} \sigma \frac{\partial}{\partial N} \log \frac{1}{R} ds = - \int_0^{2\pi} \sigma(x + \epsilon \cos \theta, y + \epsilon \sin \theta) d\theta,$$

which tends to $-2\pi\sigma(x,y)$ as $\epsilon\to0$ since $\sigma$ is continuous. Also the double integral over $D_1$ is equal to

$$\int_0^\epsilon dR \int_0^{2\pi} d\theta\,\{\cos\theta\,\sigma_x(x+R\cos\theta,y+R\sin\theta)$$
$$+\sin\theta\,\sigma_y(x+R\cos\theta,y+R\sin\theta)\},$$

which tends to zero as $\epsilon\to0$, since $\sigma_x$ and $\sigma_y$ are continuous.

But $\nabla^2u$ does not depend on $\epsilon$, and so is equal to the limit of $\nabla^2u_1$ as $\epsilon\to0$. Therefore

$$\nabla^2u = -2\pi\sigma(x,y).$$

## 8.4   Some other logarithmic potentials

We have already met two other forms of logarithmic potential which are of importance in mathematical physics. As we shall not make much use of them, their properties will be stated without proof.

Let $C$ be a regular arc with parametric equations $x=f(s)$, $y=g(s)$, $s$ being the arc length. If $\mu(s)$ is continuous, the logarithmic potential

$$u(x,y) = \int_C \mu(s)\log\frac{1}{R}ds,$$

where

$$R^2 = (x-\xi)^2+(y-\eta)^2,$$

$(\xi,\eta)$ being the coordinates of the integration point, is continuous everywhere. At points not on $C$, $u$ has continuous derivatives of all orders and satisfies Laplace's Equation.

Suppose, in addition, that $\mu(s)$ has a bounded derivative and that $C$ has bounded curvature. Let the unit tangent and normal vectors at a typical point $P_0(x_0,y_0)$ of $C$ be $T_0$ and $N_0$. If $P$ crosses the surface along the normal at $P_0$, $\partial u(x,y)/\partial T_0$ is continuous, but $\partial u(x,y)/\partial N_0$ is not. As $P$ moves to $P_0$ along the normal,

$$\frac{\partial u(x,y)}{\partial N_0} \to \pm\mu(P_0)+\int_C \mu(s)\frac{\eta-y_0}{(\xi-x_0)^2+(\eta-y_0)^2}ds,$$

the sign being $+$ or $-$ according as $P$ approaches $P_0$ in the direction $N_0$ or $-N_0$. Thus the tangential component of $\mathrm{grad}\,u$ is continuous, but the normal component is not.

The logarithmic potential of a particle of mass $-m$ at the origin and a particle of mass $+m$ at $(h,0)$, where $h>0$, is

$$m\log r-m\log(r^2-2hx+h^2)^{\frac{1}{2}}$$

at the point $(x, y)$. If we use polar coordinates, this is

$$-\tfrac{1}{2} m \log \left(1 - 2\frac{h}{r}\cos\theta + \frac{h^2}{r^2}\right) = \frac{mh}{r}\cos\theta + mO\left(\frac{h^2}{r^2}\right)$$

if $h/r$ is small. If we make $h \to 0$, $mh \to \mu$, we obtain

$$\mu\frac{\cos\theta}{r},$$

which is the potential of a doublet at $O$, directed in the positive direction of the $x$-axis; it can be written as

$$\mu\frac{\partial}{\partial x}\log r.$$

If the doublet is in the direction of the unit vector $N$, its potential is

$$\mu\frac{\partial}{\partial N}\log r.$$

The second type of logarithmic potential in Green's equivalent layer is that due to a normally directed distribution of doublets.

If we have a normally directed distribution of doublets on a regular arc $C$, the strength per unit length being $\mu(s)$, the resulting potential is

$$u(x, y) = \int_C \mu(s)\frac{\partial}{\partial N}\log R\, ds.$$

The gradient of $u$ is continuous across $C$, but $u$ itself has a jump $2\pi\mu$. The value of $u$ at a point of $C$ is the mean of the limits as we approach $C$ from the two sides.

## 8.5 Harmonic functions

A function $u(x, y)$ is said to be *harmonic* in a domain $D$ if it has continuous second derivatives and satisfies Laplace's equation. In this section, we discuss the properties of harmonic functions by means of Green's theorem.

We assume that $D$ is bounded but not necessarily simply connected. The boundary $\partial D$ of the domain is assumed to consist of a finite number of non-intersecting regular closed curves, in the terminology of Note 4 of the Appendix. If $D$ is simply connected, $\partial D$ consists of one regular closed curve; $D$ is then the domain inside $\partial D$. The closure of $D$, denoted by $\bar{D}$, is the union of $D$ and $\partial D$.

*If u is harmonic in a bounded domain $D_1$ containing $\bar{D}$, then*

$$\int_{\partial D} \frac{\partial u}{\partial N} \, ds = 0.$$

$\partial/\partial N$ denotes differentiation along the normal to $\partial D$ drawn out of $D$. If the direction cosines of the normal are $(l, m)$, the integral is equal to

$$\int_{\partial D} (lu_x + mu_y) \, ds = \iint_D (u_{xx} + u_{yy}) \, dx \, dy = 0.$$

Green's theorem is applicable because $u$ has continuous second derivatives in $D_1$.

*If u has continuous second derivatives in a bounded domain D and if*

$$\int_{C_0} \frac{\partial u}{\partial N} \, ds = 0$$

*for every regular closed curve $C_0$ bounding a domain $D_0$ contained in D, then u is harmonic in D.*

The condition implies that

$$\iint_{D_0} \nabla^2 u \, dx \, dy = 0$$

for every domain $D_0$. Hence $u$ satisfies Laplace's equation everywhere in $D$.

*If u is harmonic in a disc D with centre $(x_0, y_0)$ and radius R and if u is continuous in $\bar{D}$, then*

$$u(x_0, y_0) = \frac{1}{2\pi} \int_0^{2\pi} u(x_0 + R\cos\theta, y_0 + R\sin\theta) \, d\theta.$$

If $0 < r < R$,
$$\int_C \frac{\partial u}{\partial N} \, ds = 0,$$

where $C$ is the circle $(x - x_0)^2 + (y - y_0)^2 = r^2$. Hence, using polar coordinates, we have

$$\int_0^{2\pi} \frac{\partial}{\partial r} u(x_0 + r\cos\theta, y_0 + r\sin\theta) \, d\theta = 0,$$

and so, if $0 < R' < R$,

$$\int_0^{R'} dr \int_0^{2\pi} \frac{\partial}{\partial r} u(x_0 + r\cos\theta, y_0 + r\sin\theta) \, d\theta = 0.$$

Since $u$ is continuous in $\bar{D}$, we may invert the order of integration and obtain
$$\int_0^{2\pi} \{u(x_0 + R'\cos\theta, y_0 + R'\sin\theta) - u(x_0, y_0)\} \, d\theta = 0.$$

Therefore

$$u(x_0, y_0) = \frac{1}{2\pi} \int_0^{2\pi} u(x_0 + R' \cos\theta, y_0 + R' \sin\theta)\, d\theta$$

for every $R' < R$. Since $u(x_0 + r \cos\theta, y_0 + r \sin\theta)$ is uniformly continuous when $0 \leqslant r \leqslant R$, we may proceed to the limit as $R' \to R - 0$ under the sign of integration and obtain the desired result, which is called *Gauss's mean value theorem*.

A corollary of this result is obtained as follows. If $0 < R' < R$,

$$u(x_0, y_0) \int_0^{R'} r\, dr = \frac{1}{2\pi} \int_0^{R'} r\, dr \int_0^{2\pi} u(x_0 + r\cos\theta, y_0 + r\sin\theta)\, d\theta,$$

and so

$$u(x_0, y_0) = \frac{1}{\pi R'^2} \iint u(x, y)\, dx\, dy,$$

where integration is over the disc $(x - x_0)^2 + (y - y_0)^2 < R'^2$. Making $R' \to R - 0$, we obtain

$$u(x_0, y_0) = \frac{1}{\pi R^2} \iint_D u(x, y)\, dx\, dy,$$

where $D$ is the disc with centre $(x_0, y_0)$ and radius $R$.

*Let $u$ be harmonic in a bounded domain $D$ and continuously differentiable in $\bar{D}$. If $u$ vanishes everywhere on $\partial D$, $u$ vanishes identically in $D$. If $\partial u/\partial N$ vanishes everywhere on $\partial D$, $u$ is constant in $D$.*

This follows from the identity

$$\iint_D (u_x^2 + u_y^2)\, dx\, dy = \int_{\partial D} u\, \frac{\partial u}{\partial N}\, ds.$$

If $u$ vanishes everywhere on $\partial D$, or if $\partial u/\partial N$ vanishes everywhere on $\partial D$, the double integral vanishes. Since $u_x$ and $u_y$ are real and continuous, $u_x$ and $u_y$ vanish everywhere on $D$, and so $u$ is constant. The two results follow.

*Let $u_1$ and $u_2$ be harmonic in a bounded domain $D$ and continuously differentiable in $\bar{D}$. If $u_1 = u_2$ everywhere on $\partial D$, $u_1 = u_2$ everywhere on $D$. If $\partial u_1/\partial N = \partial u_2/\partial N$ everywhere on $\partial D$, $u_1$ differs from $u_2$ by a constant.*

This result, known as the *uniqueness theorem*, follows from the previous one if we put $u = u_1 - u_2$. A different proof of this result, which uses the maximum principle, is given at the end of this section.

*If $u$ is harmonic in a bounded domain $D$, it possesses derivatives of all orders, themselves harmonic in $D$.*

If $(x_0, y_0)$ is a point of $D$, it is the centre of a disc $K$ whose closure lies in $D$. If $(x, y)$ is any point of $K$, by formula (2) of §8.2,

$$u(x, y) = \frac{1}{2\pi} \int_{\partial K} \log \frac{1}{R} \frac{\partial u}{\partial N} ds - \frac{1}{2\pi} \int_{\partial K} u \frac{\partial}{\partial N} \log \frac{1}{R} ds, \qquad (1)$$

where                         $R^2 = (x - \xi)^2 + (y - \eta)^2,$

$(\xi, \eta)$ being a typical point of $\partial K$. Since any number of differentiations can be carried out under the sign of integration, $u$ has derivatives of all orders on $K$, and thence throughout $D$.

Since

$$\nabla^2 u_x(x, y) = \frac{\partial}{\partial x} \nabla^2 u(x, y),$$

$u_x$ is harmonic in $D$ – similarly for all the derivatives.

Let $u_{mn}$ be the value of $\partial^{m+n} u / \partial x^m \partial y^n$ at $(x_0, y_0)$. Then

$$u(x, y) = \sum_0^\infty \frac{u_{mn}}{m! \, n!} (x - x_0)^m (y - y_0)^n$$

is the formal Taylor series expansion of $u$. It can be shown that the series is uniformly and absolutely convergent on a square

$$|x - x_0| + |y - y_0| < k,$$

and hence that *a function harmonic on a bounded domain $D$ is analytic there.*

Suppose that $\partial K$ has radius $a$, so that

$$\xi = x_0 + a \cos \phi, \quad \eta = y_0 + a \sin \phi.$$

On $\partial K$,                    $u = f(\phi), \quad \dfrac{\partial u}{\partial N} = g(\phi),$

where $f(\phi)$ and $g(\phi)$ are differentiable as often as we please. If

$$x = x_0 + r \cos \theta, \quad y = y_0 + r \sin \theta$$

we have                    $R^2 = a^2 + r^2 - 2ar \cos(\theta - \phi).$

Hence                    $\log R = \log a - \sum_1^\infty \dfrac{r^n}{na^n} \cos n(\theta - \phi),$

$$\frac{\partial}{\partial N} \log R = \frac{1}{a} + \sum_1^\infty \frac{r^n}{a^{n+1}} \cos n(\theta - \phi),$$

the series being uniformly and absolutely convergent when

$$0 \leqslant r \leqslant a' < a.$$

Substituting these series in (1) and integrating term-by-term as we may when $r \leqslant a'$, we have

$$u(x, y) = a_0 + \sum_1^\infty (a_x \cos n\theta + b_n \sin n\theta) \frac{r^n}{a^n}, \tag{2}$$

where $a_0 = \frac{1}{2\pi} \int_0^{2\pi} f(\phi) \, d\phi,$

$$a_n = \frac{1}{2\pi} \int_0^{2\pi} f(\phi) \cos n\phi \, d\phi + \frac{a}{2\pi n} \int_0^{2\pi} g(\phi) \cos n\phi \, d\phi,$$

$$b_n = \frac{1}{2\pi} \int_0^{2\pi} f(\phi) \sin n\phi \, d\phi + \frac{a}{2\pi n} \int_0^{2\pi} g(\phi) \sin n\phi \, d\phi.$$

The coefficient $a_0$ does not involve $g(\phi)$ since

$$\int_0^{2\pi} g(\phi) \, d\phi = 0.$$

Now $r^n \cos n\theta = (x - x_0)^n - {}^nC_2(x - x_0)^{n-2}(y - y_0)^2 + \ldots,$

$$r^n \sin n\theta = {}^nC_1(x - x_0)^n - {}^nC_3(x - x_0)^{n-3}(y - y_0)^3 + \ldots.$$

The general term of (2) then has the absolute value

$$|a_n \cos n\theta + b_n \sin n\theta| \frac{r^n}{a^n}$$

$$\leqslant \frac{M}{a^n} \sum_{p=0}^n {}^nC_p \, |x - x_0|^{n-p} \, |y - y_0|^p$$

$$= \frac{M}{a^n} \{|x - x_0| + |y - y_0|\}^n.$$

We can therefore substitute in (2) for $r^n \cos n\theta$ and $r^n \sin n\theta$ in terms of $x$ and $y$, and obtain a double series which converges uniformly and absolutely with respect to $x$ and $y$ in any square

$$|x - x_0| + |y - y_0| \leqslant k,$$

whenever $k < a$. Hence $u(x, y)$ is analytic in $D$.

*If $u$ is harmonic on a bounded domain $D$ and continuous on $\bar{D}$ (the closure of $D$), it is bounded on $\bar{D}$ and attains its supremum on the boundary $\partial D$. If it also attains its supremum at a point of $D$, it is constant. (The maximum principle).*

Let $M$ be the supremum of $u$ on $\bar{D}$. Since $u$ is continuous, there is at least one point of $\bar{D}$ at which $u = M$. We have to show that if $u = M$ at a point inside $\partial D$, then $u$ is a constant; it will then follow that, if $u$ is not constant, it must attain its supremum only at a point of $\partial D$.

Suppose that $u$ attains the supremum $M$ at a point $(x_0, y_0)$ of the bounded open connected set $D$. Since

$$u(x_0, y_0) = \frac{1}{2\pi} \int_0^{2\pi} u(x_0 + R\cos\theta, y_0 + R\sin\theta)\, d\theta$$

for every circle $(x - x_0)^2 + (y - y_0)^2 = R^2$ lying in $D$, $u$ is equal to $M$ on every such circle. It follows that $u = M$ on the largest disc with centre $(x_0, y_0)$ contained in $D$.

Let $F$ be the set of points of $\bar{D}$ at which $u = M$. Since $u$ is continuous, $F$ is a closed set. We have to show that $F$ is $\bar{D}$. If this is not so, there is at least one point $P$ of $\partial F$ which belongs to $D$. But since $P$ belongs to $F$, $u$ is equal to $M$ at $P$. Hence there is a disc with centre $P$ which lies in $D$ and on which $u$ is equal to $M$. This disc is therefore part of $F$, and $P$ is an interior point of $F$ and so does not belong to $\partial F$, a contradiction. Hence $F$ is $\bar{D}$, and $u$ is constant on $\bar{D}$.

In particular, *if $u$ is harmonic on a bounded domain $D$ and continuous on $\bar{D}$, and if $u$ is identically zero on $\partial D$, then $u$ is zero everywhere on $\bar{D}$.* For the suprema of $u$ and $-u$ are both zero on $\bar{D}$. From this follows a uniqueness theorem. *If $u_1$ and $u_2$ are harmonic in a bounded domain $D$ and continuous on $\bar{D}$, and if $u_1 = u_2$ everywhere on $\partial D$, then $u_1 = u_2$ everywhere on $\bar{D}$.* This is a consequence of the preceding result; take $u = u_1 - u_2$.

## 8.6 Dirichlet's principle

Let $D$ be a bounded domain whose boundary $\partial D$ consists of a finite number of non-intersecting regular closed curves. The problem of Dirichlet is to show that there exists a function $u(x, y)$, harmonic in $D$ and continuous in $\bar{D}$, which satisfies on $\partial D$ the boundary condition $u = f$, where $f$ is a given continuous function. If there is such a function, it is, as we have seen, unique.

Let $\mathscr{D}_f$ be the family of all functions $u$, which are continuous in $\bar{D}$ and have continuous second derivatives in $D$, and which satisfy the boundary condition $u = f$ on $\partial D$. If $u$ belongs to $\mathscr{D}_f$,

$$I(u) = \iint_D (u_x^2 + u_y^2)\, dx\, dy$$

exists and is not negative. The set of all such integrals has a non-negative infimum $m$. But it is not necessarily the case that there exists a function $U$, belonging to $\mathscr{D}_f$ for which $I(U) = m$.

Assuming that there is such a function, let us consider $u = U + \epsilon V$ where $V$ belongs to $\mathscr{D}_0$ and $\epsilon$ is a constant. Since $V = 0$ on $\partial D$, $U + \epsilon V$

belongs to $\mathscr{D}_f$, and $I(U+\epsilon V)$ has the minimum value $m$ when $\epsilon = 0$, no matter what function $V$ of $\mathscr{D}_0$ we use. Then, by the ordinary calculus rule,

$$\iint_D (U_x V_x + U_y V_y)\,dx\,dy = 0.$$

But

$$\int_{\partial D} V\frac{\partial U}{\partial N}\,ds = \iint_D (U_x V_x + U_y V_y)\,dx\,dy + \iint_D V\nabla^2 U\,dx\,dy.$$

The integral on the left is zero. Hence

$$\iint_D V\nabla^2 U\,dx\,dy = 0$$

for every function $V$ belonging to $\mathscr{D}_0$. This implies that $\nabla^2 U = 0$ throughout $D$. For suppose $\nabla^2 U$ is positive, say, at some point $P_0$ of $D$. By continuity, there exists a disc $K$ with centre $P_0$ on which $\nabla^2 U$ is positive. But we can choose a function $V$ of $\mathscr{D}_0$ which is zero outside $K$ and positive inside $K$; and then

$$\iint_D V\nabla^2 U\,dx\,dy > 0.$$

So $\nabla^2 U$ is not positive anywhere on $D$; similarly it is not negative. Hence
$$\nabla^2 U = 0.$$

All we have proved is that, if there is a function $U$ belonging to $\mathscr{D}_f$ for which $I(u)$ attains the infimum $m$, then $U$ is harmonic. The assumption that there is a function which minimises the integral $I(u)$ was called by Riemann *Dirichlet's principle*. The principle remained suspect until, in 1899, Hilbert showed that it can be used under proper conditions on the domain, the boundary values and the class of admissable functions.

### 8.7  A problem in electrostatics

Consider an infinity long perfectly conducting cylinder with generators parallel to the z-axis. Suppose that the plane $z = 0$ cuts the cylinder in a regular closed curve bounding a domain $D$. An infinitely long line charge of density $e$ per unit length parallel to the z-axis through the point $(x_0, y_0, 0)$ of $D$ has electrostatic potential

$$2e\log\frac{1}{R},$$

where
$$R^2 = (x-x_0)^2 + (y-y_0)^2.$$

If the conducting cylinder is earthed, a surface charge of density $\sigma$ is induced on the conductor. This surface charge produces an electrostatic field of potential $u$ inside the cylinder; $u$ does not depend on $z$, is harmonic in $D$ and continuous in $\bar{D}$. The total field inside the cylinder has potential

$$U = u + 2e\log\frac{1}{R}.$$

The electric force is $\operatorname{grad} U$. On first sight, we should expect on physical grounds, that $U$ would have continuous first derivatives on $\bar{D}$. But since $U$ is zero on the conductor, the electric force is normal to the conductor and has magnitude $\partial U/\partial N$. As we go round $\partial D$, the direction of the normal vector changes suddenly at each corner, if such exist. The direction of the electric force would then seem to change suddenly at a corner. This is impossible unless $\partial U/\partial N$ either tends to zero or to infinity as we approach the corner.

Since the charge density $\sigma$ is equal to $(1/4\pi)\,\partial U/\partial N$, the total charge per unit length is

$$\frac{1}{4\pi}\int_{\partial D}\frac{\partial U}{\partial N}\,ds$$

and this is equal to $-e$. Thus, although $\partial U/\partial N$ may be infinite at a finite number of points of $\partial D$, it is integrable.

The potential $U$, with $e = \frac{1}{2}$, is called the *Green's function*, and is denoted usually by $G(x,y;x_0,y_0)$. No doubt it is physically obvious that Green's function exists; to prove mathematically that it exists is equivalent to proving the existence of the solution of a particular case of the problem of Dirichlet.

The same physical argument can be applied when $\partial D$ consists of a finite number of non-intersecting regular closed curves.

### 8.8 Green's function and the problem of Dirichlet

In this section, we show how a knowledge of Green's function enables one to solve the problem of Dirichlet for a bounded domain $D$. We assume that $D$ is simply connected and that its frontier is a regular closed curve $\partial D$. If $u$ is continuously differentiable in $\bar{D}$, and if $u$ has continuous second derivatives in $D$, then

$$u(x,y) = \frac{1}{2\pi}\int_{\partial D}\left(u\frac{\partial\log R}{\partial N} - \frac{\partial u}{\partial N}\log R\right)ds + \frac{1}{2\pi}\iint_{D}\Delta u(\xi,\eta)\log R\,d\xi\,d\eta$$

when $(x,y)$ is a point of $D$, provided that the double integral exists; here

$$R^2 = (x-\xi)^2 + (y-\eta)^2.$$

If $v$ is harmonic in $D$ and continuously differentiable in $\bar{D}$, we also have

$$0 = \frac{1}{2\pi} \int_{\partial D} \left( u \frac{\partial v}{\partial N} - v \frac{\partial u}{\partial N} \right) + \frac{1}{2\pi} \iint_D \Delta u(\xi, \eta) . v(\xi, \eta) \, d\xi \, d\eta.$$

Subtracting,

$$u(x, y) = -\frac{1}{2\pi} \int_{\partial D} \left\{ u \frac{\partial}{\partial N} (v - \log R) - (v - \log R) \frac{\partial u}{\partial N} \right\} ds$$

$$- \frac{1}{2\pi} \iint_D \Delta u(\xi, \eta) \{ v(\xi, \eta) - \log R \} \, d\xi \, d\eta.$$

Now suppose that $D$ has a Green's function

$$G(\xi, \eta; x, y) = v(\xi, \eta; x, y) - \log R$$

which vanishes when $(\xi, \eta)$ is a point of $\partial D$. With this value of $v$,

$$u(x, y) = -\frac{1}{2\pi} \int_{\partial D} u(\xi, \eta) \frac{\partial G}{\partial N} \, ds - \frac{1}{2\pi} \iint_D \Delta u(\xi, \eta) \, G(\xi, \eta; x, y) \, d\xi \, d\eta. \tag{1}$$

This holds if $G$ is continuously differentiable with respect to $\xi$ and $\eta$ in $\bar{D}$, or, more generally, if $\partial G/\partial N$ is integrable round $\partial D$.

If $u(x, y)$ satisfies Poisson's equation $\nabla^2 u = -2\pi\sigma(x, y)$, this becomes

$$u(x, y) = -\frac{1}{2\pi} \int_{\partial D} u(\xi, \eta) \frac{\partial G}{\partial N} \, ds + \iint_D \sigma(\xi, \eta) \, G(\xi, \eta; x, y) \, d\xi \, d\eta.$$

Presumably this would give the solution of Poisson's equation in $D$ when $u$ is known on $\partial D$. It is necessary to prove the existence of Green's function and to show that

$$u(x, y) = -\frac{1}{2\pi} \int_{\partial D} f(\xi, \eta) \frac{\partial G}{\partial N} \, ds + \iint_D \sigma(\xi, \eta) \, G(\xi, \eta; x, y) \, d\xi \, d\eta \tag{2}$$

tends to $f(x_0, y_0)$ as $(x, y)$ tends, on $D$, to any point $(x_0, y_0)$ of $\partial D$. The corresponding solution for Laplace's equation is

$$u(x, y) = -\frac{1}{2\pi} \int_{\partial D} f(\xi, \eta) \frac{\partial G}{\partial N} \, ds. \tag{3}$$

## 8.9   Properties of Green's function

Let $D$ be a bounded domain whose frontier consists of a finite number of non-interesting regular closed curves and which possesses a Green's function

$$G(x, y; x_0, y_0) = v(x, y; x_0, y_0) - \log R,$$

where

$$R^2 = (x - x_0)^2 + (y - y_0)^2,$$

$(x_0, y_0)$ being a fixed point of $D$. Regarded as a function of the co-ordinates $(x, y)$ of a variable point of $D$, $v$ is assumed to be harmonic in $D$, continuous in $\bar{D}$ and continuously differentiable in $\bar{D}$.

*The integral of $\partial G/\partial N$ round $\partial D$ is $-2\pi$.*

This follows from equation (1) of §8.8 if we put $u \equiv 1$. The physical argument suggests that it holds even if $\partial D$ has corners.

*For every pair of points $(x, y)$ and $(x_0, y_0)$ of $D$, $G(x, y; x_0, y_0) > 0$.*

For values of $\epsilon$ which are not too large, the closed disc $R \leqslant \epsilon$ lies in $D$. Let $C_0$ be the circle $R = \epsilon$, $D_0$ the domain obtained by deleting the disc from $D$. Regarded as a function of $(x, y)$, $G$ is harmonic in $D_0$, continuous in $\bar{D}_0$. On $\partial D$, $G$ is zero; on $C_0$, $G$ is positive since $G \to +\infty$ as $R \to 0$. By the maximum principle, $G$ attains its supremum and infimum in $\bar{D}_0$ on the boundary of $\bar{D}_0$. Hence $G$ is strictly positive when $(x, y)$ is any point of $D_0$. Since $\epsilon$ can be as small as we please, this proves the result.

*The Green's function possesses the symmetry property*

$$G(x, y; x_0, y_0) = G(x_0, y_0; x, y)$$

*for every pair of points $(x, y)$ and $(x_0, y_0)$ of $D$.*

Write

$$G(x, y; x_0, y_0) = G_0(x, y), \quad G(x, y; x_1, y_1) = G_1(x, y).$$

Let $R_0$ and $R_1$ be the distances of $(x, y)$ from $(x_0, y_0)$ and $(x_1, y_1)$ respectively. Let $C_0$ be the circle $R_0 = \delta$, $C_1$ the circle $R_1 = \epsilon$, where $\delta$ and $\epsilon$ are such that $C_0, C_1$ and $\partial D$ do not intersect; then the discs $R_0 < \delta, R_1 < \epsilon$ lie in $D$. Since $G_0$ and $G_1$ are harmonic in the domain bounded by $C_0, C_1$ and $\partial D$,

$$\int_{C_0} \left( G_0 \frac{\partial G_1}{\partial N} - G_1 \frac{\partial G_0}{\partial N} \right) ds + \int_{C_1} \left( G_0 \frac{\partial G_1}{\partial N} - G_1 \frac{\partial G_0}{\partial N} \right) ds$$

$$+ \int_{\partial D} \left( G_0 \frac{\partial G_1}{\partial N} - G_1 \frac{\partial G_0}{\partial N} \right) ds = 0.$$

The third integral vanishes since $G_0$ and $G_1$ vanish when $(x, y)$ is on $\partial D$. Near $(x_0, y_0)$, $G_1(x, y)$ is continuous, but $G_0$ is of the form

$$v(x, y; x_0, y_0) - \log R_0,$$

where $v$ is harmonic. It follows that the first integral tends to

$$-2\pi G_1(x_0, y_0)$$

as $\delta \to 0$; similarly the second integral tends to $2\pi G_0(x_1, y_1)$ as $\epsilon \to 0$. Hence $G_1(x_0, y_0) = G_0(x_1, y_1)$.

### 8.10 The case of polynomial data

In continuation of §8.8, let $f(x,y)$ be any polynomial. If there is a function $u(x,y)$ harmonic in $D$ and continuously differentiable in $\bar{D}$, which takes on the boundary $\partial D$ the same values as $f(x,y)$, then
$$U = u - f(x,y)$$
vanishes on $\partial D$ and satisfies Poisson's equation
$$\nabla^2 u = -2\pi\sigma(x,y),$$
where $\sigma = \nabla^2 f/4\pi$ is also a polynomial. By (2) of §8.8, we have
$$U(x,y) = \iint_D \sigma(\xi,\eta)\, G(\xi,\eta;x,y)\, d\xi\, d\eta. \tag{1}$$
Hence, instead of (3) of §8.8, we have a different formula
$$u(x,y) = f(x,y) + \iint_D \sigma(\xi,\eta)\, G(\xi,\eta;x,y)\, d\xi\, d\eta \tag{2}$$
for the harmonic function.

Since $G(\xi,\eta;x,y)$ vanishes when $(x,y)$ is a point of $\partial D$ by the symmetry property, the function defined by (2) does take the prescribed value $f(x,y)$ when $(x,y)$ is a point of $D$. This is not enough; we have to show that (1) gives a function harmonic in $D$ and continuous in $\bar{D}$; for, if this were so, $u(x,y)$ would tend to $f(x,y)$ as $(x,y)$ moves up to the boundary.

The Green's function
$$G(\xi,\eta;x,y) = v(\xi,\eta;x,y) - \log R$$
we write as
$$G(\xi,\eta;x,y) = \log\frac{a}{R} + \{v(\xi,\eta;x,y) - \log a\}$$
where $a$ is a constant greater than the diameter of $D$. On $\partial D$, $v - \log R$ is zero, and so $v - \log a$ is negative. By the maximum principle, $v - \log a$ is negative on $\bar{D}$. Therefore
$$0 \leqslant G(\xi,\eta;x,y) \leqslant \log\frac{a}{R}$$
for all points $(\xi,\eta)$ of $\bar{D}$.

Let $(x_0,y_0)$ be any point of $\partial D$. Let $D_0$ be the part of $D$ for which
$$(\xi-x_0)^2 + (\eta-y_0)^2 < \epsilon^2$$
and let $D_1$ be the rest of $D$. Then
$$U(x,y) = \iint_{D_0} \sigma(\xi,\eta)\, G(\xi,\eta;x,y)\, d\xi\, d\eta + \iint_{D_1} \sigma(\xi,\eta)\, G(\xi,\eta;x,y)\, d\xi\, d\eta.$$

Since $\sigma$ is a polynomial, it is bounded on $\bar{D}$. There exists a constant $K$ such that $|\sigma| \leqslant K$ there, and so

$$|U(x,y)| \leqslant K \iint_{D_0} G(\xi,\eta;x,y)\,d\xi\,d\eta + K \iint_{D_1} G(\xi,\eta;x,y)\,d\xi\,d\eta.$$

Suppose that $(x,y)$ is a point of $D_0$. The integral over $D_1$ tends to zero as $(x,y)$ moves to $(x_0,y_0)$ on $D$. Hence

$$\limsup_{(x,y)\to(x_0,y_0)} |U(x,y)| \leqslant \limsup K \iint_{D_0} G(\xi,\eta;x,y)\,d\xi\,d\eta.$$

But, if $D_2$ is the disc $R < 2\epsilon$, $D_2$ contains $D_0$ and so

$$\iint_{D_0} G(\xi,\eta;x,y)\,d\xi\,d\eta \leqslant \iint_{D_0} \log\frac{a}{R}\,d\xi\,d\eta < \iint_{D_2} \log\frac{a}{R}\,d\xi\,d\eta$$

$$= 2\pi\epsilon^2(1 + 2\log a - 2\log 2\epsilon).$$

As $\epsilon$ can be as small as we please, $U(x,y)$ tends to zero as $(x,y)$ moves to $(x_0,y_0)$ on $\partial D$. By (2), $u(x,y)$ tends to $f(x_0,y_0)$.

We can write (1) in the form

$$U(x,y) = \iint_D \sigma(\xi,\eta)\log\frac{1}{R}\,d\xi\,d\eta + \iint_D \sigma(\xi,\eta)v(\xi,\eta;x,y)\,d\xi\,d\eta.$$

The first term is a logarithmic potential of the type considered in §8.2. It is continuous and continuously differentiable everywhere. Moreover, if $\sigma$ is continuously differentiable (as is certainly the case here since it is polynomial), the first term has continuous second derivatives on $D$ and satisfies Poisson's equation there.

By the symmetry relation $v(\xi,\eta;x,y)$ is a harmonic function of $(x,y)$ in $D$. Regarded as a function of $(\xi,\eta)$ it is harmonic in $D$ and continuous differentiable in $\bar{D}$ save possibly at corners of $\partial D$. So the second term is harmonic in $D$. Hence $\nabla^2 U = -2\pi\sigma$.

The restrictions on $f(x,y)$ could be lightened. For example, it would suffice if $f(x,y)$ had continuous derivatives of the third order on some domain containing $D$.

## 8.11   Some examples of Green's function

The first example is Green's function for a disc whose boundary is the circle $C$, with equation $r = a$ in polar coordinates. If $P_0$ is the fixed point of the disc which is the singularity of the Green's function and $P$ is a variable point of the disc,

$$G(P;P_0) = \log\frac{1}{P_0P} + v,$$

where $v$ is a harmonic function. Guided by known results in electro-statics, we expect that

$$G(P; P_0) = \log \frac{1}{P_0 P} - \log \frac{1}{P_1 P} + \text{constant},$$

where $P_1$ is the point inverse to $P_0$ with respect to the circle $C$.

Now if $Q$ is any point on the circle $C$, the triangles $OP_0 Q$ and $OQP_1$ are similar, and so

$$P_0 Q / P_1 Q = OP_0 / OQ.$$

Therefore $$G(Q; P_0) = \log \frac{P_1 Q}{P_0 Q} + \text{constant},$$

which is zero if the constant is $\log OP_0 / OQ$. Hence

$$G(P; P_0) = \log \frac{P_1 P \cdot OP_0}{P_0 P \cdot OQ}.$$

Let $P, P_0$ have polar coordinates $(\rho, \phi)$, $(r, \theta)$; then $P_1$ has polar coordinates $(a^2/r, \theta)$. It follows that

$$G(P; P_0) = \tfrac{1}{2} \log \frac{a^4 - 2a^2 \rho r \cos(\theta - \phi) + \rho^2 r^2}{a^2 \{ r^2 - 2\rho r \cos(\theta - \phi) + \rho^2 \}}.$$

It can be readily checked that this function has all the required properties.

On $C$, $\partial G / \partial N$ is the limit of $\partial G / \partial \rho$ as $\rho \to a$. Hence at the point $Q(a, \phi)$

$$\frac{\partial}{\partial N} G(Q; P_0) = - \frac{a^2 - r^2}{a(a^2 - 2ar \cos(\theta - \phi) + r^2)}.$$

The function, harmonic on the disc, which takes the value $f(\phi)$ at the point $(a \cos \phi, a \sin \phi)$ of $C$, is therefore

$$u(r \cos \theta, r \sin \theta) = \frac{1}{2\pi} (a^2 - r^2) \int_0^{2\pi} \frac{f(\phi) \, d\phi}{a^2 - 2ar \cos(\theta - \phi) + r^2}.$$

This is known as Poisson's integral. It reduces to Gauss's mean value formula when $r = 0$.

Now suppose that $P_0$ lies in the semicircle $0 < r < a, 0 < \theta < \pi$. As before, let $P_1$ be the point inverse to $P_0$ with respect to the circle $C$; and let $P_0', P_1'$ be the images of $P_0$ and $P_1$ with respect to the $x$-axis. The Green's function for the semicircle is

$$G(P; P_0) = \log \frac{P_1 P \cdot P_0' P}{P_0 P \cdot P_1' P}$$

or, in polar coordinates,

$$G(\rho \cos \phi, \rho \sin \phi; r \cos \theta, r \sin \theta)$$

$$= \tfrac{1}{2} \log \frac{\{r^2 - 2r\rho \cos (\theta + \phi) + \rho^2\} \{a^4 - 2a^2 r\rho \cos (\theta - \phi) + r^2 \rho^2\}}{\{r^2 - 2r\rho \cos (\theta - \phi) + \rho^2\} \{a^4 - 2a^2 r\rho \cos (\theta + \phi) + r^2 \rho^2\}}.$$

This function is a harmonic function of $(\rho \cos \phi, \rho \sin \phi)$ in any domain which contains none of the points $P_0, P_0', P_1, P_1'$, and so is continuously differentiable. In particular, $\partial G/\partial N$ is continuous on the boundary of the semicircle and vanishes at the corners.

The Green's function for the half-plane $\eta > 0$ with singularity $(x, y)$ where $y > 0$, can be found by the method of images; it is

$$G(\xi, \eta; x, y) = \tfrac{1}{2} \log \frac{(\xi - x)^2 + (\eta + y)^2}{(\xi - x)^2 + (\eta - y)^2}.$$

From this, we should expect that, if $u$ is harmonic in $y > 0$,

$$u(x, y) = \frac{1}{\pi} \int_{-\infty}^{\infty} u(\xi, 0) \frac{y}{(x - \xi)^2 + y^2} d\xi$$

under suitable conditions. The discussion must be deferred until after we have considered functions harmonic on an unbounded domain.

Similarly, the Green's function for the quadrant $\xi > 0, \eta > 0$ with singularity $(x, y)$ is

$$G(\xi, \eta; x, y) = \tfrac{1}{2} \log \frac{(\xi - x)^2 + (\eta + y)^2}{(\xi - x)^2 + (\eta - y)^2} \cdot \frac{(\xi + x)^2 + (\eta - y)^2}{(\xi + x)^2 + (\eta + y)^2}.$$

again by the method of images. $G$ is a harmonic function of $(\xi, \eta)$ on any domain which does not contain any of the points $(\pm x, \pm y)$, and so is continuously differentiable. $\partial G/\partial N$ is continuous on the boundary of the quadrant and vanishes at the origin which is a corner of the boundary.

The method of images can be used to find the Green's function for other domains, such as an infinite strip or a rectangle, but the results are complicated since an infinity of images arise. Care must be taken that the process of taking images does not introduce unwanted singularities. A simple instance of this is when the domain $D$ is the whole plane cut along the positive part of the $x$-axis. If we simply took the image of $(x, y)$ in the $x$-axis, we should get a singularity at $(x, -y)$ which lies in $D$; we should get the Green's function for the half-plane. In the cut plane, the angle variables are restricted to lie between 0 and $2\pi$. It can be shown, by using Sommerfield's multiple-valued potential,† that the Green's function for the cut plane is

$$G(\rho \cos\phi, \rho \sin\phi; r\cos\theta, r\sin\theta) = \tfrac{1}{2}\log\frac{\rho+r-2\sqrt{(\rho r)}\cos\tfrac{1}{2}(\theta+\phi)}{\rho+r-2\sqrt{(\rho r)}\cos\tfrac{1}{2}(\theta-\phi)}.$$

This vanishes when $\phi = 0$ and when $\phi = 2\pi$. It has the correct logarithmic singularity when the point of polar coordinates $(\rho, \phi)$ moves to $(r, \theta)$ since

$$\{\rho+r-2\sqrt{(\rho r)}\cos\tfrac{1}{2}(\theta-\phi)\}\{\rho+r+2\sqrt{(\rho r)}\cos\tfrac{1}{2}(\theta-\phi)\}$$
$$= \rho^2+r^2-2\rho r\cos(\theta-\phi).$$

It also has a logarithmic singularity when $\rho = r$, $\cos\tfrac{1}{2}(\theta+\phi) = 1$. Now $\cos\tfrac{1}{2}(\theta+\phi) = 1$ when $\phi = 4n\pi - \theta$ where $n$ is an integer or zero. As $\theta$ and $\phi$ are restricted to lie between 0 and $2\pi$, this is impossible; the Green's function has but one singularity in the cut plane.

The origin is a 'corner' of the boundary of angle $2\pi$. When $\rho$ is small, $\partial G/\partial\rho$ and $\rho^{-1}\partial G/\partial\theta$ are both of the order $O(\rho^{-\frac{1}{2}})$ and so tend to infinity as the point of polar coordinates $(\rho, \phi)$ moves up to the corner.

### 8.12 Poisson's integral

Let $u(x,y)$ be harmonic in a domain $D$. If $K$ is an open disc with centre $(x_0, y_0)$ and radius $a$, which is contained in $D$, then if $(x,y)$ is any point of $K$,

$$u(x,y) = \frac{1}{2\pi}\int_0^{2\pi} u(x_0+a\cos\phi, y_0+a\sin\phi)\,P(r,\theta-\phi)\,d\phi,$$

where $\quad x = x_0+r\cos\theta, \quad y = y_0+r\sin\theta \quad (r<a)$

and $\quad\quad P(r,\theta) = \dfrac{a^2-r^2}{a^2-2ar\cos\theta+r^2}.$    (1)

Here $u$ is known to be harmonic in a domain containing $K$; the formula merely states a relation connecting the known value of $u$ at any point of the disc $K$ and the known values of $u$ on $\partial K$. We now ask what properties

$$u(x,y) = \frac{1}{2\pi}\int_0^{2\pi} f(\phi)\,P(r,\theta-\phi)\,d\phi \quad\quad (2)$$

has when $f$ is an arbitrary function. By a shift of origin, we may suppose $x_0$ and $y_0$ to be zero.

The simplest case is when $f(\phi)$ is a continuous periodic function of

---

† A brief account of Sommerfeld's multiple-valued potentials is given in Jeans, *The Mathematical Theory of Electricity and Magnetism*, 5th edn (Cambridge, 1925), pp. 279–283.

period $2\pi$. We need the following properties of the Poisson kernel $P(r,\theta)$:

(i) for every value of $\phi$, $P(r,\theta-\phi)$ is harmonic in $r < a$,

(ii) $$P(r,\theta) > 0 \quad \text{when} \quad r < a,$$

(iii) $$\int_0^{2\pi} P(r,\theta-\phi)\,d\phi = 2\pi.$$

It is evident that we may differentiate (2) under the sign of integration as often as we please with respect to $x$ and $y$, provided that $r < a$. It follows, using (i) that *the function $u(x,y)$, defined by* (2), *is harmonic in $r < a$, and has partial derivatives of all orders, which are also harmonic in $r < a$*. We next show that $u(x,y)$ *tends to $f(\alpha)$ as $(x,y)$ tends to $(a\cos\alpha, a\sin\alpha)$ in any manner in $r < a$.*

Now

$$|u(r\cos\theta, r\sin\theta)-f(\alpha)| \leqslant |u(r\cos\theta, r\sin\theta)-f(\theta)| + |f(\theta)-f(\alpha)|.$$

Since $f$ is continuous, the second term can be made as small as we please by taking $\theta$ near enough to $\alpha$. Hence it suffices to consider radial approach to the boundary. By a rotation, we can make $\alpha$ equal to zero. Then

$$u(r,0)-f(0) = \frac{1}{2\pi}\int_0^{2\pi} \{f(\phi)-f(0)\}\,P(r,\phi)\,d\phi,$$

and so $$|u(r,0)-f(0)| \leqslant \frac{1}{2\pi}\int_0^{2\pi} |f(\phi)-f(0)|\,P(r,\phi)\,d\phi.$$

Since $f$ is continuous, it is bounded; there exists a constant $M$ such that $|f(\phi)| < M$. Hence, for any positive value of $\delta$,

$$\int_\delta^{2\pi-\delta} |f(\phi)-f(0)|\,P(r,\phi)\,d\phi \leqslant 2M(a^2-r^2)\int_\delta^{2\pi-\delta} \frac{d\phi}{a^2+r^2-2ar\cos\phi}$$

$$< \frac{2M(a^2-r^2)}{a^2+r^2-2ar\cos\delta}\int_\delta^{2\pi-\delta} d\phi$$

$$< \frac{4\pi M(a^2-r^2)}{a^2+r^2-2ar\cos\delta}.$$

Given any positive value of $\epsilon$, we can choose $\delta$ so that $|f(\phi)-f(0)| < \epsilon$ when $-\delta \leqslant \phi \leqslant \delta$. Therefore

$$\int_{-\delta}^{\delta} |f(\phi)-f(0)|\,P(r,\phi)\,d\phi < \epsilon\int_{-\delta}^{\delta} P(r,\phi)\,d\phi$$

$$< \epsilon\int_{-\delta}^{2\pi-\delta} P(r,\phi)\,d\phi = 2\pi\epsilon.$$

Hence $$|u(r,0)-f(0)| < \epsilon + \frac{2M(a^2-r^2)}{a^2+r^2-2ar\cos\delta}.$$

Keeping $\delta$ fixed, $\displaystyle\limsup_{r\to a-0}|u(r,0)-f(0)| \leqslant \epsilon.$

As $\epsilon$ is arbitrary, $u(r,0)$ tends to $f(0)$ as $r \to a-0$, which was to be proved.

Another method of justifying Poisson's integral depends on the use of Fourier series. (See Notes 8 and 9.) Suppose that $\{a_n\}$ and $\{b_n\}$ are bounded sequences. Then if $x = r\cos\theta$, $y = r\sin\theta$, the series

$$u(x,y) = \tfrac{1}{2}a_0 + \sum_1^\infty (a_n\cos n\theta + b_n\sin n\theta)\frac{r^n}{a^n}$$

is absolutely convergent in $r \leqslant ka$, for any positive value of $k$ less than unity, and converges uniformly; so also do the series obtained by differentiating with respect to $r$ and $\theta$ (or $x$ and $y$) any number of times. Since $r^n\cos n\theta$ and $r^n\sin n\theta$ satisfy Laplace's equation, it follows by differentiating under the sign of integration that $u(x,y)$ is harmonic in $r < a$.

Now suppose that $a_n$ and $b_n$ are the coefficients in the Fourier series of a function $f(\theta)$ which is integrable in Lebesgue's sense, so that

$$a_n = \frac{1}{\pi}\int_0^{2\pi} f(\phi)\cos n\phi\, d\phi, \quad b_n = \frac{1}{\pi}\int_0^{2\pi} f(\phi)\sin n\phi\, d\phi.$$

Since $\{a_n\}$ and $\{b_n\}$ are null-sequences, they are certainly bounded. Hence

$$u(x,y) = \frac{1}{2\pi}\left\{\int_0^{2\pi} f(\phi)\,d\phi + 2\sum_1^\infty \frac{r^n}{a^n}\int_0^{2\pi} f(\phi)\cos n(\theta-\phi)\,d\phi\right\}$$

is harmonic in $r < a$.

A Fourier series is summable by Abel's method; hence

$$\lim_{r\to a-0} u(x,y) = f(\theta)$$

if $f(\phi)$ is continuous when $\phi = \theta$; the limit is $f(\theta)$ almost everywhere.

If $0 \leqslant r \leqslant ka$, where $0 < k < 1$, we may alter the order of integration and summation in the series solution to obtain

$$u(x,y) = \frac{1}{2\pi}\int_0^{2\pi} f(\phi)\left\{1 + 2\sum_1^n \frac{r^n}{a^n}\cos n(\theta-\phi)\right\}d\phi$$

$$= \frac{1}{2\pi}\int_0^{2\pi} f(\phi)\frac{a^2-r^2}{a^2+r^2-2ar\cos(\theta-\phi)}\,d\phi,$$

which is Poisson's integral. Thus, if $f(\phi)$ is integrable in Lebesgue's sense, Poisson's integral gives a function $u(r\cos\theta, r\sin\theta)$, harmonic in $r < a$, which tends to $f(\theta)$ as $r \to a - 0$, provided that $f$ is continuous when $\phi = \theta$. In particular, if $f(\phi)$ is continuous, $u(r\cos\theta, r\sin\theta)$ tends to $f(\theta)$ as $r \to a - 0$, for all values of $\theta$.

### 8.13   The problem of Neumann

The problem of Neumann is to find a function $u$ harmonic in a bounded domain $D$ given the values taken by $\partial u/\partial N$ on the frontier of $D$. There are two points to note. Firstly, the problem does not have a unique solution, since the addition of a constant to $u$ does not alter the boundary values of $\partial u/\partial N$; if there is a solution, it is unique only up to an additive constant. Secondly, the boundary values cannot be assigned completely arbitrarily, since

$$\int_{\partial D} \frac{\partial u}{\partial N} \, ds = 0.$$

For simplicity, we assume that the domain $D$ is simply connected, bounded by a regular closed curve $C$. We have to try to find a function $u$ which is harmonic in $D$ and continuously differentiable in $\bar{D}$ and for which $\partial u/\partial N$ takes given continuous values on $C$ satisfying the above condition.

If $(x, y)$ is a point of $D$, we have, as in §8.8,

$$u(x,y) = -\frac{1}{2\pi} \int_C \left[ u(\xi,\eta) \frac{\partial}{\partial N} \{v(\xi,\eta) - \log R\} \right.$$
$$\left. - \frac{\partial u(\xi,\eta)}{\partial N} \{v(\xi,\eta) - \log R\} \right] ds, \quad (1)$$

where $v$ is any function, harmonic in $D$ and continuously differentiable in $\bar{D}$, and
$$R^2 = (x-\xi)^2 + (y-\eta)^2.$$

We cannot eliminate the unknown boundary values of $u$ from this equation by choosing $v$ so that the normal derivative of $v - \log R$ vanishes everywhere on $D$. For, if we could so so, we should have

$$\int_C \frac{\partial v}{\partial N} \, ds = \int_C \frac{\partial}{\partial N} \log R \, ds,$$

which is impossible since the left-hand side is equal to zero but the expression on the right has the value $2\pi$.

There are two ways of avoiding this difficulty suggested by the theory of two-dimensional flow of a perfect incompressible fluid. The function

$$V = v - \log R$$

is the velocity potential at the point $(\xi, \eta)$ of irrotational flow due to a source at the point $(x, y)$. If $C$ is a rigid boundary across which there is no flow of the fluid, there must be an equal sink. If there is no sink, there must be a flow across $C$.

Suppose that there is no sink and that $\partial V/\partial N = c$ on $C$, where $c$ is a constant. Then, if $l$ is the length of $C$,

$$cl = \int_C \frac{\partial V}{\partial N}\, ds = \int_C \frac{\partial v}{\partial N}\, ds - \int_C \frac{\partial}{\partial N} \log R\, ds.$$

The first term on the right is zero since $v$ is harmonic, the second is $-2\pi$. The constant is thus

$$c = -\frac{2\pi}{l}.$$

With this choice of $V$, (1) becomes

$$u(x, y) = \frac{1}{2\pi} \int_C V \frac{\partial u}{\partial N}\, ds + \frac{1}{l} \int_C u\, ds,$$

or

$$u(x, y) = \frac{1}{2\pi} \int_C V \frac{\partial u}{\partial N}\, ds + K,$$

where $K$ is a constant. The function $V$, called the Green's function of the second kind, we denote by $\Gamma(\xi, \eta; x, y)$. Then

$$u(x, y) = \frac{1}{2\pi} \int_C \Gamma(\xi, \eta; x, y) \frac{\partial}{\partial N} u(\xi, \eta)\, ds + K. \tag{2}$$

If the problem of Neumann has a solution, it is given by (2) where $K$ is an arbitrary constant.

In the alternative method, we make $C$ a rigid boundary and introduce an equal sink at the fixed point $(x_0, y_0)$ of $D$. If

$$R_0^2 = (\xi - x_0)^2 + (\eta - y_0)^2,$$

we have

$$u(x, y) - u(x_0, y_0) = -\frac{1}{2\pi} \int_C \left\{ u \frac{\partial}{\partial N} (v - \log R + \log R_0) \right.$$
$$\left. - (v - \log R + \log R_0) \frac{\partial u}{\partial N} \right\} ds.$$

We now choose $v$ so that

$$\frac{\partial v}{\partial N} = \frac{\partial}{\partial N} \log \frac{R}{R_0}$$

on $C$, as we may since it gives

$$\int_C \frac{\partial v}{\partial N} ds = \int_C \frac{\partial}{\partial N} \log R \, ds - \int_C \frac{\partial}{\partial N} \log R_0 \, ds = 0.$$

If $\qquad N(\xi, \eta; x, y; x_0, y_0) = v - \log R + \log R_0,$
we have

$$u(x,y) = \frac{1}{2\pi} \int_C N(\xi, \eta; x, y; x_0, y_0) \frac{\partial}{\partial N} u(\xi, \eta) \, ds + K,$$

where the constant $K$ is $u(x_0, y_0)$.

## 8.14   Harnack's first theorem on convergence

*Let $\{u_n(x,y)\}$ be a sequence of functions, each harmonic in a bounded domain $D$ and continuous on $\bar{D}$, the closure of $D$. Let $u_n(x,y)$ be equal to $f_n(x,y)$ when $(x,y)$ is any point of $\partial D$. If the sequence $\{f_n(x,y)\}$ converges to $f(x,y)$ uniformly on $\partial D$, the sequence $\{u_n(x,y)\}$ converges uniformly on $\bar{D}$ to a function $u(x,y)$ which is harmonic on $D$. Moreover, $u(x,y)$ is continuous on $\bar{D}$ and is equal to $f(x,y)$ on $\partial D$. Also any derivative of $u_n(x,y)$ converges uniformly to the corresponding derivative of $u$ on any closed subset of $D$.*

For any positive value of $\epsilon$, there exists an integer $n_0$ such that, for all points $(x,y)$ on $\partial D$,

$$-\epsilon < f_m(x,y) - f_n(x,y) < \epsilon,$$

wherever $m > n \geqslant n_0$. Hence, by the maximum principle,

$$-\epsilon < u_m(x,y) - u_n(x,y) < \epsilon$$

for all $(x,y)$ on $\bar{D}$, whenever $m > n \geqslant n_0$. Hence the sequence $\{u_n(x,y)\}$ converges uniformly on $\bar{D}$ to a function $u(x,y)$; $u$ is continuous on $\bar{D}$ and is equal to $f$ on $\partial D$.

Use polar coordinates, with any point $(x_0, y_0)$ of $D$ as origin. We can choose $a$ so that $r \leqslant a$ lies in $D$. If $x = x_0 + r \cos\theta, y = y_0 + r \sin\theta$ where $r < a$, $u_n(x,y)$ is equal to

$$\frac{1}{2\pi} \int_0^{2\pi} u_n(x_0 + a\cos\phi, y_0 + a\sin\phi) \frac{a^2 - r^2}{a^2 - 2ar\cos(\theta - \phi) + r^2} d\phi. \quad (1)$$

When $m > n \geqslant n_0$, we have

$$\left| \{u_m(x_0 + a \cos \phi, y_0 + a \sin \phi) - u_n(x_0 + a \cos \phi, y_0 + a \sin \phi)\} \right.$$
$$\left. \times \frac{a^2 - r^2}{a^2 - 2ar \cos(\theta - \phi) + r^2} \right| < \epsilon \, \frac{a + r}{a - r},$$

and so the integrand in (1) converges uniformly with respect to $\theta$ and $\phi$ for any fixed $r < a$. Hence

$$u(x, y) = \frac{1}{2\pi} \int_0^{2\pi} u(x_0 + a \cos \phi, y_0 + a \sin \phi) \frac{a^2 - r^2}{a^2 - 2ar \cos(\theta - \phi) + r^2} \, d\phi,$$

and so $u(x, y)$ is harmonic on the arbitrary disc $r < a$ and hence everywhere on $D$.

Next

$$\frac{\partial u_n(x, y)}{\partial x} = \frac{1}{2\pi} \int_0^{2\pi} u_n(x_0 + a \cos \phi, y_0 + a \sin \phi) \frac{\partial}{\partial x} P(r, \theta - \phi) \, d\phi, \quad (2)$$

where $P$ is Poisson's kernel. It can be shown that

$$\left| \frac{\partial}{\partial x} P(r, \theta - \phi) \right| \leqslant \frac{2a(a + 3r)}{(a - r)^3} \leqslant \frac{2(1 + 3k)}{a(1 - k)^3},$$

where $r \leqslant ka < a$. The integrand in (2) converges uniformly on the closed disc $r \leqslant ka$, and so $\partial u_n / \partial x$ converges uniformly to $\partial u / \partial x$. Similarly for derivatives of any order.

If $F$ is any closed set contained in $D$, for every point $(x_1, y_1)$ of $F$ we can find a closed disc $(x - x_1)^2 + (y - y_1)^2 \leqslant a_1^2$ contained in $D$ such that the sequence $\{\partial u_n / \partial x\}$ converges uniformly to $\partial u / \partial x$ on this disc. The corresponding set of open discs is an infinite open cover of $F$. By the Heine–Borel theorem, we can choose a finite number of these open discs which also cover $F$. Hence $F$ is covered by a finite number of closed discs on which the convergence is uniform. Hence $\{\partial u_n / \partial x\}$ converges uniformly to $\partial u / \partial x$ on $F$.

## 8.15 Harnack's inequality

*Let $u(x, y)$ be a non-negative function, harmonic in a bounded domain $D$. Let*

$$(x - x_0)^2 + (y - y_0)^2 \leqslant a^2$$

*be a closed disc contained in $D$. If $(x, y)$ is an interior point of the disc at a distance $r$ from $(x_0, y_0)$, then*

$$\frac{a - r}{a + r} u(x_0, y_0) \leqslant u(x, y) \leqslant \frac{a + r}{a - r} u(x_0, y_0). \quad (1)$$

This follows at once from Poisson's integral

$$u(x,y) = \frac{1}{2\pi} \int_0^{2\pi} u(x_0 + a\cos\phi, y_0 + a\sin\phi) \, P(r, \theta - \phi) \, d\phi,$$

since $u \geqslant 0$ and $\qquad \dfrac{a-r}{a+r} \leqslant P(r, \theta) \leqslant \dfrac{a+r}{a-r}.$

From this follows an analogue of Liouville's theorem. *A non-negative function, harmonic in every bounded domain, is a constant.* For such a function is harmonic on every disc $(x-x_0)^2 + (y-y_0)^2 < a^2$. If we make $a \to \infty$ in (1), we get

$$u(x,y) = u(x_0, y_0),$$

the desired result.

A more general result is as follows. *Let $u(x,y)$ be harmonic and non-negative in a domain $D$, and let $F$ be a bounded closed subset of $D$. If $(x_0, y_0)$ is a fixed point of $F$, there exist positive constants $c_0$ and $c_1$, independent of $u$, such that*

$$c_0 u(x_0, y_0) \leqslant u(x,y) \leqslant c_1 u(x_0, y_0)$$

*at every point $(x, y)$ of $F$.*

Since $\partial D$ and $F$ are disjoint closed sets, they are at a positive distance $4R$ apart. Consider the set of all open discs of radius $R$ with centres at points of $F$. By the Heine–Borel theorem, we can choose a finite number, $n$ say, of these discs which cover $F$. We may suppose that the disc with centre $(x_0, y_0)$ belongs to this family, which we call $\Sigma$.

The function $u(x, y)$ is harmonic and non-negative in the closed disc with centre $(x_0, y_0)$ and radius $4R$, since every point of this disc is an interior point of $D$. If $K_0$ is the open disc with centre $(x_0, y_0)$ and radius $2R$,

$$\tfrac{1}{3} u(x_0, y_0) \leqslant u(x,y) \leqslant 3u(x_0, y_0)$$

everywhere on $K_0$.

There exists at least one point $(x_1, y_1)$, other than $(x_0, y_0)$, which lies in $K_0$ and is the centre of a disc of the family $\Sigma$. Again, $u$ is harmonic and non-negative on the closed disc with centre $(x_1, y_1)$ and radius $4R$. If $K_1$ is the open disc with centre $(x_1, y_1)$ and radius $2R$,

$$\tfrac{1}{3} u(x_1, y_1) \leqslant u(x,y) \leqslant 3u(x_1, y_1)$$

everywhere on $K_1$. But, since $(x_1, y_1)$ is a point of $K_0$,

$$\tfrac{1}{3} u(x_0, y_0) \leqslant u(x_1, y_1) \leqslant 3u(x_0, y_0).$$

Hence, at every point $(x, y)$ of $K_1$,

$$\tfrac{1}{9}u(x_0, y_0) \leqslant u(x, y) \leqslant 9u(x_0, y_0).$$

This also holds on $K_0$.

Proceeding in this way, we find after $n$ steps, that

$$\frac{1}{3^n}u(x_0, y_0) \leqslant u(x, y) \leqslant 3^n u(x_0, y_0)$$

on all the discs of the family $\Sigma$, and hence everywhere on $\Sigma$.

## 8.16   Harnach's second theorem on convergence

*Let $\{u_n(x, y)\}$ be a sequence of functions, each harmonic in a bounded domain $D$, such that*

$$u_n(x, y) \leqslant u_{n+1}(x, y)$$

*at every point of $D$. If the sequence is bounded at one point $(x_0, y_0)$ of $D$, it converges on $D$ to a harmonic function; and the convergence is uniform on every closed subset of $D$.*

The sequence $\{u_n(x_0, y_0)\}$ is monotonic and bounded, and so is convergent. For every positive value of $\epsilon$, there exists, therefore, an integer $n_0$ such that

$$0 \leqslant u_m(x_0, y_0) - u_n(x_0, y_0) < \epsilon$$

whenever $m > n \geqslant n_0$. If $F$ is a closed subset of $D$ containing $(x_0, y_0)$, we have, by the extension of Harnack's inequality,

$$0 \leqslant u_m(x, y) - u_n(x, y) \leqslant c_1\{u_m(x_0, y_0) - u_n(x_0, y_0)\}$$

and so $$0 \leqslant u_m(x, y) - u_n(x, y) < c_1\epsilon$$

everywhere on $F$, whenever $m > n \geqslant n_0$. Hence the sequence $\{u_n(x, y)\}$ converges uniformly on $F$; let its limit be $u(x, y)$.

By Harnack's first convergence theorem, $u(x, y)$ is harmonic in the interior of $F$. As $F$ was any closed subset of $D$, $u(x, y)$ is harmonic on $D$.

## 8.17   Functions harmonic in an annulus

If $u$ is harmonic in a domain containing an annulus $0 < a \leqslant r \leqslant b$ in polar coordinates, its value at a point $(x, y)$ of the annulus is

$$u(x, y) = \frac{1}{2\pi}\left\{\int_{C_1} + \int_{C_2}\right\}\left(u\frac{\partial \log R}{\partial N} - \log R\frac{\partial u}{\partial N}\right)ds,$$

where $C_1$ is $r = a$, $C_2$ is $r = b$, and

$$R^2 = (x - \xi)^2 + (y - \eta)^2.$$

On $C_2$, $\xi = b \cos \phi$, $\eta = b \sin \phi$, and so

$$R^2 = r^2 - 2br \cos(\phi - \theta) + b^2,$$

where $(r, \theta)$ are the polar coordinates of $(x, y)$. As in §8.5,

$$\frac{1}{2\pi} \int_{C_2} = a_0 + \sum_1^\infty (a_n \cos n\theta + b_n \sin n\theta) \frac{r^n}{b^n}.$$

But on $C_1$, $\xi = a \cos \phi$, $\eta = a \sin \phi$, and

$$R^2 = a^2 - 2ar \cos(\phi - \theta) + r^2,$$

where $r > a$. Then

$$\log R = \log r - \sum_1^\infty \frac{a^n}{nr^n} \cos n(\theta - \phi)$$

and

$$\frac{\partial \log R}{\partial N} = -\frac{\partial \log R}{\partial a} = \sum_1^\infty \frac{a^{n-1}}{r^n} \cos n(\theta - \phi).$$

It follows that

$$\frac{1}{2\pi} \int_{C_1} = a_0' \log r + \sum_1^\infty (a_n' \cos n\theta + b_n' \sin n\theta) \frac{a^n}{r^n}.$$

Here

$$a_0' = -\frac{1}{2\pi} \int_{C_1} \frac{\partial u}{\partial N} ds;$$

but this is not necessarily zero since we are not given that $u$ is harmonic in $r \leqslant a$. A function $u(x, y)$, harmonic in a domain containing the annulus $0 < a \leqslant r \leqslant b$, is of the form

$$u(x, y) = a_0 + a_0' \log r + \sum_1^\infty (a_n \cos n\theta + b_n \sin n\theta) \frac{r^n}{b^n}$$
$$+ \sum_1^\infty (a_n' \cos n\theta + b_n' \sin n\theta) \frac{a^n}{r^n}.$$

The first infinite series is convergent when $r < b$ and uniformly convergent in $r$ and $\theta$ when $r \leqslant b' < b$. The second is convergent when $r > a$ and uniformly convergent when $r \geqslant a' > a$. This is the analogue of Laurent's theorem for harmonic functions.

*If the one-valued function $u(x, y)$ is harmonic in a neighbourhood of a point $P_0$, except at $P_0$ itself, and is bounded in the neighbourhood of $P_0$, then we can define $u$ at $P_0$ so that $u$ is harmonic in the whole neighbourhood.*

We may take $P_0$ to be the origin and the neighbourhood to contain $r \leqslant b$. Then, changing the notation,

$$u(x, y) = a_0 + a_0' \log r + \sum_1^\infty (a_n \cos n\theta + b_n \sin n\theta) r^n$$
$$+ \sum_1^\infty (a_n' \cos n\theta + b_n' \sin n\theta) r^{-n},$$

the series being uniformly convergent on any closed set contained in $0 < r < b$. Suppose that $|u| \leqslant K$ in $r \leqslant b$. By integrating term-by-term we have

$$2\pi(a_0 + a_0' \log r) = \int_0^{2\pi} u(r\cos\theta, r\sin\theta)\,d\theta$$

and so

$$|a_0 + a_0' \log r| \leqslant K.$$

As this holds no matter how small $r$ may be, we must have $a_0' = 0$. Similarly, if $n > 0$,

$$|a_n r^n + a_n' r^{-n}| = \left| \frac{1}{\pi} \int_0^{2\pi} u(r\cos\theta, r\sin\theta)\cos n\theta\,d\theta \right| \leqslant 2K.$$

Again, as this holds no matter how small $r$ may be, $a_n' = 0$. Similarly $b_n' = 0$. Hence

$$u(x,y) = a_0 + \sum_1^\infty (a_n \cos n\theta + b_n \sin n\theta)\, r^n$$

for $0 < r < b$, and also for $r = 0$ if we define $u(0,0)$ to be $a_0$. Since $\Sigma(|a_n| + |b_n|)c^n$ is convergent for $c < b$, $u$ is harmonic in $r < b$. $P_0$ is called a *removable singularity*.

*If the one-valued function $u(x,y)$ is harmonic in a neighbourhood of a point $P_0$ except at $P_0$ itself, and $u$ tends to $+\infty$ as $(x,y)$ tends to $P_0$, then $u = a_0' \log r + v$, where $a_0'$ is negative and $v$ is harmonic in a neighbourhood of $P_0$.*

Since $u \to +\infty$, for any positive value of $K$, no matter how large, we can choose $r_0$ so that $u > K$ when $0 < r < r_0$. Using the Laurent expansion, we have

$$2\pi(a_0 + a_0' \log r) + \pi(a_n r^n + a_n' r^{-n}) = \int_0^{2\pi} u(r\cos\theta, r\sin\theta)\{1 + \cos n\theta\}\,d\theta$$

$$\geqslant K \int_0^{2\pi} (1 + \cos n\theta)\,d\theta = 2\pi K,$$

since $1 + \cos n\theta \geqslant 0$. Similarly

$$2\pi(a_0 + a_0' \log r) - \pi(a_n r^n + a_n' r^{-n}) \geqslant 2\pi K.$$

These inequalities hold in $0 < r < r_0$, no matter how small $r$ may be. If $a_n' \neq 0$, these inequalities cannot both be true since $\pm \pi a_n' r^{-n}$ are the dominant terms on the left. Hence $a_n' = 0$; similarly $b_n' = 0$. The inequalities also imply that $a_0' \log r \to +\infty$ as $r \to 0$; hence $a_0' < 0$.

Hence

$$u(x,y) = a_0 + a_0' \log r + \sum_1^\infty (a_n \cos n\theta + b_n \sin n\theta)\, r^n$$

which was to be proved.

## 8.18   Unbounded domains

So far, we have considered only bounded domains. If $D$ is unbounded, we say that $u(x, y)$ is harmonic on $D$ if it is bounded and is harmonic on every bounded domain contained in $D$.

We may suppose that the origin is not a point of $D$, so that $D$ lies outside a circle $x^2 + y^2 = a^2$. The point $(x', y')$ inverse to $(x, y)$ with respect to this circle has coordinates

$$x' = a^2 x/(x^2 + y^2), \quad y' = a^2 y/(x^2 + y^2),$$

or, in polar coordinates,

$$r' = a^2/r, \quad \theta' = \theta.$$

The inverse of the domain $D$ is a domain $D'$, which lies inside the circle. The origin is not a point of $D'$; but $D'$ does contain a 'punctured' disc

$$0 < x'^2 + y'^2 < b^2.$$

Since $\qquad (x^2 + y^2)\left(\dfrac{\partial^2 u}{\partial x^2} + \dfrac{\partial^2 u}{\partial y^2}\right) = (x'^2 + y'^2)\left(\dfrac{\partial^2 u}{\partial x'^2} + \dfrac{\partial^2 u}{\partial y'^2}\right),$

the function $\qquad u_1(x', y') = u\left(\dfrac{a^2 x'}{x'^2 + y'^2}, \dfrac{a^2 y'}{x'^2 + y'^2}\right)$

is harmonic on the punctured disc, and is bounded there. The origin is therefore a removable singularity of $u_1(x', y')$; as $r' \to 0$, $u_1(x', y')$ then tends to a limit $a_0$. The function which is equal to $u_1(x', y')$ on $0 < r' < b$ and is equal to $a_0$ at the origin is harmonic on $r' < b$.

Let $C$ be a regular closed curve. The exterior problem of Dirichlet is to find a function which is harmonic on the domain $D$ exterior to $C$ and which takes given continuous values on $C$. By the definition, we require the function $u(x, y)$ to be bounded and to be harmonic on every domain contained in $D$. If the boundedness condition is dropped, the problem does not have a unique solution.

Choose as origin a point inside $C$. The inverse of $D$ with respect to the circle $r = a$ is a bounded domain, punctured at the origin. The inverse of $C$ is a regular closed curve $\Gamma$. If $u_1(x', y')$ is the unique function which is harmonic inside $\Gamma$ and takes the corresponding continuous boundary values on $\Gamma$, then

$$u_1\left(\frac{a^2 x}{x^2 + y^2}, \frac{a^2 y}{x^2 + y^2}\right)$$

is the required solution of the exterior problem of Dirichlet. The existence and uniqueness of the solution of the interior problem of Dirichlet implies the existence and uniqueness of the solution of the exterior problem of Dirichlet in the class of bounded functions.

A simple example is the solution of the exterior problem of Dirichlet when the boundary is a circle $r = a$. The solution of the interior problem is

$$u(r, \theta) = \frac{1}{2\pi} \int_0^{2\pi} \frac{a^2 - r^2}{a^2 - 2ar\cos(\theta - \phi) + r^2} f(\phi)\, d\phi.$$

If we invert with respect to $r = a$, the interior domain becomes the unbounded domain $r > a$. Replacing $r$ by $a^2/r$, we get

$$u(r, \theta) = \frac{1}{2\pi} \int_0^{2\pi} \frac{r^2 - a^2}{a^2 - 2ar\cos(\theta - \phi) + r^2} f(\phi)\, d\phi,$$

a result which can also be obtained from the fact that a function which is bounded and harmonic when $r > a$ is of the form

$$a_0 + \sum_1^\infty (a_n \cos n\theta + b_n \sin n\theta)\frac{a^n}{r^n}.$$

More generally, the solution of the exterior problem of Dirichlet in the class of bounded functions can be deduced from the Green's function solution of the interior problem obtained by inversion. This implies that we can define the Green's function in the ordinary way for an unbounded domain, provided that we require it to be bounded at infinity. For example, the Green's function for the half-plane $\eta > 0$ with singularity $(x, y)$ where $y > 0$ is

$$G(\xi, \eta; x, y) = \tfrac{1}{2} \log \frac{(\xi - x)^2 + (\eta + y)^2}{(\xi - x)^2 + (\eta - y)^2}.$$

This satisfies the condition of vanishing when $\eta = 0$; it has the correct logarithmic singularity at $(x, y)$ and is bounded at infinity.

## 8.19   Connexion with complex variable theory

Let $w = f(z)$ be an analytic function of the complex variable $z = x + iy$, regular in a bounded domain $D$. If $w = u + iv$, where $u$ and $v$ are real, $u$ and $v$ are harmonic in $D$. For both have continuous derivatives of all orders; and since $u_x = v_y, u_y = -v_x$, both $u$ and $v$ satisfy Laplace's equation.

Conversely, if $u$ is harmonic in a simply connected domain $D$, whose boundary is a regular closed curve $C$, then $u$ is the real part of an analytic function, regular in $D$. By Green's theorem,

$$\int_\Gamma (u_x\, dy - u_y\, dx)$$

vanishes for every regular closed curve $\Gamma$ lying in $D$. Hence

$$v(x, y) = \int_{(x_0, y_0)}^{(x, y)} (u_x \, dy - u_y \, dx)$$

is independent of the path of integration; $v(x, y)$ is a one-valued function which has continuous derivatives in $D$, and $v_x = -u_y, v_y = u_x$. Therefore $u + iv$ is an analytic function of $z$, regular in $D$. The harmonic function $v$ is called the conjugate of $u$.

If $(x_0, y_0)$ is a point of $D$, $u$ is the real part of an analytic function $f(z)$, regular in a disc $|z - z_0| < R$ contained in $D$. By Taylor's theorem,

$$f(z) = \sum_0^\infty c_n (z - z_0)^n$$

is convergent on the disc. Hence $u$ can be expressed, either as

$$u = \sum_0^\infty |c_n| \, r^n \cos(n\theta + \alpha_n),$$

where $c_n = |c_n| \, e^{i\alpha_n}$ and $z = z_0 + r e^{\theta i}$, or as

$$u = \sum_{m, n=0}^m a_{mn} (x - x_0)^m (y - y_0)^n.$$

By the sort of argument used in §8.5, we can show that this double series is absolutely convergent on the square

$$|x - x_0| + |y - y_0| < R,$$

and hence that $u(x, y)$ is analytic.

## 8.20 Conformal mapping

If $\zeta = F(z)$ is an analytic function, regular in a bounded domain $D$, it maps $D$ into a domain $\Delta$ in the $\zeta$ plane. Such a mapping is conformal, since the angle between the tangents at the point of intersection of two curves is unaltered by the mapping. If the mapping is a bijection, that is, to every point of $D$ there corresponds one point of $\Delta$ and to every point of $\Delta$ there corresponds one point of $D$, $\zeta = F(z)$ is said to map $D$ conformally on $\Delta$; such a mapping is sometimes said to be simple or *schlicht*.

If $\zeta = F(z)$ maps $D$ conformally onto $\Delta$, for any function $u(x, y)$, we have

$$\frac{\partial^2 u}{\partial x^2} + \frac{\partial^2 u}{\partial y^2} = |F'(z)|^2 \left( \frac{\partial^2 u}{\partial \xi^2} + \frac{\partial^2 u}{\partial \eta^2} \right),$$

where $z = x + iy$, $\zeta = \xi + i\eta$. Hence if $u$ is harmonic on $D$, the conformal mapping turns it into a function harmonic on $\Delta$.

Suppose that $D$ is bounded by a regular closed curve $C$. Then, by Riemann's theorem on conformal mapping, there exists a unique analytic function $F(z)$, regular in $D$, such that $\zeta = F(z)$ maps $D$ conformally on $|\zeta| < 1$, transforms any point $a$ within $C$ into the origin, and a given direction at $a$ into the positive direction of the real axis. It would seem, therefore, that all we have to do to solve the problem of Dirichlet for $D$ is to map $D$ conformally on the unit circle $|\zeta| < 1$ and solve the corresponding problem in the $\zeta$ plane. If we know an explicit formula for the mapping, this is satisfactory. The argument fails to prove the general existence theorem since Riemann's theorem is equivalent to the theorem concerning the existence and uniqueness of the solution of Dirichlet's problem.

As an example, we deduce Poisson's formula from Gauss's mean value theorem. The conformal transformation

$$\zeta = \frac{a(z - z_0)}{a^2 - z\bar{z}_0},$$

where $|z_0| < a$ and $\bar{z}_0$ is the conjugate of $z_0$, maps $|z| \leqslant a$ onto $|\zeta| \leqslant 1$, the image of $z_0$ being the origin of the $\zeta$ plane. This mapping turns a function $u(x, y)$ harmonic in the $z$ plane into a function $U(\xi, \eta)$ harmonic in $|\zeta| \leqslant 1$; and $u(x_0, y_0) = U(0, 0)$. But

$$U(0, 0) = \frac{1}{2\pi} \int_\Gamma U(\xi, \eta) \, d\sigma,$$

where $\sigma$ is the arc length of the unit circle $\Gamma$ whose equation is $|\zeta| = 1$. Hence

$$u(x_0, y_0) = \frac{1}{2\pi} \int_C U(\xi, \eta) \frac{d\sigma}{ds} \, ds,$$

where $s$ is the arc length of $C$, $|z| = a$. On $C$, $z = ae^{\phi i}$; and $z_0 = re^{\theta i}$ where $r < a$. Now

$$\frac{d\sigma}{ds} = \left| \frac{d\zeta}{dz} \right| = \left| \frac{a(a^2 - |z_0|^2)}{a(a^2 - z\bar{z}_0)^2} \right| = \frac{a^2 - r^2}{a(a^2 - 2ar\cos(\theta - \phi) + r^2)}.$$

Therefore

$$u(x_0, y_0) = \frac{1}{2\pi} \int_0^{2\pi} u(r\cos\phi, r\sin\phi) \frac{a^2 - r^2}{a^2 - 2ar\cos(\theta - \phi) + r^2} \, d\phi.$$

## 8.21 The problem of Neumann

Let $C$ be a regular closed curve bounding a domain $D$. Position on $C$ is defined by the arc length $s$, measured from some fixed point on $C$; $s$ is chosen to increase as the point of parameter $s$ describes $C$ in the positive sense. The problem of Neumann is to find a function $u(x,y)$, harmonic in $D$ and continuously differentiable in $\bar{D}$, which is such that $\partial u/\partial N$ takes given continuous values on $C$. Thus, on $C$,

$$\frac{\partial u}{\partial N} = g(s),$$

where $g(s)$ is periodic, of period $l$, where $l$ is the length of $C$. This problem can be transformed into the Dirichlet problem, by using conjugate functions.

Let $v(x,y)$ be the harmonic function conjugate to $u$. By the Cauchy-Riemann conditions,

$$\frac{\partial u}{\partial N} = \frac{\partial v}{\partial s}$$

on $C$. Hence, on $C$, $\quad v = \int_0^s g(s)\,ds + A = f(s)$,

where $A$ is an arbitrary constant. Since $g(s)$ is periodic of period $l$ and

$$\int_a^{l+a} g(s)\,ds = 0$$

for all $a$, $f(s)$ is a continuous function of period $l$. If $v(x,y)$ is the solution of the problem of Dirichlet for $D$ with boundary data $f(s)$, the solution $u$ of the problem of Neumann is the function whose conjugate is $v$, or, alternatively, is the conjugate of $v$ with the sign changed.

As an example, consider the problem of Neumann for the disc $r < a$. The solution is to be harmonic in $D$, continuously differentiable in $\bar{D}$; on $r = a$, $\partial u/\partial r = g(\phi)$, with a change of parameter to $\phi = s/a$. The conjugate function $v$ takes on $C$ the values defined by

$$f(\phi) = a \int_0^\phi g(\phi)\,d\phi,$$

dropping the constant of integration.

Since

$$v(r\cos\theta, r\sin\theta) = \frac{1}{2\pi}\int_0^{2\pi} f(\phi)\,P(r,\theta-\phi)\,d\phi,$$

the solution of the problem of Neumann is

$$u(r\cos\theta, r\sin\theta) = -\frac{1}{2\pi}\int_0^{2\pi} f(\phi)\,Q(r,\theta-\phi)\,d\phi,$$

where $Q$ is the conjugate of the Poisson kernel. But

$$\frac{a e^{\phi i} + r e^{\theta i}}{a e^{\phi i} - r e^{\theta i}} = \frac{a^2 - r^2}{a^2 - 2ar\cos(\theta-\phi) + r^2} + \frac{2iar\sin(\theta-\phi)}{a^2 - 2ar\cos(\theta-\phi) + r^2}.$$

The second term on the right is $iQ(r,\theta-\phi)$.

We thus have

$$u(r\cos\theta, r\sin\theta) = -\frac{1}{2\pi}\int_0^{2\pi} f(\phi)\,\frac{2ar\sin(\theta-\phi)}{a^2 - 2ar\cos(\theta-\phi) + r^2}\,d\phi.$$

To express this in terms of the given function $g(\phi)$, we integrate by parts: this gives

$$u(r\cos\theta, r\sin\theta) = \frac{1}{2\pi}\int_0^{2\pi} f(\phi)\,\frac{\partial}{\partial\phi}\log(a^2 - 2ar\cos(\theta-\phi) + r^2)\,d\phi$$

$$= -\frac{a}{2\pi}\int_0^{2\pi} g(\phi)\log(a^2 - 2ar\cos(\theta-\phi) + r^2)\,d\phi.$$

An arbitrary constant can be added to this solution.

## 8.22    Green's function and conformal mapping

In the complex $z$-plane, let $D$ be a simply connected domain bounded by a regular closed curve $C$. By Riemann's mapping theorem, there exists a unique analytic function $f(z)$ such that $\zeta = f(z)$ maps $D$ conformally on the disc $|\zeta| < 1$, maps a given point $z_0$ of $D$ on the origin and turns any given direction at $z_0$ into the positive direction of the real axis in the $\zeta$ plane. When $z$ is on $C$, $|\zeta| = 1$. The function $f(z)$ has a simple zero at $z_0$ and so is of the form $(z - z_0)\,g(z)$ where $g(z)$ is analytic and does not vanish on $D$; for only one point of $D$ is mapped into the origin.

If we put $\qquad\qquad f(z) = e^{-U-iV},$

then $\qquad\qquad U = -\log|f(z)| = u - \log|z - z_0|,$

where $\qquad\qquad u = -\operatorname{re}\log g(z)$

is harmonic on $D$. On $C$, $U$ vanishes. Hence $U$ is the Green's function for $D$. If we know a conformal mapping of $D$ of the desired type, the Green's function can be found.

An interesting case is the Green's function for a closed polygon. Suppose that the corners of the polygon are $z_1, z_2, \ldots, z_n$ taken in order in the positive sense. As we go through the corner $z_k$ in the positive sense, the direction of the side turns through an angle $\pi\alpha_k$. At an ordinary corner $0 < \alpha_k < 1$, but at a re-entrant corner

$$-1 < \alpha_k < 0;$$

and $\Sigma\alpha_k = 2$. The Schwarz–Christoffel transformation which maps the polygon onto $|\zeta| < 1$ and turns a given point $z_0$ inside the polygon into the origin is given by

$$z - z_0 = K\int_0^\zeta \prod_1^n (\zeta - \zeta_k)^{-\alpha_k}\,d\zeta.$$

Since
$$z - z_1 = K \int_{\zeta_1}^{\zeta} (\zeta - \zeta_1)^{-\alpha_1} \prod_2^{\infty} (\zeta - \zeta_k)^{-\alpha_k} d\zeta,$$

the mapping near the corner $z$, has the form

$$z - z_1 = (\zeta - \zeta_1)^{1-\alpha_1} \sum_0^{\infty} A_p (\zeta - \zeta_1)^p,$$

where the infinite series has a finite radius of convergence. Hence, by the formula for the reversion of series

$$\zeta - \zeta_1 = \sum_0^{\infty} a_p (z - z_1)^{p/(1-\alpha_1)},$$

where the series again has a finite radius of convergence.

If we write $\zeta = e^{-U-iV}$, $U$ is the Green's function for $D$. And

$$|\text{grad } U| = |U_x - iU_y| = |U_x + iV_x| = \left| \frac{1}{\zeta} \frac{d\zeta}{dz} \right|.$$

Since $\zeta_1$ is not zero, $|\text{grad } U|$ near the corner $z_1$, behaves like

$$\frac{|a_1|}{|\zeta_1|} \cdot \frac{1}{1 - \alpha_1} |z - z_1|^{1/(1-\alpha_1)-1} = K_1 \rho^{1/(1-\alpha_1)-1},$$

where $K_1$ is a non-zero constant and $\rho$ is the distance of $z$ from the corner $z_1$.

If $z_1$ is an ordinary corner, $0 < \alpha_1 < 1$. Hence as $z$ moves up to $z_1$ on the interior of the polygon, grad $U$ tends to zero like a multiple of $\rho^{\alpha_1/(1-\alpha_1)}$. But if $z_1$ is a re-entrant corner, $-1 < \alpha_1 < 0$. If we write $\alpha_1 = -\beta$, then as $z$ moves up to $z_1$ on the interior of the polygon, grad $U$ tends to infinity like a multiple of $\rho^{-\beta/(1+\beta)}$. In either case, $1/(1-\alpha_1)$ is positive, and so grad $U$ is integrable round the polygon.

If we know the Green's function for a domain $D$, we can deduce the Green's function for a domain $\Delta$ if we can map $D$ conformally on $\Delta$. For example, the Green's function for the half-plane $\eta > 0$ with singularity $(x, y)$ where $y > 0$ is

$$G(\xi, \eta; x, y) = \tfrac{1}{2} \log \frac{(x-\xi)^2 + (y+\eta)^2}{(x-\xi)^2 + (y-\eta)^2}.$$

If we write $z = x + iy$, $\zeta = \xi + i\eta$, this becomes

$$G = \log \frac{|z - \bar{\zeta}|}{|z - \zeta|},$$

where $\bar{\zeta}$ is the conjugate of $\zeta$. When $z$ and $\zeta$ lie in the upper half-plane $0 < \text{ph } z < \pi$, $0 < \text{ph } \zeta < \pi$. If $\zeta$ lies on the boundary, $\text{ph } \zeta = 0$ or $\text{ph } \zeta = \pi$.

If $0 < \alpha \leqslant 2$, the relation $w = z^\alpha$ maps the half-plane $0 < \mathrm{ph}\, z < \pi$ onto the angle $0 < \mathrm{ph}\, w < \pi\alpha$ (the cut plane if $\alpha = 2$), and the mapping is conformal except at the origin. Hence the Green's function for the angle is

$$G = \log \frac{|w^{1/\alpha} - (\overline{\varpi})^{1/\alpha}|}{|w^{1/\alpha} - \varpi^{1/\alpha}|}.$$

If we put

$$w = r e^{i\theta}, \quad \varpi = \rho e^{i\phi},$$

where $0 < \theta < \pi\alpha, 0 < \phi < \pi\alpha$, this becomes

$$G = \tfrac{1}{2}\log \frac{r^{2/\alpha} - 2r^{1/\alpha}\rho^{1/\alpha}\cos\dfrac{1}{\alpha}(\theta + \phi) + \rho^{2/\alpha}}{r^{2/\alpha} - 2r^{1/\alpha}\rho^{1/\alpha}\cos\dfrac{1}{\alpha}(\theta - \phi) + \rho^{2/\alpha}},$$

the result we stated in §8.11 for the case $\alpha = 2$. When $\rho$ is small,

$$\frac{\partial G}{\partial \rho} \doteqdot \frac{2}{\alpha}\frac{\rho^{1/\alpha - 1}}{r^{1/\alpha}}\sin\frac{\theta}{\alpha}\sin\frac{\phi}{\alpha}, \quad \frac{1}{\rho}\frac{\partial G}{\partial \phi} \doteqdot \frac{2}{\alpha}\frac{\rho^{1/\alpha - 1}}{r^{1/\alpha}}\sin\frac{\theta}{\alpha}\cos\frac{\phi}{\alpha}.$$

Hence, if $0 < \alpha < 1$, when the angle is an ordinary acute or obtuse angle, $\partial G/\partial \rho$ and $\rho^{-1}\partial G/\partial \phi$ tend to zero as $\rho \to 0$ like multiples of $\rho^{1/\alpha - 1}$. But if $1 < \alpha \leqslant 2$, when the angle is a reflex angle, $\partial G/\partial \rho$ and $\rho^{-1}\partial G/\partial \phi$ tend to infinity as $\rho \to 0$. This agrees with the result obtained for a polygon.

Since the problem of what happens at a corner is a local problem we should expect that the Green's function for a domain whose boundary consists of a finite number of regular arcs would behave in a similar way at corners.

### Exercises

**1.** Prove that, if $D$ is the disc $\xi^2 + \eta^2 < a^2$, the logarithmic potential

$$\iint_D \log\frac{1}{R}\, d\xi\, d\eta$$

is equal to $\tfrac{1}{2}\pi(a^2 - r^2) - \pi a^2\log a$ when $r^2 = x^2 + y^2 < a^2$, and is equal to $-\pi a^2 \log r$ when $r > a$.

**2.** Prove that the logarithmic potential

$$\int_{-a}^{a} \log\frac{1}{R}\, d\xi,$$

where $R^2 = (x-\xi)^2 + y^2$, is equal to

$$\tfrac{1}{2}(x-a)\log\{(x-a)^2+y^2\} - \tfrac{1}{2}(x+a)\log\{(x+a)^2+y^2\} + 2a$$

$$-y\tan^{-1}\frac{x+a}{y} + y\tan^{-1}\frac{x-a}{y}$$

provided that $(x,y)$ is not a point of the segment $-a \leqslant x \leqslant a$, $y = 0$. Show that, on the segment, the potential is

$$2a - (a-x)\log(a-x) - (a+x)\log(a+x).$$

Verify that this potential has the properties stated in §8.4.

Show also that

$$\int_{-a}^{a} \frac{\partial}{\partial N} \log\frac{1}{R}\, d\xi,$$

where $N$ is parallel to the axis of $y$, is equal to

$$\tan^{-1}\frac{x+a}{y} - \tan^{-1}\frac{x-a}{y},$$

when $y \neq 0$. What happens when $y = 0$?

**3.** Prove that the potential at a point $P$ of a normally directed distribution of doublets of uniform density $\sigma$ on a regular arc $C$ is $\sigma\psi$ where $\psi$ is the angle subtended by $C$ at $P$.

**4.** $u(x,y)$ is harmonic on the disc $x^2+y^2 < a^2$ and vanishes on the diameter $y = 0$. Prove that $u$ is an odd function of $y$.

**5.** Find the function $u(x,y)$, harmonic on the disc $r^2 = x^2+y^2 < a^2$ given that, on $r = a$, $u = 1$ when $y > 0$ and $u = 0$ when $y < 0$. Find also the function, harmonic and bounded when $r > a$, which takes the same values on $r = a$.

**6.** Find the function $u(x,y)$, harmonic on $r < a$, which takes the values $|y|$ on $r = a$.

**7.** The function $f(\xi)$ is bounded and continuous. Prove that

$$u(x,y) = \frac{1}{\pi}\int_{-\infty}^{\infty} f(\xi)\frac{y}{(x-\xi)^2+y^2}\, d\xi$$

is harmonic in $y > 0$ and that as $(x,y)$ tends to $(a,0)$ on any path in $y > 0$, $u(x,y)$ tends to $f(a)$. Show that $u(x,y)$ is also harmonic in $y < 0$, but is discontinuous across the $x$-axis.

Deduce that

$$u(x,y) = \frac{1}{2\pi}\int_{-\infty}^{\infty} g(\xi)\log\{(x-\xi)^2+y^2\}\, d\xi$$

is the solution of the problem of Neumann for the half-plane $y > 0$.

**8.** $u(x, y)$ is harmonic on the disc $x^2 + y^2 < a^2$ and is identically zero on the smaller disc $x^2 + y^2 < b^2$. Show that $u$ is identically zero. Hence prove that, if $u_1$ and $u_2$ are harmonic on a domain $D$ and are equal on a domain contained in $D$, then $u_1 = u_2$ everywhere on $D$.

**9.** $u(x, y)$ is harmonic in a domain containing the closed disc $K$,

$$x^2 + y^2 \leqslant a^2.$$

If

$$\iint_K \{u(x, y)\}^2 \, dx \, dy = M,$$

prove that

$$|u(0, 0)| \leqslant \frac{1}{a} \left( \frac{M}{\pi} \right)^{\frac{1}{2}}$$

by using Gauss's mean value theorem. Deduce that if $(x, y)$ is an interior point of $K$,

$$|u(x, y)| \leqslant \frac{1}{a - r} \left( \frac{M}{\pi} \right)^{\frac{1}{2}},$$

where $r^2 = x^2 + y^2$. Hence show that, if $u$ is harmonic everywhere and is not identically zero, the integral of $u^2$ over the whole plane is divergent.

**10.** Deduce from Harnack's Inequality that if $u(x, y)$ is non-negative and harmonic everywhere, it is constant.

**11.** Prove that, if $\Gamma(\xi, \eta; x, y)$ is the Green's function of the second kind,

$$\Gamma(x_0, y_0; x_1, y_1) + \frac{1}{l} \int_C \Gamma(\xi, \eta; x_1, y_1) \, ds$$

$$= \Gamma(x_1, y_1; x_0, y_0) + \frac{1}{l} \int_C \Gamma(\xi, \eta; x_0, y_0) \, ds.$$

**12.** Prove that, for the disc $x^2 + y^2 \leqslant a^2$, the Green's function of the second kind is

$$\Gamma(\xi, \eta; x, y) = -\tfrac{1}{2} \log (r^2 - 2\rho r \cos (\theta - \phi) + \rho^2)$$
$$- \tfrac{1}{2} \log (a^4 - 2a^2 \rho r \cos (\theta - \phi) \rho^2 r^2),$$

where $(\rho, \phi)$ and $(r, \theta)$ are the polar coordinates of $(\xi, \eta)$ and $(x, y)$. Deduce the solution of the problem of Neumann for the disc.

**13.** Show that $w = e^z$ maps the infinite strip $0 \leqslant y \leqslant \pi$, where $z = x + iy$, conformally on the half-plane $\operatorname{im} w \geqslant 0$. Hence show that the Green's function for the strip is

$$G(\xi, \eta; x, y) = \tfrac{1}{2} \log \frac{e^{2x} - 2e^{x+\xi} \cos (y + \eta) + e^{2\xi}}{e^{2x} - 2e^{x+\xi} \cos (y - \eta) + e^{2\xi}}.$$

Deduce that, if $u$ is harmonic in the strip,

$$u(x, y) = \frac{\sin y}{2\pi} \int_{-\infty}^{\infty} \left\{ \frac{u(\xi, 0)}{\cosh (\xi - x) - \cos y} + \frac{u(\xi, \pi)}{\cosh (\xi - x) + \cos y} \right\} d\xi.$$

**14.** $u(x, y)$ is bounded and harmonic in the whole plane cut along the positive part of the $x$-axis. As $(x, y)$ moves to the point $(\xi, 0)$ where $\xi > 0$ by a path in the upper half plane, $u(x, y)$ tends to $f_+(\xi)$ where $f_+$ is bounded and continuous. As $(x, y)$ moves to the same point by a path in the lower half-plane, $u(x, y)$ tends to $f_-(\xi)$ where $f_-$ is also bounded and continuous. Prove that if $(x, y)$ has polar coordinates $(r, \theta)$ when $0 < \theta < 2\pi$,

$$u(x, y) = \frac{1}{2\pi} \int_0^\infty f_+(\xi) \sqrt{\left(\frac{r}{\xi}\right)} \frac{\sin \frac{1}{2}\theta}{r + \xi - 2\sqrt{(r\xi)} \cos \frac{1}{2}\theta} \, d\xi$$

$$+ \frac{1}{2\pi} \int_0^{2\pi} f_-(\xi) \sqrt{\left(\frac{r}{\xi}\right)} \frac{\sin \frac{1}{2}\theta}{r + \xi + 2\sqrt{(r\xi)} \cos \frac{1}{2}\theta} \, d\xi.$$

**15.** $u(x, y)$ has continuous second derivatives on a bounded domain $D$ where it satisfies $\nabla^2 u = -2\pi\sigma(x, y)$ where $\sigma$ is non-negative. If $(x_0, y_0)$ is any point of $D$, there exists a disc with centre $(x_0, y_0)$ and radius $a$ which lies in $D$. Prove that

$$\int_0^{2\pi} u(x_0 + r \cos \theta, y_0 + r \sin \theta) \, d\theta$$

is a decreasing function of $r$ when $r < a$. Hence show that

$$u(x_0, y_0) \geqslant \frac{1}{2\pi} \int_0^{2\pi} u(x_0 + r \sin \theta, y_0 + r \sin \theta) \, d\theta.$$

# 9

# SUBHARMONIC FUNCTIONS AND THE PROBLEM OF DIRICHLET

## 9.1 The enunciation of the problem

If $D$ is a bounded domain, the problem is to show that there exists a function, harmonic in $D$ and continuous in $\bar{D}$, which takes given continuous values on the frontier $\partial D$ of $D$. A domain is an open connected set; its frontier is the intersection of its closure and the closure of its complement. There are several proofs of this existence theorem. We give here the proof, associated with the names of Perron and Remak, which uses the theory of subharmonic functions.

No assumption is made initially about the connectivity of $D$ or the structure of $\partial D$. A function is found which is harmonic in $D$ and is specially related to the boundary values. Conditions are found on the structure of $\partial D$, under which this harmonic function is continuous in $\bar{D}$ and takes the given values on $\partial D$. These conditions are satisfied when $D$ is simply connected and is bounded by a regular closed curve; more generally, when $D$ is multiply connected and is bounded by a finite number of non-intersecting regular closed curves.

Although the function which solves the problem of Dirichlet is 'found', the method does not provide a practical method of solving particular problems.

## 9.2 Subharmonic functions

A function $u(x, y)$, continuous in a domain $D$, is said to be subharmonic in $D$ if, for every closed disc $(\xi - x)^2 + (\eta - y)^2 \leqslant r^2$ contained in $D$

$$u(x, y) \leqslant \frac{1}{2\pi} \int_0^{2\pi} u(x + r\cos\theta, y + r\sin\theta) \, d\theta.$$

If $u$ and $-u$ are both subharmonic, $u$ is harmonic, by the converse of Gauss's mean value theorem.

*If $u_1, u_2, \ldots, u_n$ are subharmonic on a domain $D$, so also are*

$$c_1 u_1 + c_2 u_2 + \ldots + c_n u_n,$$

[ 175 ]

*where $c_1, c_2, \dots, c_n$ are positive constants, and*

$$\max (u_1, u_2, \dots, u_n).$$

The first result is obvious. The second can be proved by induction since

$$\max (u_1, u_2, \dots, u_n) = \max (u_n, v),$$

where

$$v = \max (u_1, u_2, \dots, u_{n-1}).$$

So it suffices to prove the second result in the case $n = 2$.

Let

$$u = \max (u_1, u_2).$$

This means that, at each point $(x, y)$, the value of $u(x, y)$ is the greater of $u_1(x, y)$ and $u_2(x, y)$. Since

$$u = \tfrac{1}{2}(u_1 + u_2 + |u_1 - u_2|)$$

$u$ is continuous on $D$. If $u = u_1$ at the point $(x, y)$, we have

$$u(x, y) = u_1(x, y) \leqslant \frac{1}{2\pi} \int_0^{2\pi} u_1(x + r\cos\theta, y + r\sin\theta)\, d\theta$$

$$\leqslant \frac{1}{2\pi} \int_0^{2\pi} u(x + r\cos\theta, y + r\sin\theta)\, d\theta.$$

We get the same result if $u = u_2$ at the point considered. Hence $u$ is subharmonic.

*If $u$ is subharmonic on a bounded domain $D$ and has a relative maximum at a point of $D$, there is a disc, which lies in $D$ and has the point as centre, on which $u$ is constant.*

Take the point as origin. Then there is a disc $r < a$ whose closure lies in $D$. Since $u$ is subharmonic,

$$u(0, 0) \leqslant \frac{1}{2\pi} \int_0^{2\pi} u(r\cos\theta, r\sin\theta)\, d\theta$$

for all $r < a$. Suppose that there is no disc with centre $O$ on which $u$ is constant. Since $u$ has a relative maximum at $O$, we can find a circle $r = b$ on which the supremum $M$ of $u$ is less than $u(0, 0)$.

Let $w(x, y)$ be the function, given by Poisson's integral, which takes on $r = b$ the same values as $u(x, y)$. Then

$$w(0, 0) = \frac{1}{2\pi} \int_0^{2\pi} u(b\cos\theta, b\sin\theta)\, d\theta \geqslant u(0, 0),$$

and so

$$w(0, 0) > M.$$

This is impossible since $w(x,y)$, being harmonic in $r < b$ and continuous in $r \leqslant b$, cannot take at the origin a value greater than its supremum on $r = b$. It follows that there must be a disc with centre $O$ on which $u$ is constant.

*If $u$ is subharmonic on a bounded domain $D$ and continuous on $\overline{D}$, it attains its supremum on $\overline{D}$ at a point or points of $\partial D$. If it also attains this supremum at a point of $D$, it is a constant.*

Since $u$ is continuous on $\overline{D}$, it is bounded and attains its supremum at a point of $\overline{D}$. We have to show that, if $u$ is not constant, it cannot attain its supremum at a point of $D$.

Suppose this is not so. Let $A$ be the set of points of $D$ at which $u = M$, where $M$ is the supremum of $u$ on $\overline{D}$. If $(x,y)$ is a point of $A$, $u$ has a relative maximum at that point, and so there is an open disc with centre $(x,y)$ on which $u = M$. The set $A$ is therefore the union of a family of open discs and hence is an open set.

Let $B$ be the set of points of $D$ at which $u < M$. Since $u$ is continuous, $B$ is an open set. But $D$ is the union of the disjoint open sets $A$ and $B$, which is impossible since $D$, being a domain, is connected. Hence either $A$ or $B$ is an empty set.

If $B$ is empty, $A$ is $D$ and $u = M$ everywhere on $D$. This is impossible if $u$ is not constant. Hence $A$ is empty, and $u$ does not attain its supremum on $\overline{D}$ at any point of the interior of $\overline{D}$.

Let $\Delta$ be an open disc whose closure is contained in $D$. We can associate with any function $u$ which is continuous in $\overline{D}$ a function $u_\Delta$ with the following properties:

(i) $u_\Delta$ is continuous in $\overline{D}$;

(ii) $u_\Delta = u$ on the complement of $\Delta$ with respect to $\overline{D}$;

(iii) $u_\Delta$ is harmonic on $\Delta$.

In fact, $u_\Delta$ on $\Delta$ is given by Poisson's integral.

*If $u \leqslant u_\Delta$ for every open disc $\Delta$ where closure is contained in $D$, then $u$ is subharmonic in $D$. Conversely, if $u$ is subharmonic in $D$, so also is $u_\Delta$ for every disc $\Delta$ whose closure is contained in $D$; and $u \leqslant u_\Delta$.*

If $(x,y)$ is any point of $D$, there exists a positive number $a$ such that, if $r < a$, the closure of the disc $\Delta$ with centre $(x,y)$ and radius $r$ lies in $D$. Then, if $r < a$,

$$u(x,y) \leqslant u_\Delta(x,y) = \frac{1}{2\pi}\int_0^{2\pi} u_\Delta(x+r\cos\theta, y+r\sin\theta)\, d\theta$$

$$= \frac{1}{2\pi}\int_0^{2\pi} u(x+r\cos\theta, y+r\sin\theta)\, d\theta,$$

since $u_\Delta = u$ on $\partial\Delta$. Hence $u$ is subharmonic.

We now consider the converse. Since $u_\Delta$ is harmonic on $\Delta$, it is subharmonic on $\Delta$. Since $u_\Delta = u$ on the part of $D$ outside $\bar\Delta$, it is subharmonic there. It remains to discuss points on $\partial\Delta$.

Since $-u_\Delta$ is harmonic on $\Delta$, it is subharmonic. Hence $u - u_\Delta$ is subharmonic on $\Delta$, and so attains its supremum in $\bar\Delta$ at points of $\partial\Delta$. But $u - u_\Delta$ vanishes on $\partial\Delta$; hence $u - u_\Delta \leqslant 0$ on $\bar\Delta$. But $u_\Delta = u$ on the part of $D$ outside $\Delta$. Therefore $u \leqslant u_\Delta$ everywhere on $D$.

If $(x, y)$ is a point of $\partial\Delta$,

$$u_\Delta(x, y) = u(x, y) \leqslant \int_0^{2\pi} u(x + \rho\cos\theta, y + \rho\sin\theta)\, d\theta$$

for all sufficiently small values of $\rho$. It follows that

$$u_\Delta(x, y) \leqslant \int_0^{2\pi} u_\Delta(x + \rho\cos\theta, y + \rho\sin\theta)\, d\theta,$$

so that $u_\Delta$ is subharmonic at every point of $\partial\Delta$.

## 9.3  Lower functions

Let $f$ be continuous on the frontier $\partial D$ of a bounded domain $D$. A function $v$, which is continuous on $\bar D$ and subharmonic on $D$ and which satisfies $v \leqslant f$ on $\partial D$, is called a lower function or subfunction associated with $f$ on $D$. We denote the family of all lower functions associated with $f$ on $D$ by $\mathscr{S}_f$. The family is not empty; the infimum $m$ of $f$ on $\partial D$ belongs to $\mathscr{S}_f$. By the maximum principle, every function $v$ belonging to $\mathscr{S}_f$ satisfies on $\bar D$ the inequality $v \leqslant M$, where $M$ is the supremum of $f$ on $\partial D$.

If $v_1, v_2, \ldots, v_n$ belong to $\mathscr{S}_f$, so also does $\max(v_1, v_2, \ldots, v_n)$. For $\max(v_1, v_2, \ldots, v_n)$ is continuous on $\bar D$, subharmonic on $D$ and does not exceed $f$ on $\partial D$.

If $v$ belongs to $\mathscr{S}_f$, then $v_\Delta$ belongs to $\mathscr{S}_f$ for every disc $\Delta$ whose closure lies in $D$.

$v_\Delta$ is continuous in $\bar D$, subharmonic on $D$ and, on $\partial D$, $v_\Delta = v \leqslant f$.

## 9.4  Perron's function

If $v$ belongs to $\mathscr{S}_f$, $v \leqslant M$ at every point of $\bar D$. The values taken by the functions of the family of $\mathscr{S}_f$ at a given point $(x, y)$ of $D$ have, therefore, a finite supremum which we denote by $u(x, y)$. Perron's function

$$u(x, y) = \sup_{v \in \mathscr{S}_f} v(x, y)$$

is therefore defined everywhere on $\bar D$, where it satisfies $u \leqslant M$. On $\partial D$, $u \leqslant f$. This is all we know about $u(x, y)$; we prove that it is harmonic on $D$.

Let $\Delta$ be a disc whose closure is contained in $D$. If $P_0$ is a point of $\Delta$, there exist functions of $\mathscr{S}_f$ which take values as near as we please to $u$ at $P_0$. Hence we can find a function $v_1'$ of $\mathscr{S}_f$ such that

$$v_1'(P_0) \geqslant u(P_0) - 1.$$

If we write $\qquad v_1 = [v_1']_\Delta,$

then $v_1$ also belongs to $\mathscr{S}_f$ and is harmonic on $\Delta$. Since $v_1' \leqslant v_1$ on $\bar{D}$, we have

$$v_1(P_0) \geqslant u(P_0) - 1.$$

Again, there exists a function $v_2'$ of $\mathscr{S}_f$ such that

$$v_2'(P_0) \geqslant u(P_0) - \tfrac{1}{2}.$$

The function $\qquad v_2 = [\max(v_1, v_2')]_\Delta$

also belongs to $\mathscr{S}_f$ and is harmonic on $\Delta$. Since

$$v_2' \leqslant \max(v_1, v_2') \leqslant v_2,$$

we have $\qquad v_2(P_0) \geqslant u(P_0) - \tfrac{1}{2}.$

Proceeding in this way, we can construct two sequences $\{v_n'\}$ and $\{v_n\}$ of functions of $\mathscr{S}_f$ such that

$$v_n'(P_0) \geqslant u(P_0) - \frac{1}{n},$$

$$v_n = [\max(v_{n-1}, v_n')]_\Delta,$$

$$v_n(P_0) \geqslant u(P_0) - \frac{1}{n}.$$

Since $\qquad v_n'(P_0) \leqslant u(P_0), \quad v_n(P_0) \leqslant u(P_0),$

both sequences converge to $u(P_0)$ at $P_0$.

Now
$$v_{n-1} \leqslant \max(v_{n-1}, v_n') \leqslant [\max(v_{n-1}, v_n')]_\Delta = v_n \leqslant M$$

everywhere on $\bar{D}$. Hence $\{v_n\}$, being a bounded increasing sequence, converges everywhere on $\bar{D}$ to a function $v$. Since each function $v_n$ is harmonic on $\Delta$, the limit $v$ is harmonic on $\Delta$, by Harnack's second convergence theorem. The convergence of $\{v_n\}$ to $v$ is uniform on every closed subset of $\Delta$. We know that $v = u$ at $P_0$; we shall now show that $v = u$ everywhere on $\Delta$.

By definition, $v_n \leqslant u$ and so $v \leqslant u$ on $\Delta$. If $v \neq u$ everywhere on $\Delta$, there is a point $P_1$ of $\Delta$ such that $v(P_1) < u(P_1)$. We carry out the same construction starting with the point $P_1$ instead of $P_0$. We get two sequences $\{w_n'\}$ and $\{w_n\}$ of functions of the family $\mathcal{S}_f$ such that

$$u(P_1) \geqslant w_n'(P_1) \geqslant u(P_1) - \frac{1}{n},$$

$$w_n = [\max(w_{n-1}, w_n')]_\Delta,$$

$$u(P_1) \geqslant w_n(P_1) \geqslant u(P_1) - \frac{1}{n}.$$

Both sequences converge to $u(P_1)$ at $P_1$. And, as before, $\{w_n\}$ converges to a function $w$ which is harmonic in $\Delta$.

Now consider the sequences $\{W_n'\}$, $\{W_n\}$, where

$$W_n' = \max(v_n, w_n), \quad W_n = [W_n']_\Delta.$$

All these functions belong to the family $\mathcal{S}_f$, and $W_n$ is harmonic on $\Delta$. $\{W_n'\}$ is a bounded increasing sequence since

$$W_n' = \max(v_n, w_n) \leqslant \max(v_{n+1}, w_{n+1}) = W_{n+1}'.$$

The sequence $\{W_n\}$ is also a bounded increasing sequence. For, outside $\Delta$, $W_n = W_n'$, $W_{n+1} = W_{n+1}'$, and so $\{W_n\}$ is increasing outside $\Delta$. On $\Delta$ $W_{n+1} - W_n$ is harmonic and takes non-negative values $W_{n+1}' - W_n'$ on $\partial\Delta$; hence $W_{n+1} - W_n \geqslant 0$ on $\Delta$. The sequence $\{W_n\}$ converges, therefore, to a function $W$ which is harmonic on $\Delta$. Moreover, $W \geqslant v$, $W \geqslant w$.

Since $u$ is the supremum of the functions of the family $\mathcal{S}_f$,

$$W_n(P_0) \leqslant u(P_0)$$

and so $$W(P_0) \leqslant u(P_0) = v(P_0) \leqslant W(P_0).$$

Therefore the function $W - v$, which is harmonic on $\Delta$, vanishes at $P_0$. But $W - v \geqslant 0$. Hence $W - v$ attains its minimum at an interior point of $\Delta$, and so is identically zero on $\Delta$. In particular,

$$v(P_1) = W(P_1) \geqslant w(P_1) = u(P_1).$$

This is a contradiction, since we assumed $V(P_1) < u(P_1)$. Hence $v = u$ everywhere on $\Delta$, and $v$ is harmonic on $\Delta$. Hence Perron's function $u$ is harmonic on any disc $\Delta$ where closure is contained in $D$. As $D$ is an open set, the union of a family of open discs, it follows that Perron's function is harmonic in $D$.

## 9.5   Barriers

In this section, $P$ denotes a point of $D$, but $Q$ and $Q_0$ are points of $\partial D$. We have to find conditions under which Perron's function $u(P)$ tends to $f(Q_0)$ as $P$ tends to $Q_0$ by a path in $D$. To do this, we introduce the concept of a barrier. *A barrier is a function $\omega(P; Q)$ which is continuous in $\bar{D}$, harmonic in $D$, whose boundary-values $\omega(Q; Q_0)$ are strictly positive except at the point $Q_0$; and $\omega(Q_0; Q_0) = 0$.* If there is such a barrier at $Q_0$, $u(P)$ does tend to $f(Q_0)$ as $P$ tends to $Q_0$. If every point of $\partial D$ has a barrier, Perron's function is the required solution of the problem of Dirichlet, which we know is unique.

The boundary function $f(Q)$ is continuous on the closed set $\partial D$ and so is bounded; there exists a constant $K$ such that $|f(Q)| \leqslant K$. The result will follow if we can show that, for every positive value of $\epsilon$,

$$\limsup_{P \to Q_0} u(P) \leqslant f(Q_0) + \epsilon, \quad \liminf_{P \to Q_0} u(P) \geqslant f(Q_0) - \epsilon,$$

when there exists a barrier $\omega(P; Q_0)$.

Let $\Delta$ be a disc with centre $Q_0$ where radius is so small that $\Delta$ does not contain the whole of the domain $D$; $\Delta'$ is the complement of $\Delta$. For every positive value of $\epsilon$, we can choose the radius of $\Delta$ so that

$$|f(Q) - f(Q_0)| < \epsilon$$

for all points $Q$ in $\Delta \cap \partial D$. On $D \cap \Delta'$, $\omega(P; Q_0)$ is harmonic and strictly positive, and so has a positive minimum $\omega_0$ which is attained on the boundary of $D \cap \Delta'$.

Consider the harmonic function

$$W(P) = f(Q_0) + \epsilon + \frac{\omega(P; Q_0)}{\omega_0} [K - f(Q_0)].$$

Everywhere on $\bar{D}$,      $W(P) \geqslant f(Q_0) + \epsilon$.

Since $f(Q)$ lies between $f(Q_0) \pm \epsilon$ on $\partial D \cap \Delta$, we have

$$W(Q) > f(Q)$$

there. But on $\partial D \cap \Delta'$

$$W(Q) \geqslant f(Q_0) + \epsilon + [K - f(Q_0)] = K + \epsilon > f(Q).$$

Thus $W - f > 0$ on $\partial D$.

If $v$ belongs to the family $\mathscr{S}_f$, $v - W$ is continuous in $\bar{D}$, subharmonic in $D$, and

$$v - W \leqslant f - W < 0$$

on $\partial D$. Hence, by the maximum principle,

$$v(P) < W(P)$$

everywhere on $D$. Since $u$ is the supremum of all functions $v$ belonging to $\mathscr{S}_f$,

$$u(P) \leqslant W(P),$$

and so
$$\limsup_{P \to Q_0} u(P) \leqslant \limsup_{P \to Q_0} W(P) = \lim_{P \to Q_0} W(P) = f(Q_0) + \epsilon.$$

Next consider the harmonic function
$$V(P) = f(Q_0) - \epsilon - \frac{\omega(P, Q_0)}{\omega_0} [K + f(Q_0)].$$

Everywhere on $\bar{D}$,        $V(P) \leqslant f(Q_0) - \epsilon,$

and, in particular on $\partial D \cap \Delta$,
$$V(Q) \leqslant f(Q_0) - \epsilon < f(Q).$$

But on $\partial D \cap \Delta'$,
$$V(Q) \leqslant f(Q_0) - \epsilon - [K + f(Q_0)] = -K - \epsilon < f(Q).$$

Therefore $V$ belongs to $\mathscr{S}_f$, and so
$$V(P) \leqslant u(P).$$

Hence    $\displaystyle \liminf_{P \to Q_0} u(P) \geqslant \liminf_{P \to Q_0} V(P) = \lim_{P \to Q_0} V(P) = f(Q_0) - \epsilon.$

We have thus proved that, for every positive value of $\epsilon$,
$$f(Q_0) - \epsilon \leqslant \liminf_{P \to Q_0} u(P) \leqslant \limsup_{P \to Q_0} u(P) \leqslant f(Q_0) + \epsilon.$$

Hence $u(P)$ tends to $f(Q_0)$ as $P \to Q_0$, provided that there is a barrier at $Q_0$.

## 9.6   Some examples of barriers

The simplest case of a barrier arises when $\bar{D}$ lies in an open half-plane except for one point $Q_0$ of $\partial D$ which lies on the line which is the frontier of the half-plane. Take $Q_0$ as origin, $y = 0$ as the line and $y > 0$ as the half-plane. Then $\omega = y$ is a barrier for $Q_0$. For example, if the frontier of $D$ is an ellipse, this geometrical condition is satisfied at each point of $\partial D$. Hence the problem of Dirichlet has a solution for an ellipse.

A somewhat similar test occurs when there exists a disc $\Delta$ such that $\Delta$ and $\bar{D}$ are disjoint, but $\bar{\Delta}$ and $\bar{D}$ have a single point $Q_0$ in common. If $O$ is the centre of the disc,
$$\omega(P) = \log \frac{OP}{OQ_0}$$

is a barrier at $Q_0$. This condition is again satisfied at each point of $\partial D$ if the frontier of $D$ is an ellipse. It follows again that the problem of Dirichlet has a unique solution for an ellipse.

In another simple example, the point $Q_0$ is the end-point of a straight segment $Q_0 A$ every point of which except $Q_0$ lies in the complement of $\overline{D}$. If we choose the axes so that $Q_0$ is the origin and $A$ is $(a, 0)$ where $a > 0$, we can use polar coordinates $(r, \theta)$ and $(r_1, \theta_1)$ with $Q_0$ and $A$ as origins, so that

$$x = r \cos \theta = a + r_1 \cos \theta_1, \quad y = r \sin \theta = r_1 \sin \theta_1,$$

where $\theta$ and $\theta_1$ lie between $\pm \pi$. Then

$$\sqrt{\left(\frac{r}{r_1}\right)} \cos \tfrac{1}{2}(\theta - \theta_1)$$

is a barrier at $Q_0$. It is harmonic, since it is the real part of $\sqrt{\{z/(z-a)\}}$, it vanishes on $y = 0$, $0 \leqslant x < a$ and is positive elsewhere. If this geometrical condition is satisfied at each point of $\partial D$, the problem of Dirichlet has a unique solution for $D$. A simple example is the doubly-connected domain bounded by two concentric circles.

If $D$ is a simply-connected domain bounded by a regular closed curve, $\partial D$ may have a finite number of corners. If $Q_0$ is not a corner, we may take $Q_0 A$ along the outward normal. If $Q_0$ is a corner which is not re-entrant, we may take $Q_0 A$ along either of the outward normals at $Q_0$ to the two arcs which meet there. So, if $\partial D$ is a regular closed curve with no re-entrant corners, the problem of Dirichlet has a solution for $D$. The situation is more complicated at a re-entrant corner $Q_0$. If the two arcs which join at $Q_0$ do not touch, we can still find a segment $Q_0 A$ there. Even if the arcs touch but form a keratoid cusp at which the two arcs lie on opposite sides of the tangent at $Q_0$, the test is applicable. But if the two touching arcs form a ramphoid cusp at which the two arcs lie on the same side of the tangent at $Q_0$, the test fails.

There is another barrier which enables us to take account of re-entrant corners. Suppose that, at a point $Q_0$ of $\partial D$, there is a regular arc $\Gamma$ which lies, apart from its end point $Q_0$, in $\overline{D}'$, the complement of $\overline{D}$. Let $\Delta$ be a disc with centre $Q_0$ of so small a radius that its circumference intersects $\partial D$ and $\Gamma$. As we go along $\Gamma$ from $Q_0$, there will be a first point $A$ which lies on $\partial \Delta$. Cut $\Delta$ along the arc $Q_0 A$. Call the cut disc $\Delta_0$. Take $Q_0$ as origin, the tangent to $\Gamma$ at $Q_0$ as axis of $x$, and write $x = r \cos \theta, y = r \sin \theta$. Then

$$\omega = -\frac{\log(r/a)}{\{\log(r/a)\}^2 + \theta^2} \quad (r \neq 0),$$

$$= 0 \quad (r = 0)$$

is a barrier at $Q_0$. $\omega$ is one-valued and continuous on $\bar{D}$ and on $\Delta_0$. Since, when $z \neq 0$,

$$\omega = -\operatorname{re} \frac{1}{\log(z/a)},$$

$\omega$ is harmonic on $D$. And, if we choose $a$ large enough, $\omega > 0$ on $\partial D$ except at $Q_0$. Hence the result.

This enables us to complete the discussion where $D$ is simply-connected and $\partial D$ is a regular closed curve. If $Q_0$ is not a re-entrant corner, we take as the regular arc the straight segments of the previous test. If $Q_0$ is a re-entrant corner, take $Q_0$ as origin and choose the $x$-axis so that, near $Q_0$, $\bar{D}'$ lies in $x > 0$. Near $Q_0$, the two arcs of $\partial D$ have equations $y = \phi(x)$ and $y = \phi(x)$ where $\phi$ and $\psi$ are differentiable. Since the two arcs intersect only at $Q_0$, there is an interval $0 < x < a$ on which $\phi(x) < \psi(x)$; and $\phi(x) < y < \psi(x)$ belongs to $\bar{D}'$. Then, for $0 \leqslant x < a$,

$$y = \tfrac{1}{2}\{\phi(x) + \psi(x)\}$$

is a regular arc which lies, apart from its end point $Q_0$, in $\bar{D}'$. The problem of Dirichlet thus has a unique solution when $\partial D$ is a regular closed curve. The same argument applies when $D$ is a multiply-connected domain bounded by a finite number of non-intersecting regular closed curves.

## 9.7 Discontinuous boundary data

In the case when the domain $D$ is bounded by a regular closed curve, the requirement that the boundary data be continuous can be lightened; we can allow for a finite number of ordinary discontinuities. It suffices to consider one such discontinuity.

Suppose that the boundary function $f$ has one ordinary discontinuity at a point $(a, b)$ which is not a cusp of $\partial D$. Since $\partial D$ is regular, there is either a tangent to $\partial D$ at $(a, b)$ or the two regular arcs which join these have tangents in different directions. We may suppose that the axes are chosen so that the approach to $(a, b)$ on the two arcs is as $x \to a - 0$ and as $x \to a + 0$. As $x \to a + 0$ on $\partial D$,

$$\frac{y-b}{x-a} \to \tan\alpha \quad (-\tfrac{1}{2}\pi < \alpha < \tfrac{1}{2}\pi),$$

and as $x \to a - 0$, $\quad \dfrac{y-b}{x-a} \to \tan\beta \quad (\tfrac{1}{2}\pi < \beta < \tfrac{3}{2}\pi).$

If there is a unique tangent at $(a, b)$, $\beta = \pi + \alpha$. As $x \to a + 0$ on $\partial D$, $f \to f_+$; as $x \to a - 0$, $f \to f_-$.

If
$$F(x,y) = f(x,y) + K\tan^{-1}\frac{y-b}{x-a},$$

$F \to f_+ + K\alpha$ as $x \to a+0$, $F \to f_- + K\beta$ as $x \to a-0$. Hence $F$ is continuous at $(a,b)$ if $K = (f_+ - f_-)/(\beta - \alpha)$.

Let $U(x,y)$ be the function which is harmonic in $D$, continuous on $\bar{D}$ and which takes the continuous boundary values $F$ on $\partial D$. Then

$$u(x,y) = U(x,y) - K\tan^{-1}\frac{y-b}{x-a},$$

with an appropriate choice of the branch of the inverse tangent, is harmonic in $D$ and takes the continuous boundary values $f$ on $\partial D$ except at $(a,b)$.

Since $U$ is continuous on $\bar{D}$, $U$ tends to a limit $L$ as $(x,y)$ tends to $(a,b)$ on a path in $D$ not tangent to the regular arcs which meet at $(a,b)$; and the inverse tangent tends to a limit $\gamma$ definitely between $\alpha$ and $\beta$. Since $F$ is continuous,

$$L = f_+ + K\alpha = f_- + K\beta.$$

Hence     $$u \to f_+ - K(\gamma - \alpha) = \frac{1}{\beta-\alpha}\{(\beta-\gamma)f_+ + (\gamma-\alpha)f_-\}.$$

Hence $u$ tends to a limit, depending on the direction in which $(x,y)$ approaches $(a,b)$; this limit lies between $f_+$ and $f_-$.

# 10

# EQUATIONS OF ELLIPTIC TYPE IN THE PLANE

## 10.1 The linear equation

Suppose that a linear equation of the second order with two independent variables is of elliptic type in some region in which the coordinates $(x, y)$ can be chosen so that the characteristics are $x \pm iy = $ constant. Then the equation is of the form

$$\frac{\partial^2 u}{\partial x^2} + \frac{\partial^2 u}{\partial y^2} + 2a \frac{\partial u}{\partial x} + 2b \frac{\partial u}{\partial y} + cu = F, \qquad (1)$$

where $a, b, c$ and $F$ are functions of $x$ and $y$ alone. For brevity we write this equation as
$$L(u) = F.$$

Associated with $L$ is the adjoint operator $L^*$ defined by

$$L^*(u) \equiv \frac{\partial^2 u}{\partial x^2} + \frac{\partial^2 u}{\partial y^2} - 2 \frac{\partial}{\partial x}(au) - 2 \frac{\partial}{\partial y}(bu) + cu. \qquad (2)$$

Then
$$vL(u) - uL^*(v) = \frac{\partial H}{\partial x} + \frac{\partial K}{\partial y},$$

where
$$H = vu_x - uv_x + 2auv, \quad K = vu_y - uv_y + 2buv.$$

$L$ is said to be self-adjoint if the operators $L$ and $L^*$ are identical; this happens if and only if $a$ and $b$ are zero. If $a$ and $b$ are constants, the operator $M$ defined by

$$M(u) = e^{ax+by} L(e^{-ax-by}u)$$

is self-adjoint.

The problem of Dirichlet for the linear equation (1) is to find a solution which is continuously differentiable in the closure of a bounded domain $D$, which has continuous second derivatives in $D$ and which takes given continuous boundary values on the frontier $\partial D$ of $D$. We have seen that, in the particular case of Laplace's equation or Poisson's equation when $a, b, c$ are identically zero, the problem of Dirichlet has a solution and that it is unique, provided that the boundary satisfies certain conditions. This is not necessarily the case for the general linear equation.

If the linear equation (1) had two solutions $u_1$ and $u_2$, the function $u = u_1 - u_2$ would satisfy $L(u) = 0$ and would vanish on $\partial D$. For Laplace's equation, this implies that $u$ is identically zero on the closure of $D$, so that, if the problem of Dirichlet has a solution, it is unique. It can happen that the problem of Dirichlet for $L(u) = 0$ with zero boundary values on $\partial D$ possesses solutions which are not identically zero on $\bar{D}$, so that the problem of Dirichlet for $L(u) = F$ does not necessarily possess a unique solution.

The other problem associated with a bounded domain is the problem of Neumann, in which the normal derivative is assigned on $\partial D$.

In the case of an unbounded domain, we have to impose in both these problems conditions on the behaviour of the solution at infinity, just as we did for Laplace's equation.

## 10.2   The reduced wave equation

In the theory of small vibrations of a uniform stretched plane membrane with a fixed rim, the small normal displacement $U$ of the membrane at the point $(x, y)$ at the instant $t$ satisfies the equation of wave motions

$$\frac{\partial^2 U}{\partial x^2} + \frac{\partial^2 U}{\partial y^2} - \frac{\partial^2 U}{\partial t^2} = 0, \tag{1}$$

with an appropriate choice of units of length and time. Since the equation is linear, the general solution is the sum of solutions corresponding to the vibrations in normal modes of oscillation. In a normal mode, $U$ is of the form $u(x, y) \cos{(kt + \alpha)}$ where $k$ and $\alpha$ are constants, and $u$ satisfies the reduced wave equation

$$\frac{\partial^2 u}{\partial x^2} + \frac{\partial^2 u}{\partial y^2} + k^2 u = 0. \tag{2}$$

If the boundary of the membrane is a regular closed curve $C$, we require $u$ to satisfy the usual differentiability conditions inside and on $C$ and to vanish on $C$.

In the simplest case, $C$ is the square with sides $x = 0$, $x = a$, $y = 0$, $y = a$. Then (2) has the solution

$$u = \sin{\frac{m\pi x}{a}} \sin{\frac{n\pi y}{a}},$$

which vanishes on $C$ provided that $m$ and $n$ are positive integers and that

$$k^2 = (m^2 + n^2)\, \pi^2/a^2. \tag{3}$$

In the problem of the vibrations of a square membrane, this equation gives the frequencies of the normal modes of vibration, and the general solution of equation (1) is of the form

$$\sum_{m,n} a_{mn} \sin\frac{m\pi x}{a} \sin\frac{n\pi y}{a} \cos\left(k_{mn}t + \alpha_{mn}\right),$$

where
$$k_{mn} = \sqrt{(m^2 + n^2)}\,\pi/a.$$

Now forget the membrane problem and consider the reduced wave equation (2) as an elliptic equation with a given $k$. If $a = \sqrt{2\pi/k}$, equation (2) has the solution

$$u = \sin\frac{kx}{\sqrt{2}} \sin\frac{ky}{\sqrt{2}}, \tag{4}$$

which vanishes on $C$, but does not vanish identically inside $C$. Therefore for this square, the problem of Dirichlet does not have a unique solution; we can always add to any solution a multiple of the solution (4). But if $a < \sqrt{2\pi/k}$, there is no non-zero solution of (2) which vanishes on $C$; hence, if the problem of Dirichlet for (2) has a solution, it seems likely that it is unique if the square is sufficiently small.

The problem of the normal modes of vibration of a circular membrane with fixed rim $r = a$, using polar coordinates, can be discussed in a similar way. In a normal mode of vibration, $U$ is of the form $u\cos(kt + \alpha)$ where $u$ satisfies the reduced wave equation. Now, in polar coordinates, equation (2) is

$$\frac{\partial^2 u}{\partial r^2} + \frac{1}{r}\frac{\partial u}{\partial r} + \frac{1}{r^2}\frac{\partial^2 u}{\partial \theta^2} + k^2 u = 0.$$

By the method of separation of variables, we find that this equation has the particular solution

$$u = J_m(kr)\cos\left(m\theta + \epsilon_m\right),$$

where $m$ is a positive integer or zero, to ensure that the solution is finite and one-valued when $r \leqslant a$, and $J_m$ is the Bessel function of order $m$. In the membrane problem, we get the normal modes by choosing $k$ so that $J_m(ka)$ vanishes, and the general solution is the sum of solution corresponding to the various normal modes.

Now $z^{-m}J_m(z)$ is an even function of $z$, whose zeros are all real and simple. Denote the positive zeros by $j_{m,1}, j_{m,2}, \ldots$ in increasing order of magnitude. Then the normal modes of vibration of the circular membrane with fixed rim $r = a$ are given by

$$U = J_m(j_{m,n}r/a)\cos\left(m\theta + \epsilon_{m,n}\right)\cos\left(k_{m,n}t + \alpha_{m,n}\right),$$

where
$$k_{m,n}a = j_{m,n}.$$

Again, forget about the membrane problem and consider the reduced wave equation (2) in its own right. If $k = j_{m,n}/a$, there exists a solution

$$u = J_m(j_{m,n} r/a) \cos(m\theta + \epsilon),$$

which vanishes on $r = a$. Hence for the disc $r \leqslant a$, the problem of Dirichlet does not have a unique solution. The least of the zeros $j_{m,n}$ is $j_{0,1}$. If $a < j_{0,1}/k$, there is no non-zero solution of (2) which vanishes on $r = a$. Hence, if the problem of Dirichlet for (2) has a solution, it seems likely that it is unique if the disc is sufficiently small.

The argument in this section is not entirely rigorous, but it does suggest the sort of result we should expect.

## 10.3 The elementary solution

The theory of harmonic functions in the plane depends on the existence of the solution $\log R$ where $R$ is the distance from the singularity $(x_0, y_0)$ to the variable point $(x, y)$. A homogeneous linear equation

$$L(u) \equiv \nabla^2 u + 2au_x + 2bu_y + cu = 0 \tag{1}$$

also possesses a solution with a similar logarithmic singularity; it was called by Hadamard the *elementary solution*. Its existence for $\nabla^2 u + cu = 0$ was first proved by Picard. In his *Lectures on Cauchy's Problem*, Hadamard deals with the general linear homogeneous equation which is not a parabolic type. He discusses the problem in a space of any number of dimensions; in $n$-dimensional space, the elementary solution behaves near the singularity like $1/R^{n-2}$.

The reduced wave equation

$$u_{xx} + u_{yy} + k^2 u = 0$$

has two solutions which depend only on $R$, viz. $J_0(kR)$ and $Y_0(kR)$. $J_0(z)$ is an integral function of $z$ in the complex plane, but $Y_0(z)$ has a logarithmic singularity at $z = 0$; in fact,

$$Y_0(z) = \frac{2}{\pi} J_0(z) \log z + V(z),$$

where $V(z)$ is an integral function. Hence $Y_0(kR)$ is the required elementary solution of the reduced wave equation. It is not uniquely determined; we could add to it any constant multiple of $J_0(kR)$.

The discussion here is based on Hadamard's proof and the method of Frobenius for solving by series an ordinary linear homogeneous differential equation.

We choose the singular point as origin, and we assume that there is a neighbourhood of the origin in which the coefficients $a, b, c$ in the equation

$$L(u) \equiv \nabla^2 u + 2a\frac{\partial u}{\partial x} + 2b\frac{\partial u}{\partial y} + cu = 0 \qquad [(1)]$$

are analytic functions. Follow Hadamard, we write $\Gamma$ for $r^2$, where $(r, \theta)$ are the polar coordinates of the point $(x, y)$. We ask whether $L(u) = 0$ has a solution of the form

$$u = \sum_0^\infty U_n \Gamma^{n+\nu}, \qquad (2)$$

where $\nu$ is a constant at our disposal and where the coefficients $U_n$ are analytic.

Since
$$L(U\Gamma^\nu) = \Gamma^\nu L(U) + 4\nu\Gamma^{\nu-1}\left(r\frac{\partial U}{\partial r} + (\nu + \Psi)\, U\right),$$

where
$$\Psi = ax + by,$$

we have
$$L(u) = \sum_0^\infty \Gamma^{n+\nu} L(U_n) + 4\sum_0^\infty \Gamma^{n+\nu-1}(n+\nu)\left(r\frac{\partial U_n}{\partial r} + (n+\nu+\Psi)\, U_n\right),$$

assuming that term-by-term differentiation is valid. It follows that

$$L(u) = 4\nu\Gamma^{\nu-1}\left(r\frac{\partial U_0}{\partial r} + (\nu + \Psi)\, U_0\right) \qquad (3)$$

if the other coefficients satisfy the recurrence relation

$$r\frac{\partial U_n}{\partial r} + (n+\nu+\Psi)\, U_n = -\frac{1}{4(n+\nu)}L(U_{n-1}). \qquad (4)$$

If $U_0$ is given, this differential equation gives the other coefficients successively; it is an ordinary differential equation in which the variable $\theta$ is only a parameter.

If we choose $U_0$ so that

$$r\frac{\partial U_0}{\partial r} + \Psi U_0 = 0,$$

(3) becomes
$$L(u) = 4\nu^2\Gamma^{\nu-1}U_0,$$

the resemblance to the case of equal exponents in Frobenius's method being evident. The equation satisfied by $U_0$ is an ordinary differential equation with solution

$$U_0 = A\exp\left(-\int_0^r \Psi\frac{dr'}{r}\right),$$

where $A$ may depend on $\theta$.

Since $a$ and $b$ are analytic near the origin, there is a disc on which

$$\Psi = ax + by = \Sigma' \alpha_{mn} x^m y^n,$$

where the $\Sigma'$ indicate summation over all non-negative integers except

$$m = n = 0.$$

Since $\qquad \displaystyle\int_0^r \frac{x^m y^n}{r}\, dr = \cos{}^m\theta \sin{}^n\theta \int_0^r r^{m+n-1} dr = \frac{x^m y^n}{m+n},$

we have $\qquad \displaystyle\int_0^r \Psi \frac{dr}{r} = \Sigma' \alpha_{mn} \frac{x^m y^n}{m+n}$

on the disc where the series for $\Psi$ is uniformly and absolutely convergent. Hence

$$U_0 = A \exp\left(-\Sigma' \alpha_{mn} \frac{x^m y^n}{m+n}\right).$$

If, in (4), we substitute $\quad \Psi = -\dfrac{r}{U_0} \dfrac{\partial U_0}{\partial r},$

we obtain $\quad r \dfrac{\partial}{\partial r}\left(\dfrac{U_n}{U_0}\right) + (n+\nu)\dfrac{U_n}{U_0} = -\dfrac{1}{4(n+\nu) U_0} L(U_{n-1}).$

It follows that

$$U_n = -\frac{U_0}{4(n+\nu) r^{n+\nu}} \int_0^r L(U_{n-1}) \frac{r^{n+\nu-1}}{U_0}\, dr,$$

the lower limit of integration being taken to be zero to ensure that each coefficient $U_n$ is analytic near the origin.

If we write $\qquad U_n = \dfrac{\Gamma(\nu+1)}{\Gamma(\nu+n+1)} U_0 W_n,$

where $W_0 = 1$, we have

$$u = U_0 \Gamma^\nu \sum_0^\infty \frac{\Gamma(\nu+1)}{\Gamma(\nu+n+1)} W_n \Gamma^n. \tag{5}$$

The recurrence formula becomes

$$W_n = -\frac{1}{4r^{n+\nu}} \int_0^r L_1(W_{n-1}) r^{n+\nu-1} dr,$$

where $\qquad L_1(W) = \dfrac{\partial^2 W}{\partial x^2} + \dfrac{\partial^2 W}{\partial y^2} + 2a_1 \dfrac{\partial W}{\partial x} + 2b_1 \dfrac{\partial W}{\partial y} + c_1 W,$

with $\qquad a_1 = a + \dfrac{1}{U_0}\dfrac{\partial U_0}{\partial x}, \quad b_1 = b + \dfrac{1}{U_0}\dfrac{\partial U_0}{\partial y}, \quad c_1 = \dfrac{1}{U_0} L(U_0).$

Since $U_0$ is analytic and does not vanish in a neighbourhood of the origin, these coefficients are all analytic there. By the method of dominant functions, it can be shown that the series (5) is uniformly and absolutely convergent in a neighbourhood of the origin, and the formal method by which (5) was obtained is then valid. The details under more general conditions will be found in Hadamard's book.

Since $u$ defined by (5) satisfies

$$L(u) = 4\nu^2 \Gamma^{\nu-1} U_0,$$

where $\Gamma$ and $U_0$ do not depend on $\nu$, we obtain a solution $u_1$ of $L(u) = 0$ by putting $\nu = 0$; this solution

$$u_1 = U_0 \sum_0^\infty \frac{w_n}{n!} \Gamma^n,$$

where 
$$w_0 = 1, \quad w_n = -\frac{1}{4r^n} \int_0^r L_1(w_{n-1}) r^{n-1} dr,$$

is analytic in a neighbourhood of the origin.

The series (5) and the series obtained by differentiation with respect to $\nu$ can be shown to converge uniformly on any interval $\alpha \leqslant \nu \leqslant \beta$ which contains no negative integer. Since

$$L\left(\frac{\partial u}{\partial \nu}\right) = 8\nu \Gamma^{\nu-1} U_0 + 4\nu^2 \Gamma^{\nu-1} U_0 \log \Gamma,$$

a second solution of $L(u) = 0$ is

$$u_2 = \left(\frac{\partial u}{\partial \nu}\right)_{\nu=0},$$

or
$$u_2 = u_1 \log \Gamma + U_0 \sum_0^\infty \left[\frac{\partial}{\partial \nu} \frac{\Gamma(\nu+1) W_n}{\Gamma(\nu+n+1)}\right]_{\nu=0} \Gamma^n.$$

This is the required elementary solution. The first term contains the logarithmic singularity, the second is analytic in a neighbourhood of the origin; we could add to the second term any solution of $L(u) = 0$, analytic near the origin.

This discussion of the theory of Hadamard's elementary solution is incomplete – a discussion of the convergence of the series (5) is needed. All that is given here is a method of finding the form of an elementary solution. The difficulty in the method is the evaluation of the integrals which determine successively the coefficients $W_n$. In some problems it is possible to calculate explicitly the first few coefficients and to make an inspired guess at the form of the elementary solution and then to find it by direct substitution in the differential equation.

## 10.4   Boundary value problems

Let $D$ be a bounded domain whose frontier $\partial D$ consists of a finite number of non-intersecting regular closed curves. The most important boundary value problems for the linear equation

$$L(u) \equiv \nabla^2 u + 2au_x + 2bu_y + cu = F(x, y),$$

where $a, b, c, F$ are analytic, are the problems of Dirichlet and of Neumann. In the problem of Dirichlet, we have to find a solution with continuous first derivatives in $\bar{D}$ and continuous bounded second derivatives in $D$, given that $u = f(x, y)$ on $\partial D$, $f$ being continuous. In the problem of Neumann, $\partial u / \partial N$ is given on $\partial D$. We try to solve these problems by the use of Green's theorem. (See Note 5.)

Let $L^*$ be the operator adjoint to $L$. If the double integral exists

$$\iint_D \{vL(u) - uL^*(v)\} \, dx \, dy = \int_{\partial D} (lH + mK) \, ds,$$

where $(l, m)$ are the direction cosines of the normal to $\partial D$ drawn out of $D$, and

$$H = vu_x - uv_x + 2auv, \quad K = vu_y - uv_y + 2buv.$$

This is true if $u$ and $v$ have continuous bounded second derivatives on $D$ and continuous first derivatives on $\bar{D}$. In particular, if $L^*(v) = 0$,

$$\iint_D vF \, dx \, dy = \int_{\partial D} \left( v \frac{\partial u}{\partial N} - u \frac{\partial v}{\partial N} + 2(la + mb) \, uv \right) ds. \tag{1}$$

We assumed that $a, b, c$ are analytic. Then $L^*(v) = 0$ has an elementary solution

$$v = v_0 \log \frac{1}{R} + v_1,$$

where

$$R^2 = (x - x_0)^2 + (y - y_0)^2,$$

and

$$v_0(x_0, y_0) = 1.$$

If $(x_0, y_0)$ is a point of $D$, $v_0$ and $v_1$ are analytic in $D$, but $v$ has a logarithmic singularity at $(x_0, y_0)$. Equation (1) no longer holds; it is replaced by

$$u(x_0, y_0) = \frac{1}{2\pi} \int_{\partial D} \left\{ \left( v \frac{\partial u}{\partial N} - u \frac{\partial u}{\partial N} \right) + 2(la + mb) \, uv \right\} ds - \frac{1}{2\pi} \iint_D vF \, dx \, dy,$$

by the argument we used in the theory of Laplace's equation.

This formula contains the values of $u$ and $\partial u / \partial N$ on $\partial D$. In order to solve the problem of Dirichlet, we must in some way eliminate $\partial u / \partial N$. The elementary solution $v$ is not uniquely determined; we can add to $v_1$ any solution $v_2$ of $L^*(v) = 0$ which has no singularities on $\bar{D}$. Suppose that $v_1$ is fixed. Then we can take as the elementary solution

$$v = v_0 \log \frac{1}{R} + v_1 + v_2$$

and choose $v_2$ so that $\qquad v_2 = -v_0 \log \frac{1}{R} - v_1$

on $\partial D$, if this is possible. Assuming that this particular problem of Dirichlet for the adjoint equation is soluble, $v$ vanishes on $\partial D$, and we get the required solution of $L(u) = F$. The elementary solution chosen in this way is the Green's function.

In the same way, we can solve the problem of Neumann if we can find an elementary solution of $L^*(v) = 0$ of the form

$$v = v_0 \log \frac{1}{R} + v_1 + v_3$$

where the solution $v_3$ of the adjoint equation $L^*(v) = 0$ satisfies the condition

$$\frac{\partial v_3}{\partial N} - 2(la + mb) v_3 = -\frac{\partial}{\partial N} \left( v_0 \log \frac{1}{R} + v_1 \right) + 2(la + mb) \left( v_0 \log \frac{1}{R} + v_1 \right)$$

on $\partial D$. But $v_3$ is not a solution of the problem of Neumann for the adjoint equation; it has to satisfy a more general condition of the form

$$\frac{\partial v_3}{\partial N} + h v_3 = g(x, y)$$

on the boundary $\partial D$.

Even if we succeed in finding a solution of the problem of Dirichlet for the equation $L(u) = F(x, y)$, we still have to show that the solution is unique. If there were two solutions $u_1$ and $u_2$, $u = u_1 - u_2$ would satisfy $L(u) = 0$ and would vanish on $\partial D$. To prove uniqueness, we have to show that the only solution of this last problem is identically zero. As we have seen in the case of the reduced wave equation, this is not necessarily the case.

Consider the self-adjoint equation

$$\nabla^2 u + cu = 0,$$

where $c$ is now not constant on $D$. Since

$$\iint_D \{u \nabla^2 u + u_x^2 + u_y^2\} \, dx \, dy = \int_{\partial D} u \frac{\partial u}{\partial N} \, ds,$$

we have $\qquad \displaystyle\iint_D (u_x^2 + u_y^2) \, dx \, dy = \iint_D cu^2 \, dx \, dy$

if $u$ vanishes on $\partial D$. If $c \leqslant 0$ everywhere on $D$,

$$\iint_D (u_x^2 + u_y^2)\, dx\, dy \leqslant 0.$$

Therefore $u_x$ and $u_y$ vanish everywhere on $D$. Since we are considering solutions which are continuously differentiable on the closure of $D$, $u$ is constant on $D$ and so is identically zero on $\bar{D}$. It follows that, if the problem of Dirichlet for $\nabla^2 u + cu = F(x, y)$, where $c \leqslant 0$ on $D$, has a solution, the solution is unique. This result can be extended.

*If $c \leqslant 0$ on $D$, the equation*

$$L(u) \equiv \nabla^2 u + 2au_x + 2bu_y + cu = F(x, y)$$

*has at most one solution which is continuously differentiable on $\bar{D}$, which has continuous bounded second derivatives on $D$ and which takes given continuous values on the boundary $\partial D$.*

This is equivalent to proving that if $u$ satisfies the differentiability conditions and is a solution of $L(u) = 0$ which vanishes on $\partial D$, then $u$ is identically zero on $\bar{D}$.

Suppose firstly that $c < 0$ everywhere on $D$. Then no solution of $L(u) = 0$ has a positive maximum or a negative minimum at any point of $D$. For if such a solution had a relative maximum at a point $P$ of $D$, $u_x$ and $u_y$ would vanish at $P$ and we should have $u_{xx} \leqslant 0$, $u_{yy} \leqslant 0$ there. By the differential equation, $cu \geqslant 0$ at $P$ and hence $u \leqslant 0$ there. Similarly if it had a relative minimum at a point $P$ of $D$, we should have $u \geqslant 0$ there.

Since $u$ is continuous, it attains its supremum and its infimum on $\bar{D}$. A supremum attained at a point of $D$ (an interior point of $\bar{D}$) is a relative maximum and so cannot be positive. But $u$ vanishes on $\partial D$, and therefore the supremum of $u$ on $\bar{D}$ is zero. Similarly the infimum of $u$ on $\bar{D}$ is zero. Hence $u$ vanishes identically on $\bar{D}$.

The case $c \leqslant 0$ can be transformed into the case $c < 0$ by a change of dependent variable. If we put

$$u = (A - e^{\alpha x})\, U,$$

where $A$ and $\alpha$ are positive constants, the equation $L(u) = 0$ becomes

$$\nabla^2 U + 2a'U_x + 2b'U_y + c'U = 0,$$

where

$$c' = \frac{Ac - e^{\alpha x}(\alpha^2 + 2a\alpha + c)}{A - e^{\alpha x}}.$$

Now $a$ and $c$, being continuous, are bounded on $\bar{D}$. Choose $\alpha$ so large that

$$\alpha^2 + 2a\alpha + c > 0$$

on $\bar{D}$. Then choose $A$ so that $A > e^{\alpha h}$ where $h$ is the supremum of $x$ on $\bar{D}$. Then
$$A - e^{\alpha x} \geqslant A - e^{\alpha h} > 0$$

on $\bar{D}$. But
$$c' = -\frac{A|c| + e^{\alpha x}(\alpha^2 + 2a\alpha + c)}{A - e^{\alpha x}} < 0$$

on $\bar{D}$. It follows that $U_1$ and hence $u$, are identically zero on $\bar{D}$.

If the condition $c \leqslant 0$ does not hold everywhere on $D$, it suffices to require that the area of $D$ is sufficiently small. The result is stated by Petrovsky.†

## 10.5   The linear equation with constant coefficients

The homogeneous linear equation of the second order with constant coefficients can be put into the form
$$\frac{\partial^2 u}{\partial x^2} + \frac{\partial^2 u}{\partial y^2} + \lambda u = 0, \tag{1}$$

where $\lambda$ is a constant. If the problem of Dirichlet for this equation has a solution, it is unique when $\lambda \leqslant 0$ but not necessarily unique when $\lambda > 0$.

Let us consider this in more detail when the domain is a disc $D$ with the circle $C$ (equation $r = a$ in the polar coordinates) as frontier. Try to find a solution which is continuously differentiable in $r \leqslant a$, which has continuous second derivatives in $r < a$, and which satisfies the condition $u = f(\theta)$ when $r = a$, $f$ being of period $2\pi$. The function $f(\theta)$ must then have a continuous derivative $f'(\theta)$. Then $f(\theta)$ has a Fourier series
$$\tfrac{1}{2}a_0 + \sum_1^\infty (a_n \cos n\theta + b_n \sin n\theta),$$

which is uniformly convergent on any finite interval, since $na_n$ and $nb_n$ tend to zero as $n \to \infty$.

If $\lambda = k^2$, (1) has a formal solution
$$u = \tfrac{1}{2}a_0 \frac{J_0(kr)}{J_0(ka)} + \sum_1^\infty (a_n \cos n\theta + b_n \sin n\theta) \frac{J_n(kr)}{J_n(ka)}. \tag{2}$$

Now $z^{-n} J_n(z)$ is an even function of $z$ which has only real simple zeros; the positive zeros are $j_{n,1}, j_{n,2}, j_{n,3}, \ldots$ in increasing order of magnitude. If $J_m(ka) = 0$, the series (2) becomes meaningless, except in the case when $a_m = b_m = 0$. The series (2) does then give a solution of the problem of Dirichlet; but it is not unique – we can always add to it a term of the form
$$(A \cos m\theta + B \sin m\theta) J_m(kr).$$

† *Lectures on Partial Differential Equations* (New York, 1955), p. 232.

This difficulty does not arise if $a$ is small enough. If $j_{n,1}$ is the least positive zero of $z^{-n}J_n(z)$, the numbers $\{j_{n,1}: n = 0, 1, 2, ...\}$ form an increasing sequence. If $0 < ka < j_{0,1}$, none of the functions

$$J_0(ka), J_1(ka), J_2(ka), \ldots$$

vanishes, and the formal solution exists.

If $\lambda$ is negative, say $\lambda = -k^2$, equation (1) has solutions

$$I_n(kr)\cos n\theta, \quad I_n(kr)\sin n\theta,$$

where
$$I_n(z) = i^{-n}J_n(iz) = (\tfrac{1}{2}z)^n \sum_{k=0}^{\infty} \frac{(\tfrac{1}{2}z)^{2k}}{k!(n+k)!}.$$

The zeros of $I_n(z)$, apart from the zero at the origin, are all purely imaginary, being of the form $\pm ij_{n,1}, \pm ij_{n,2}, \ldots$. A formal solution is then

$$\tfrac{1}{2}a_0 \frac{I_0(kr)}{I_0(ka)} + \sum_{1}^{\infty} (a_n \cos n\theta + b_n \sin n\theta)\frac{I_n(kr)}{I_n(ka)}. \tag{3}$$

When $r = a$, its sum is $f(\theta)$. We have to discuss its convergence and continuity on $r \leqslant a$.

When $x \geqslant 0$, the function $I_n(x)$ behaves like a multiple of $x^n$ when $n$ is large, since

$$I_n(x) = \frac{(\tfrac{1}{2}x)^n}{n!}\left\{1 + \sum_{k=1}^{\infty} \frac{n!}{(n+k)!\,k!}(\tfrac{1}{4}x^2)^k\right\}.$$

$$= \frac{(\tfrac{1}{2}x)^n}{n!}\{1 + \eta_n(x)\},$$

where
$$0 \leqslant \eta_n(x) \leqslant \frac{1}{n+1}\sum_{k=1}^{\infty}\frac{1}{k!}(\tfrac{1}{4}x^2)^k < \frac{1}{n+1}\exp(\tfrac{1}{4}x^2).$$

Hence
$$\frac{I_n(kr)}{I_n(ka)} = \left(\frac{r}{a}\right)^n \{1 + \eta_n(r)\}/\{1 + \eta_n(a)\}.$$

Since $\{na_n\}$ and $\{nb_n\}$ are null sequences, the series (3) and the series obtained by differentiation as often as we please are uniformly and absolutely convergent on every disc $r \leqslant R$ when $R < a$. The sum of the series is therefore a solution of $\nabla^2 u - k^2 u = 0$ continuously differentiable as often as we please on $r < a$.

On $r \leqslant a$ and on any finite interval of values of $\theta$, the series (3) is of the form
$$\Sigma u_n(\theta)\,v_n(r),$$
where

$$u_0 = \tfrac{1}{2}a_0, \quad u_n = a_n \cos n\theta + b_n \sin n\theta, \quad v_n = I_n(kr)/I_n(ka).$$

198] ELLIPTIC EQUATIONS [10.5

The series $\Sigma u_n(\theta)$ is uniformly convergent. For any particular value of $r$ in the interval $[0,a]$, we have

$$0 \leqslant v_n(r) \leqslant 1;$$

and since $$v_n = \frac{r^n}{a^n}\{1+\omega_n(r,a)\},$$

where $\omega_n(r,a)$ tends to zero as $n \to \infty$ uniformly with respect to $r$, $\{v_n(r)\}$ is a non-increasing sequence on $[0,a]$. It follows by Abel's test that the series (3) is uniformly convergent on $x^2+y^2 \leqslant a^2$ and so is continuous there.

The series (3) thus gives a solution of $\nabla^2 u - k^2 u = 0$ which is continuous on $r \leqslant a$, which is continuously differentiable as often as we please on $r < a$ and which takes the given continuously differentiable values $f(\theta)$ on $r = a$.

The corresponding problem for $\nabla^2 u + k^2 u = 0$ can be discussed in a similar way.

## 10.6   The use of the elementary solution

Let $D$ be a bounded domain whose frontier consists of a finite number of non-intersecting regular closed curves. If $L(u) \equiv \nabla^2 u + k^2 u$ where $k$ is a positive constant, we have

$$\iint_D \{uL(v)-vL(u)\}\,dx\,dy = \int_{\partial D}\left(u\frac{\partial v}{\partial N} - v\frac{\partial u}{\partial N}\right)ds,$$

where $\partial/\partial N$ denotes differentiation along the normal to $\partial D$ drawn out of $D$. Sufficient conditions for this use of Green's theorem are that $u$ and $v$ be continuously differentiable in the closure of $D$ and possess bounded continuous second derivatives in $D$. If $u$ and $v$ are solutions of $L(w) = 0$, the left-hand side vanishes and

$$\int_{\partial D}\left(u\frac{\partial v}{\partial N} - v\frac{\partial u}{\partial N}\right)ds = 0. \tag{1}$$

If $R$ is the distance from a fixed point $(x_0,y_0)$ of $D$ to a variable point $(x,y)$, this result holds if $v$ is $J_0(kR)$, but not if $v$ is the elementary solution

$$Y_0(kR) = \frac{2}{\pi}\log R \,.\, J_0(kR) + \dots,$$

where the terms omitted have no singularity in $D$.

If $v$ is the elementary solution, we have to exclude from $D$ a disc with centre $(x_0,y_0)$ just as in the theory of Laplace's equation. The result is that, if $(x_0, y_0)$ is a point of $D$,

$$u(x_0, y_0) = \frac{1}{4} \int_{\partial D} \left\{ u \frac{\partial}{\partial N} Y_0(kR) - Y_0(kR) \frac{\partial u}{\partial N} \right\} ds. \tag{2}$$

This formula involves the boundary values of $u$ and its normal derivative, and so does not provide a solution either of the problem of Dirichlet or of the problem of Neumann. To solve them, we have to construct Green's functions.

If $\partial D$ is a circle $x^2 + y^2 = a^2$, a typical point of the boundary has polar coordinates $(a, \theta)$. If $(x_0, y_0)$ has polar coordinates $(r_0, \theta_0)$, then

$$R^2 = a^2 + r_0^2 - 2ar_0 \cos(\theta - \theta_0).$$

If we use the addition theorem

$$Y_0(kR) = Y_0(ka) J_0(kr_0) + 2 \sum_1^\infty Y_n(ka) J_n(kr_0) \cos n(\theta - \theta_0),$$

we obtain a solution

$$u(r_0, \theta_0) = \tfrac{1}{2} a_0 J_0(kr_0) + \sum_1^\infty (a_n \cos n\theta_0 + b_n \sin n\theta_0) J_n(kr_0),$$

where the coefficients $a_n$ and $b_n$ depend on the boundary values of $u$ and $\partial u/\partial N$. In particular, (2) gives the value of $u$ at the origin

$$u_0 = \tfrac{1}{4} k Y_0'(ka) \int_C u\, ds - \tfrac{1}{4} Y_0(ka) \int_C \frac{\partial u}{\partial N} ds,$$

where $C$ is the circle $r = a$. But (1), with $v = J_0(kR)$, gives

$$0 = \tfrac{1}{4} k J_0'(ka) \int_C u\, ds - \tfrac{1}{4} J_0(ka) \int_C \frac{\partial u}{\partial N} ds.$$

Hence      $J_0(ka) u_0 = \dfrac{k}{4} \{ J_0(ka)\, Y_0'(ka) - J_0'(ka)\, Y_0(ka) \} \displaystyle\int_C u\, ds$

in which the values of $\partial u/\partial N$ on the boundary do not appear. The expression

$$J_0(z)\, Y_0'(z) - J_0'(z)\, Y_0(z)$$

is called the Wronskian of $J_0$ and $Y_0$; it has the value $2/(\pi z)$. We have thus arrived at the mean-value formula

$$u_0 = \frac{1}{2\pi a J_0(ka)} \int_C u\, ds.$$

Since $J_0'(z) = -J_1(z)$, we have another mean-value formula

$$u_0 = -\frac{1}{2\pi a J_1(ka)} \int_C \frac{\partial u}{\partial N} ds$$

depending only on the values of the normal derivative on $C$.

## 10.7  Divergent waves

So far, we have considered the solution of the reduced wave equation only on a bounded domain. When the domain is unbounded, a condition at infinity is necessary. An important case arises in the theory of the reflection of sound waves of small amplitude by a rigid obstacle. If the incident waves have a velocity potential $\phi_i$, the reflected waves a velocity potential $\phi_s$, the condition on the boundary of the obstacle is

$$\frac{\partial}{\partial N}(\phi_i + \phi_s) = 0.$$

The condition at infinity is that $\phi_s$ should represent a divergent wave motion.

We assume that the motion is 'monochromatic', that the velocity potentials vary in a simple-harmonic manner with the time. It is convenient to use complex velocity potentials of the form

$$\phi = u e^{-ikct}$$

where $k > 0$, and take real parts at the end. Then $u$ satisfies the reduced wave equation $\nabla^2 u + k^2 u = 0$. In this section we consider the case when the wave motion is cylindrical; the obstacle is a cylinder with generators parallel to the $z$-axis; $\phi_i$, and hence also $\phi_s$, are independent of $z$.

We saw in §6.1 that the velocity potential of cylindrical waves due to a source of strength $2\pi F(t)$ per unit length on the $z$-axis is

$$\phi = \int_{-\infty}^{\infty} F(t - r\cosh\alpha)\,d\alpha$$

where $r$ is the distance from the axis of $z$. In particular, for monochromatic divergent cylindrical waves

$$\phi = \exp(-ikct)\int_{-\infty}^{\infty} \exp(ikr\cosh\alpha)\,d\alpha = \pi i H_0^{(1)}(kr)\exp(-ikct),$$

where $H_0^{(1)}$ denotes the Hankel function of order zero. The function of order $n$ is

$$H_n^{(1)}(kr) = J_n(kr) + iY_n(kr).$$

Using the asymptotic expansion for the Hankel function $H_0^{(1)}(kr)$, we find that

$$\phi_0 = H_0^{(1)}(kr)\exp(-ikct) \sim \left(\frac{2}{\pi kr}\right)^{\frac{1}{2}}\exp\{i(kr - kct - \tfrac{1}{4}\pi)\}$$

as $r \to \infty$. The velocity potential $\phi_0$ represents expanding harmonic cylindrical waves with amplitude decreasing like $1/\sqrt{r}$. A more general velocity potential is

$$\phi_n = H_n^{(1)}(kr) \frac{\cos}{\sin} n\theta \, \exp(-ikct),$$

which behaves like a multiple of

$$\frac{\exp\{ik(r-ct)\}}{\sqrt{r}} \frac{\cos}{\sin} n\theta,$$

as $r \to \infty$.

If $\phi_n = u_n \exp(-ikct)$, it can be shown that, for any integer $n$,

(i) as $r \to \infty$, $u_n \sqrt{r}$ is bounded, uniformly with respect to $\theta$,

(ii) $\sqrt{r}\{\partial u_n/\partial r - iku_n\} \to 0$ as $r \to \infty$, uniformly with respect to $\theta$. These two conditions were called by Sommerfeld the *finiteness condition* and the *radiation condition*. They play an important part in the theory of the solution of the reduced wave equation in an unbounded domain.

Let $D$ be the domain which is bounded internally by a regular closed curve $C$ and externally by a circle $\Gamma$ of large radius $a$. Then if $(x_0, y_0)$ is a point of $D$, we obtain, as in §10.6 the expression

$$u(x_0, y_0) = \frac{i}{4} \int_{\partial D} \left\{ \frac{\partial u}{\partial N} H_0^{(1)}(kR) - u \frac{\partial}{\partial N} H_0^{(1)}(kR) \right\} ds$$

for the value of a solution $u$ of $\nabla^2 u + k^2 u = 0$ at the point $(x_0, y_0)$. This is obtained by combining formula (2) and formula (1) with $v = J_0(kR)$. We may take $\Gamma$ to be the circle $R = a$. Hence

$$u(x_0, y_0) = \frac{i}{4} \int_{\Gamma} \left\{ \frac{\partial u}{\partial R} H_0^{(1)}(kR) - u \frac{\partial}{\partial R} H_0^{(1)}(kR) \right\} ds,$$

$$+ \frac{i}{4} \int_{C} \left\{ u \frac{\partial}{\partial n} H_0^{(1)}(kR) - \frac{\partial u}{\partial n} H_0^{(1)}(kR) \right\} ds,$$

where $n$ is the unit normal vector drawn out of the bounded domain whose frontier is $C$, this being oppositely directed to $N$.

In the integral over $\Gamma$, the integrand is

$$\left( \frac{\partial u}{\partial R} - iku \right) v - \left( \frac{\partial v}{\partial R} - ikv \right) u,$$

where $\qquad\qquad v = H_0^{(1)}(kR).$

This function $v$ satisfies Sommerfeld's finiteness and radiation conditions. There exists a constant $K_1$ such that $|v| < K_1/\sqrt{R}$; and, for every positive value of $\epsilon$, there exists a positive number $a_1$ such that

$$\left| \frac{\partial v}{\partial R} - ikv \right| < \frac{\epsilon}{\sqrt{R}},$$

when $R > a_1$. Suppose that $u$ also satisfies the finiteness and radiation conditions. Then there exists a constant $K_2$ such that $|u| < K_2/\sqrt{R}$; and, with the same value of $\epsilon$, there exists a positive number $a_2$ such that

$$\left| \frac{\partial u}{\partial R} - iku \right| < \frac{\epsilon}{\sqrt{R}},$$

when $R > a_2$.

Choose $a$ greater than max $(a_1, a_2)$. Then

$$\left| \frac{i}{4} \int_\Gamma \left\{ \frac{\partial u}{\partial R} H_0^{(1)}(kR) - u \frac{\partial}{\partial R} H_0^{(1)}(kR) \right\} ds \right| < \tfrac{1}{2}\pi (K_1 + K_2)\epsilon,$$

so that the integral round $\Gamma$ tends to zero as $a \to \infty$. Therefore, at any point $(x_0, y_0)$ outside $C$, the solution $u$ of the reduced wave equation has the value

$$u(x_0, y_0) = \frac{i}{4} \int_C \left\{ u \frac{\partial}{\partial n} H_0^{(1)}(kR) - \frac{\partial u}{\partial n} H_0^{(1)}(kR) \right\} ds$$

provided that $u$ satisfies Sommerfeld's conditions. This formula expresses a divergent wave function $u$ in terms of the values taken by $u$ and $\partial u/\partial n$ on $C$. But it cannot be used to solve scattering problems in acoustics because, although $\partial u_s/\partial n$ is known on the rigid boundary $C$, $u_s$ is unknown there. To eliminate the unknown values of $u_s$ on $C$, we should have to find a Green's function. Usually this is not possible; we give a simple example when it can be done in the next section.

We shall discuss Sommerfeld's conditions in more detail in Ch. 11 when we consider the reduced wave equation in three-dimensional space.

## 10.8   The half-plane problem

*Let $u$ be a solution of $\nabla^2 u + k^2 u = 0$ which is continuously differentiable in $y \geqslant 0$, has continuous second derivatives and satisfies Sommerfeld's conditions at infinity. Let $u = f(x)$ or $u_y = g(x)$ when $y = 0$. Then, when $y_0 > 0$,*

$$u(x_0, y_0) = -\tfrac{1}{2}i \int_{-\infty}^{\infty} g(x) H_0^{(1)}(kR_0)\, dx,$$

*or*

$$u(x_0, y_0) = -\tfrac{1}{2}i \frac{\partial}{\partial y_0} \int_{-\infty}^{\infty} f(x) H_0^{(1)}(kR_0)\, dx,$$

*where*

$$R_0^2 = (x - x_0)^2 + y_0^2.$$

These formulae express $u$ in terms of the boundary values of $u$ or of $\partial u/\partial N$.

Let $D$ be the domain defined by

$$x^2 + y^2 < a^2, \quad y > 0.$$

If $(x_0, y_0)$ is a point of $D$, so that $y_0 > 0$,

$$u(x_0, y_0) = \frac{i}{4} \int_{\partial D} \left\{ \frac{\partial u}{\partial N} H_0^{(1)}(kR) - u \frac{\partial}{\partial N} H_0^{(1)}(kR) \right\} ds,$$

where $\qquad R^2 = (x - x_0)^2 + (y - y_0)^2;$

but $\qquad 0 = \frac{i}{4} \int_{\partial D} \left\{ \frac{\partial u}{\partial N} H_0^{(1)}(k\bar{R}) - u \frac{\partial}{\partial N} H_0^{(1)}(k\bar{R}) \right\} ds,$

where $\qquad \bar{R}^2 = (x - x_0)^2 + (y + y_0)^2$

since $(x_0, -y_0)$ is in $y < 0$. As in §10.7, the integrals round the semi-circle $x^2 + y^2 = a^2, y \geqslant 0$ tend to zero as $a \to \infty$.

On $y = 0, \partial u/\partial N = -g(x)$. Also

$$\frac{\partial}{\partial y} H_0^{(1)}(kR) = -H_1^{(1)}(kR) k \frac{\partial R}{\partial y} = -H_1^{(1)}(kR) \frac{k(y - y_0)}{R},$$

so that, on $y = 0$,

$$\frac{\partial}{\partial N} H_0^{(1)}(kR) = -H_1^{(1)}(kR_0) \frac{ky_0}{R_0},$$

where $\qquad R_0^2 = (x - x_0)^2 + y_0^2.$

Similarly, on $y = 0$,

$$\frac{\partial}{\partial N} H_0^{(1)}(k\bar{R}) = H_1^{(1)}(kR_0) \frac{ky_0}{R_0}.$$

Therefore, when $y_0 > 0$,

$$u(x_0, y_0) = -\frac{i}{4} \int_{-\infty}^{\infty} g(x) H_0^{(1)}(kR_0) \, dx + \frac{i}{4} \int_{-\infty}^{\infty} f(x) H_1^{(1)}(kR_0) \frac{ky_0}{R_0} \, dx,$$

and

$$0 = -\frac{i}{4} \int_{-\infty}^{\infty} g(x) H_0^{(1)}(kR_0) \, dx - \frac{i}{4} \int_{-\infty}^{\infty} f(x) H_1^{(1)}(kR_0) \frac{ky_0}{R_0} \, dx.$$

Adding, we have

$$u(x_0, y_0) = -\tfrac{1}{2} i \int_{-\infty}^{\infty} g(x) H_0^{(1)}(kR_0) \, dx.$$

Subtracting, $\quad u(x_0, y_0) = \frac{1}{2}i \int_{-\infty}^{\infty} f(x) H_1^{(1)}(kR_0) \dfrac{ky_0}{R_0} dx,$

or $\qquad u(x_0, y_0) = -\frac{1}{2}i \dfrac{\partial}{\partial y_0} \int_{-\infty}^{\infty} f(x) H_0^{(1)}(kR_0) dx.$

To give a rigorous proof of these results, we should have to impose suitable conditions on $f(x)$ and $g(x)$ and consider the limits of these integrals as $y_0 \to +0$.

## 10.9 A boundary and initial value problem

We discussed in Ch. 7 the Cauchy problem for the equation of wave motions

$$\frac{\partial^2 u}{\partial x^2} + \frac{\partial^2 u}{\partial y^2} - \frac{\partial^2 u}{\partial t^2} = 0,$$

namely to find $u$ when $t = 0$, given that $u = f(x, y)$, $u_t = h(x, y)$ when $t = 0$. In the theory of small vibrations of a uniform stretched membrane with a fixed rim, we have to solve a boundary value and an initial value problem; we are given the values of $u$ and $u_t$ on a plane domain $D$ when $t = 0$ and also given that $u$ vanishes on the boundary of $D$ when $t \geqslant 0$. The solution is a sum of terms, each representing a normal mode of vibration, in which $u = U(x, y) \cos(\omega t + \epsilon)$. The function $U$ satisfies the reduced wave equation

$$\nabla^2 U + \omega^2 U = 0.$$

In the simplest case, $D$ is a rectangle $0 \leqslant x \leqslant a$, $0 \leqslant y \leqslant b$. By the method of separation of variables, we obtain

$$U = \sin \frac{m\pi x}{a} \sin \frac{n\pi y}{b},$$

where $m$ and $n$ are positive integers because $U$ vanishes on $\partial D$, and $\omega = \omega_{m,n}$ where

$$\omega_{m,n}^2 = \frac{m^2 \pi^2}{a^2} + \frac{n^2 \pi^2}{b^2}.$$

Hence we get a formal solution

$$u = \sum_{m,n=1} (A_{m,n} \cos \omega_{m,n} t + B_{m,n} \sin \omega_{m,n} t) \sin \frac{m\pi x}{a} \sin \frac{n\pi y}{b},$$

where the coefficients $A_{m,n}$ and $B_{m,n}$ have to be chosen so that $u = f$, $u_t = h$ when $t = 0$. For a rigorous discussion we need to know the difficult theory of double Fourier series. The calculation of the coefficients is easy; the usual Fourier rule gives

$$A_{m,n} = \frac{4}{ab} \iint_D f(x,y) \sin \frac{m\pi x}{a} \sin \frac{n\pi y}{b} \, dx \, dy$$

$$\omega_{m,n} B_{m,n} = \frac{4}{ab} \iint_D h(x,y) \sin \frac{m\pi x}{a} \sin \frac{n\pi y}{b} \, dx \, dy.$$

When $D$ is the disc $r < a$ in polar coordinates, the reduced wave equation

$$\frac{\partial^2 U}{\partial r^2} + \frac{1}{2} \frac{\partial U}{\partial r} + \frac{1}{r^2} \frac{\partial^2 U}{\partial \theta^2} + \omega^2 U = 0$$

has solutions $J_n(\omega r) \cos(n\theta + \alpha)$, $Y_n(\omega r) \cos(n\theta + \beta)$.

Since $U$ is one-valued on the disc $n$ is a positive integer or zero; and the Bessel function of the second kind cannot occur since it tends to infinity as $r \to 0$.

Since $U$ vanishes when $r = a$, $J_n(\omega a) = 0$. The equation $J_n(z) = 0$ has a multiple root at the origin and all its other roots are real and simple; they are $\pm j_{n,1}, \pm j_{n,2}, \ldots$. The positive roots from an increasing sequence, and $j_{n,k}$ tends to infinity with $k$.

Write $\omega_{n,k} = j_{n,k}/a$. Then we get a formal solution

$$u = \sum_{n=0}^{\infty} \sum_{k=1}^{\infty} (A_{n,k} \cos n\theta + B_{n,k} \sin n\theta) J_n(\omega_{n,k} r) \cos(\omega_{n,k} t + s_{n,k}),$$

where the coefficients have to be chosen so that the initial conditions on $r \leqslant a$ are satisfied. This involves the theory of Fourier–Bessel expansions, of which an account will be found in Ch. 18 of G. N. Watson's *Treatise on the Theory of Bessel Functions*.

### Exercises

**1.** Deduce from the mean value theorem of §10.6 that a solution of $\nabla^2 u + k^2 u = 0$ cannot have a positive minimum or a negative maximum.

**2.** Monochromatic sound waves of small amplitude are propagated along a cylindrical wave guide of uniform cross section $D$. Taking the $z$-axis along the guide, show that a complex velocity potential satisfying

$$\nabla^2 u = \frac{1}{c^2} \frac{\partial^2 u}{\partial t^2}$$

is of the form          $u = U e^{i(\gamma z - kct)},$

where $U$ satisfies the reduced wave equation

$$\frac{\partial^2 U}{\partial x^2} + \frac{\partial^2 U}{\partial y^2} + (k^2 - \gamma^2)\, U = 0,$$

with boundary condition $\partial U/\partial N = 0$ on the surface of the wave guide.

If the cross section of the guide is the square $0 \leqslant x \leqslant a,\, 0 \leqslant y \leqslant a$, prove that a complex velocity potential is

$$u = \cos\frac{m\pi x}{a}\cos\frac{n\pi y}{a}\, e^{i(\gamma z - kct)},$$

where

$$\gamma^2 = k^2 - (m^2 + n^2)\,\pi^2/a^2,$$

$m$ and $n$ being positive integers. Show that this mode is propagated only if $k > \pi\sqrt{(m^2+n^2)}/a$. If this mode is propagated, show that the phase velocity $V = kc/\gamma$ is greater than $c$ and that the group velocity $d(kc)/d\gamma$ is less than $c$.

# 11

# EQUATIONS OF ELLIPTIC TYPE
# IN SPACE

## 11.1  Laplace's equation

The simplest equation of elliptic type in three dimensions is Laplace's equation

$$\nabla^2 u \equiv \frac{\partial^2 u}{\partial x^2} + \frac{\partial^2 u}{\partial y^2} + \frac{\partial^2 u}{\partial z^2} = 0.$$

The theory of the solutions of this equation resembles the theory of harmonic functions in two dimensions, with some important differences.

The theory of harmonic functions in two dimensions can be made to depend on the theory of analytic functions of a complex variable $x + iy$. There is nothing corresponding to the theory of functions of a complex variable $x + iy$ in three dimensions. The nearest approach is given by Whittaker's general solution

$$u(x, y, z) = \int_0^{2\pi} f(z + ix \cos u + iy \sin u, u) \, du$$

of Laplace's equation.

The applications of Green's theorem run, in general, parallel to those in two dimensions, with the difference that the elementary solution in three dimensions is $1/R$ which vanishes at infinity, whereas in two dimensions it is $\log 1/R$.

There are coordinate systems in which Laplace's equation in three dimensions is separable, and these lead to solutions as infinite series whose terms are the various special functions of analysis.

## 11.2  Polynomial solutions

A solution of Laplace's equation, analytic in a neighbourhood of some fixed point which we may take as origin is expansible as a convergent series

$$\sum_0^\infty f_n(x, y, z),$$

where $f_n(x, y, z)$ is a homogeneous polynomial of degree $n$ which satisfies Laplace's equation. Now the most general homogeneous

208]     ELLIPTIC EQUATIONS IN SPACE          [11.2

polynomial $f_n$ of degree $n$ contains $\frac{1}{2}(n+2)(n+1)$ terms and so has the same number of independent coefficients. Since $\nabla^2 f_n$ is a homogeneous polynomial of degree $n-2$, it has $\frac{1}{2}n(n-1)$ different terms. If $\nabla^2 f_n = 0$, we have $\frac{1}{2}n(n-1)$ independent linear relations connecting the $\frac{1}{2}(n+2)(n+1)$ constants in $f_n$. Hence there are only

$$\tfrac{1}{2}(n+2)(n+1) - \tfrac{1}{2}n(n-1) = 2n+1$$

independent constants in a homogeneous polynomial of degree $n$ which satisfies Laplace's equation. Thus there are $2n+1$ independent homogeneous polynomials of degree $n$ which satisfy Laplace's equation in three dimensions. For example, if $n=1$, the three solutions are $x, y, z$; if $n=2$, the five solutions are $x^2-z^2, y^2-z^2, yz, zx, xy$. In contrast to this, in two dimensions, there are only two independent homogeneous polynomials of degree $n$ which satisfy Laplace's equation, viz. $r^n \cos n\theta$ and $r^n \sin n\theta$ in plane polar coordinates.

A particular homogeneous polynomial of degree $n$ which satisfies Laplace's equation is
$$(z + ix \cos \omega + iy \sin \omega)^n,$$

where $\omega$ is a parameter. This can be expanded as a trigonometric polynomial

$$\tfrac{1}{2}a_0(x,y,z) + \sum_{m=1}^{n} (a_m(x,y,z) \cos m\omega + b_m(x,y,z) \sin m\omega).$$

Since $\omega$ is independent of $x, y, z$, each coefficient $a_m, b_m$ is a homogeneous polynomial which satisfies Laplace's equation. The highest power of $z$ in $a_m(x,y,z)$ is $z^{n-m}$, and $a_m$ is an even function of $y$. Similarly the highest power of $z$ in $b_m(x,y,z)$ is $z^{n-m}$, but $b_m$ is an odd function of $y$. Thus the $2n+1$ polynomials $a_m$ and $b_m$ are linearly independent, and are the required $2n+1$ harmonic polynomials of degree $n$. They are given explicitly by

$$a_m(x,y,z) = \frac{1}{\pi} \int_{-\pi}^{\pi} (z + ix \cos \omega + iy \sin \omega)^n \cos m\omega \, d\omega$$

$$b_m(x,y,z) = \frac{1}{\pi} \int_{-\pi}^{\pi} (z + ix \cos \omega + iy \sin \omega)^n \sin m\omega \, d\omega.$$

Every harmonic polynomial of degree $n$ is therefore of the form

$$\int_{-\pi}^{\pi} (z + ix \cos \omega + iy \sin \omega)^n f(\omega) \, d\omega,$$

where $f(\omega)$ is a trigonometric polynomial. A harmonic polynomial of degree $n$ is usually called a *spherical harmonic*.

Now, if $V(x,y,z)$ is a harmonic function analytic in a neighbourhood of the origin, it is of the form

$$V(x,y,z) = \sum_1^\infty V_n(x,y,z),$$

where $V_n(x,y,z)$ is a harmonic polynomial of degree $n$, the series being convergent near the origin. Hence

$$V(x,y,z) = \sum_0^\infty \int_{-\pi}^\pi (z + cx\cos\omega + iy\sin\omega)^n f_n(\omega)\,d\omega,$$

where $f_n(\omega)$ is a trigonometric polynomial. From this follows Whittaker's general solution of Laplace's equation

$$V(x,y,z) = \int_{-\pi}^\pi F(z + ix\cos\omega + iy\sin\omega, \omega)\,d\omega,$$

where $F(\zeta,\omega)$ is expressible as a power series in $\zeta$, convergent in a neighbourhood of $\zeta = 0$. This solution holds under less stringent conditions. All we need is that $F$ is such that differentiation under the sign of integration is valid.

Shifting the origin, we get a solution

$$V(x,y,z) = \int_{-\pi}^\pi F(z - z_0 + ix\cos\omega - ix_0\cos\omega + iy\sin\omega - iy_0\sin\omega, \omega)\,d\omega$$

valid near $(x_0,y_0,z_0)$. The generalisation is rather trivial since this solution is merely

$$V(x,y,z) = \int_{-\pi}^\pi G(z + ix\cos\omega + iy\sin\omega, \omega)\,d\omega.$$

There is no special merit in giving $z$ priority; we can interchange $x,y,z$ as we please.

It should be noted that Whittaker's solution can represent different harmonic functions in different parts of the plane. For example

$$\frac{1}{2\pi}\int_{-\pi}^\pi \frac{d\omega}{z + ix\cos\omega + iy\sin\omega}$$

is equal to

$$\frac{1}{r} \equiv \frac{1}{\sqrt{(x^2+y^2+z^2)}},$$

when $z > 0$, but is equal to $-1/r$ when $z < 0$. When $z = 0$, the integral is not convergent, but has the principal value zero. We also have

$$\frac{1}{2\pi}\int_{-\pi}^\pi \frac{d\omega}{x + iy\cos\omega + iz\sin\omega} = \pm\frac{1}{r},$$

where the sign is $+$ or $-$ according as $x$ is positive or negative. Moreover, we have found in the region $x > 0, z > 0$ two different integral representations of $1/r$, and there does not appear to be any simple transformation of the one integral into the other.

It should be noted that, if we put $z = ict$, Whittaker's formula gives a solution

$$\int_{-\pi}^{\pi} H(ct + x\cos\omega + y\sin\omega, \omega)\, d\omega$$

of the two dimensional wave equation.

Further applications of Whittaker's solution will be found in Chapter 18 of Whittaker and Watson's *Modern Analysis*.

## 11.3   Spherical harmonics

There are $2n+1$ linearly independent spherical harmonics of degree $n$ which we denoted by

$$a_m(x, y, z) = \frac{1}{\pi}\int_{-\pi}^{\pi} (z + ix\cos\omega + iy\sin\omega)^n \cos m\omega\, d\omega$$

$$(m = 0, 1, 2, \dots, n),$$

$$b_m(x, y, z) = \frac{1}{\pi}\int_{-\pi}^{\pi} (z + ix\cos\omega + iy\sin\omega)^n \sin m\omega\, d\omega$$

$$(m = 1, 2, \dots, n).$$

In spherical polar coordinates

$$x = r\sin\theta\cos\phi, \quad y = r\sin\theta\sin\phi, \quad z = r\cos\theta,$$

these are

$$a_m(x, y, z) = \frac{r^n}{\pi}\int_{-\pi}^{\pi} \{\cos\theta + i\sin\theta\cos(\omega - \phi)\}^n \cos m\omega\, d\omega, \qquad (1)$$

and $$b_m(x, y, z) = \frac{r^n}{\pi}\int_{-\pi}^{\pi} \{\cos\theta + i\sin\theta\cos(\omega - \phi)\}^n \sin m\omega\, d\omega. \qquad (2)$$

Now

$$(\cos\theta + i\sin\theta\cos t)^n = P_n(\cos\theta) + 2\sum_{m=1}^{n} i^{-m}\frac{n!}{(n+m)!} P_n^m(\cos\theta)\cos mt,$$

$$(3)$$

where $P_n(\mu)$ is Laplace's polynomial and $P_n^m(\mu)$ is the associated function defined when $-1 < \mu < 1$ by†

$$P_n^m(\mu) = (-1)^m (1 - \mu^2)^{\frac{1}{2}m}\frac{d^m P_n(\mu)}{d\mu^m}.$$

---

† This is Hobson's definition. The factor $(-1)^m$ is sometimes omitted. Formula (3) will be found in Hobson, *The Theory of Spherical and Ellipsoidal Harmonics*, (Cambridge, 1931), p. 98; Whittaker and Watson, in *Course of Modern Analysis*, (4th edn (Cambridge, 1927), p. 392) omit the factor. For properties of the Legendre polynomials, see also Copson, *Functions of a Complex Variable* (Oxford, University Press, 1935) and Erdélyi *et al. Higher Transcendental Functions* (New York, 1953), Vol. I.

It follows that

$$a_0(x, y, z) = 2r^n P_n(\cos\theta),$$

$$a_m(x, y, z) = 2r^n_i{}^{-m} \frac{n!}{(n+m)!} P_n^m(\cos\theta)\cos m\phi,$$

$$b_m(x, y, z) = 2r^n_i{}^{-m} \frac{n!}{(n+m)!} P_n^m(\cos\theta)\sin m\phi.$$

Every spherical harmonic of degree $n$ is of the form

$$\left[ A_0 P_n(\cos\theta) + \sum_{m=1}^{n} (A_m\cos m\phi + B_m\sin m\phi)\, P_n^m(\cos\theta) \right] r^n,$$

where the coefficients $A_k$ and $B_k$ are constants. The expression inside the square brackets is often called a *surface harmonic*.

Of particular interest are solutions of Laplace's equation with symmetry about an axis, say the axis of $z$. Now such a solution which is analytic† in a neighbourhood of the origin is expansible as a series

$$\sum_0^\infty u_n(x, y, z),$$

where $u_n$ is a homogeneous harmonic polynomial of degree $n$, independent of $\phi$. Hence it is of the form

$$U(x, y, z) = \sum_0^\infty A_n r^n P_n(\mu),$$

where $\mu = \cos\theta$. If this solution takes on the sphere $r = a$ the values $f(\mu)$, we have formally

$$\sum_0^\infty A_n a^n P_n(\mu) = f(\mu).$$

Using the orthogonal properties of the Legendre polynomials, we get

$$\frac{2}{2n+1} A_n a^n = \int_{-1}^{1} f(\mu)\, P_n(\mu)\, d\mu.$$

Thus a formal solution of the problem is

$$V(x, y, z) = \sum_0^\infty \frac{2n+1}{2} \frac{r^n}{a^n} P_n(\mu) \int_{-1}^{1} f(t)\, P_n(t)\, dt.$$

The problem now is to discuss whether the series on the right converges for $r < a$ and whether its sum tends to $f(\mu)$ as $r \to a$. A full discussion of this is outside the scope of this book. We refer the reader to Ch. VII of Hobson's *The Theory of Spherical and Ellipsoidal Harmonics*.

The problem of Dirichlet for a sphere is discussed later in this chapter by the use of Poisson's integral.

† We shall see later that if a solution of Laplace's equation is harmonic in a neighbourhood of the origin, it is necessarily analytic there.

## 11.4   Green's theorem

Let $D$ be a domain whose function is a regular closed surface $S$, as defined in Note 6. $S$ is then the union of a finite number of smooth caps; on each cap, there is a continuously varying normal direction.

Let $u$, $v$, $w$ be continuous in $\bar{D}$ and have bounded continuous first derivatives in $D$. Then, if the triple integral exists

$$\iint_S (lu + mv + nw)\, dS = \iiint_D \left(\frac{\partial u}{\partial x} + \frac{\partial v}{\partial y} + \frac{\partial w}{\partial z}\right) dx\, dy\, dz, \qquad (1)$$

where $(l, m, n)$ are the direction cosines of the outward drawn normal. The fact that there may be a finite number of curves on $S$, the rims of the caps, across which the direction of the unit normal vector $N = (l, m, n)$ is discontinuous, does not affect the truth of this result.

If, in (1), we put

$$u = UV_x, \quad v = UV_y, \quad w = UV_z,$$

we obtain

$$\iint_S U\frac{\partial V}{\partial N}\, dS = \iiint_D (U_x V_x + U_y V_y + U_z V_z)\, dx\, dy\, dz + \iiint_D U\nabla^2 V\, dx\, dy\, dz. \tag{2}$$

If we interchange $U$ and $V$ and subtract, we obtain

$$\iint_S \left(U\frac{\partial V}{\partial N} - V\frac{\partial U}{\partial N}\right) dS = \iiint_D (U\nabla^2 V - V\nabla^2 U)\, dx\, dy\, dz, \qquad (3)$$

provided that $U$, $V$, $\mathrm{grad}\, U$, $\mathrm{grad}\, V$ are continuous on $\bar{D}$, that $U$ and $V$ have bounded continuous second derivatives on $D$, and that the triple integral exists.

We sometimes need the case when $\partial D$ consists of a finite number of non-intersecting regular closed surfaces. For example, $D$ might be the domain $a < r < b$ in spherical polar coordinates; then $\partial D$ is the pair of non-intersecting spheres $r = a$ and $r = b$. In such a case, $N$ is the unit normal vector drawn out of $D$.

## 11.5   Harmonic functions

A function is said to be harmonic in a bounded domain $D$ if it has continuous derivatives of the second order and satisfies Laplace's equation there. The elementary solution $1/R$, where

$$R^2 = (x-x_0)^2 + (y-y_0)^2 + (z-z_0)^2,$$

plays the same part as the logarithm does in the theory of harmonic functions in the plane.

In spherical polar coordinates

$$\nabla^2 V = \frac{\partial^2 V}{\partial r^2} + \frac{2}{r}\frac{\partial V}{\partial r} + \frac{1}{r^2}\mathscr{L}(V).$$

where $\mathscr{L}$ is a linear operator involving only differentiations with respect to the angle variables. If we put $r = a^2/s$,

$$\nabla^2 V = \frac{s^4}{a^4}\left[\frac{\partial^2 V}{\partial s^2} + \frac{1}{s^2}\mathscr{L}(V)\right].$$

Hence     $$\nabla^2(sW) = \frac{s^5}{a^4}\left[\frac{\partial^2 W}{\partial s^2} + \frac{2}{s}\frac{\partial W}{\partial s} + \frac{1}{s^2}\mathscr{L}(W)\right].$$

It follows that, if $V(x,y,z)$ is harmonic, so also is

$$\frac{1}{r}V\left(\frac{a^2x}{r^2},\frac{a^2y}{r^2},\frac{a^2z}{r^2}\right).$$

$V(x,y,z)$ is said to be harmonic at infinity if

$$\frac{1}{r}V\left(\frac{a^2x}{r^2},\frac{a^2y}{r^2},\frac{a^2z}{r^2}\right)$$

is harmonic at the origin. This enables us to extend the definition of harmonic functions to unbounded domains.

If $U = 1$ and $V$ is harmonic in $D$, equation (2) of §11.4 becomes

$$\iint_S \frac{\partial V}{\partial N}\,dS = 0. \tag{1}$$

If $V = U$ and $V$ is harmonic, the equation becomes

$$\iint_S V\frac{\partial V}{\partial N}\,dS = \iiint_D (V_x^2 + V_y^2 + V_z^2)\,dx\,dy\,dz. \tag{2}$$

Hence if $V\partial V/\partial N$ vanishes on $S$, $V_x, V_y, V_z$ all vanish on $D$ and so $V$ is a constant. From this follow the uniqueness theorems for the problems of Dirichlet and Neumann.

If $U_1$ and $U_2$ are two functions which are harmonic in $D$ and take the same continuous boundary values on $S$, $V = U_1 - U_2$, is harmonic in $D$ and vanishes on $S$. Hence $V$ vanishes on $D$, and $U_1 = U_2$ there. Again, if $U_1$ and $U_2$ are two functions which are harmonic in $D$, and if $\partial U_1/\partial N$ and $\partial U_2/\partial N$ take the same continuous boundary values on $S$, $V = U_1 - U_2$ is harmonic on $D$ and $\partial V/\partial N$ vanishes on $S$. Hence $V$ is a constant on $D$, and the functions $U_1$ and $U_2$ differ by a constant. As in the plane case, we cannot assign arbitrarily the values of the normal derivative on $S$ of a function $V$ harmonic in $D$; the boundary values must satisfy equation (1).

In equation (3) of § 11.4, suppose that $V$ is harmonic on $D$ and that $U = 1/r$. Then

$$\iint_S \left\{\frac{1}{r}\frac{\partial V}{\partial N} - V\frac{\partial}{\partial N}\left(\frac{1}{r}\right)\right\} dS = 0,$$

provided that the origin is not a point of $\bar{D}$. In particular, if $D$ is bounded by two concentric spheres $r = a$ and $r = R$, then, if $V$ is harmonic in $r \leqslant R$,

$$\iint_{r=R}\left(\frac{1}{r}\frac{\partial V}{\partial r} + \frac{V}{r^2}\right) dS = \iint_{r=a}\left(\frac{1}{r}\frac{\partial V}{\partial r} + \frac{V}{r^2}\right) dS.$$

As $a \to 0$, the expression on the right tends to $4\pi V(0,0,0)$. The expression on the left is

$$\frac{1}{R}\iint_{r=R}\frac{\partial V}{\partial r}dS + \frac{1}{R^2}\iint_{r=R} V\,dS.$$

The first term vanishes by equation (1). Hence

$$V(0,0,0) = \frac{1}{4\pi R^2}\iint_{r=R} V\,dS,$$

which is called *Gauss's mean value theorem*.

Suppose that $V$ is harmonic in a bounded domain $D$ and continuous in $\bar{D}$. If $V$ is not constant it follows from the mean value theorem that $V$ cannot have a relative maximum or a relative minimum at a point of $D$. Hence $V$ attains its supremum and its infimum on $\bar{D}$ at points of $\partial D$. In particular if $V$ is constant on $\partial V$, $V$ is constant on $\bar{D}$.

This provides an alternative proof of the uniqueness of the solution of the problem of Dirichlet, if such a solution exists. If $U_1$ and $U_2$ are two functions, harmonic in $D$ and continuous in $\bar{D}$, which take the same continuous boundary values on $\partial D$, $V = U_1 - U_2$ is also harmonic in $D$ and continuous in $\bar{D}$, and vanishes on $\partial D$. Hence $V$ is zero on $\bar{D}$.

## 11.6   Green's equivalent layer

Let $S$ be a regular closed surface bounding a domain $D$. If $U$ and $V$ have continuous first derivatives in $\bar{D}$ and have continuous second derivatives in $D$ where they satisfy Laplace's equation, it follows from (3) of §11.4 that

$$\iint_S \left( U \frac{\partial V}{\partial N} - V \frac{\partial U}{\partial N} \right) dS = 0. \tag{1}$$

If $R$ is the distance from a fixed point $(x_0, y_0, z_0)$ to the variable point $(x, y, z)$, we can put $V = 1/R$ provided that $(x_0, y_0, z_0)$ lies outside $\bar{D}$. If $(x_0, y_0, z_0)$ is a point of $D$, let $S_0$ be the sphere $R = \epsilon$, where $\epsilon$ is so small that $S_0$ lies in $D$. Then

$$\iint_S \left( U \frac{\partial}{\partial N} \frac{1}{R} - \frac{1}{R} \frac{\partial U}{\partial N} \right) dS + \iint_{S_0} \left( U \frac{\partial}{\partial N} \frac{1}{R} - \frac{1}{R} \frac{\partial U}{\partial N} \right) dS = 0.$$

Now on $S_0$, $\partial/\partial N$ is $-\partial/\partial R$. Hence

$$\iint_S \left( U \frac{\partial}{\partial N} \frac{1}{R} - \frac{1}{R} \frac{\partial U}{\partial N} \right) dS = -\iint_{S_0} \left( \frac{1}{R^2} U + \frac{1}{R} \frac{\partial U}{\partial R} \right) dS$$

$$= -\iint_{S_0} \left( U + \epsilon \frac{\partial U}{\partial R} \right) d\omega,$$

where $d\omega$ is the element of solid angle subtended at $(x_0, y_0, z_0)$ by the surface element $dS$ of $S_0$. Since $U$ and $\partial U/\partial R$ are continuous in $D$, we obtain $-4\pi U(x_0, y_0, z_0)$ as the limit of the right-hand side as $\epsilon \to 0$. Hence

$$U(x_0, y_0, z_0) = \frac{1}{4\pi} \iint_S \left( \frac{1}{R} \frac{\partial U}{\partial N} - U \frac{\partial}{\partial N} \frac{1}{R} \right) dS. \tag{2}$$

This result is known as the Green's equivalent layer theorem. It expresses $U$ on $D$ as the sum of potentials of a simple layer of density $(\partial U/\partial N)/4\pi$ on $S$ and a double layer of density $-U/4\pi$ on $S$.

Combining equations (1) and (2), we have

$$U(x_0, y_0, z_0) = \frac{1}{4\pi} \iint_S \left\{ \left( V + \frac{1}{R} \right) \frac{\partial U}{\partial N} - U \frac{\partial}{\partial N} \left( V + \frac{1}{R} \right) \right\} dS.$$

If we could choose the harmonic function $V$ so that

$$V + \frac{1}{R}$$

vanished at all points of $S$, the derivative $\partial U/\partial N$ would not occur in this integral. Such a function $V + 1/R$ is simply the electrostatic potential of a unit point charge at $(x_0, y_0, z_0)$ inside an earthed conductor $S$, and is called the *Green's function*. This physical argument suggests that there should be a Green's function.

If we denote the Green's function by $G(x_0, y_0, z_0; x, y, z)$ or, more briefly, by $G(P_0, P)$, a function harmonic in $D$ has the representation

$$U(P_0) = -\frac{1}{4\pi} \iint_S U(P) \frac{\partial}{\partial N} G(P_0, P) \, dS. \tag{3}$$

Thus the existence of a solution of the problem of Dirichlet is equivalent to the existence of a Green's function.

By the argument used in the plane case, we can prove the symmetry property

$$G(P_0, P) = G(P, P_0).$$

Lastly we assumed that the frontier of $D$ was a regular closed surface. It could equally well have been a finite number of non-intersecting regular closed surfaces.

## 11.7 Green's function for a sphere

The image of a unit point charge at a point $P_0$ inside a perfectly conducting sphere $S$ of radius $a$ and centre $O$ is a charge $-a/OP_0$ at the point $P_1$ which is the inverse of $P_0$ with respect to the sphere. For if $P$ is any point of the sphere, the triangles $OP_0P$ and $OPP_1$ are similar, and so

$$\frac{1}{P_0P} = \frac{a}{OP_0 \cdot P_1P}.$$

Hence the Green's function for the sphere is

$$G(P_0, P) = \frac{1}{P_0P} - \frac{a}{OP_0 \cdot P_1P}.$$

If $U$ is harmonic in the sphere,

$$U(P_0) = -\frac{1}{4\pi} \iint_S u(P) \frac{\partial}{\partial N} \left\{ \frac{1}{P_0P} - \frac{a}{OP_0 \cdot P_1P} \right\} dS.$$

It is convenient to use spherical polar coordinates. If $P_0$ is $(r, \theta, \phi)$ where $r < a$, then $P_1$ is $(a^2/r, \theta, \phi)$. If $P$ is $(\rho, \alpha, \beta)$

$$P_0P^2 = R_0^2 = r^2 + \rho^2 - 2r\rho \cos \gamma,$$

$$P_1P^2 = R_1^2 = \frac{a^4}{r^2} + \rho^2 - 2\frac{a^2}{r}\rho \cos \gamma,$$

where     $\cos \gamma = \cos \theta \cos \alpha + \sin \theta \sin \alpha \cos (\phi - \beta).$

Since     $\dfrac{\partial}{\partial \rho} \dfrac{1}{R_0} = \dfrac{r \cos \gamma - \rho}{R_0^3}, \quad \dfrac{\partial}{\partial \rho} \dfrac{1}{R_1} = \dfrac{a^2 \cos \gamma - r\rho}{rR_1^3},$

it follows that, when $P$ is on the surface of the sphere,

$$\frac{\partial}{\partial N} G(P_0, P) = -\frac{a^2 - r^2}{aR^3},$$

where     $R^2 = r^2 + a^2 - 2ar \cos \gamma.$

Hence     $U(r, \theta, \phi) = \dfrac{1}{4\pi a} (a^2 - r^2) \iint_S U(a, \alpha, \beta) \dfrac{dS}{R^3},$

or $$U(r,\theta,\phi) = \frac{a(a^2 - r^2)}{4\pi} \iint_\Omega \frac{U(a,\alpha,\beta)\,d\Omega}{(a^2 + r^2 - 2ar\cos\gamma)^{\frac{3}{2}}},$$

where $d\Omega$ is $\sin\alpha\,d\alpha\,d\beta$, the element of solid angle. Thus if $U(r,\theta,\phi)$ is equal to $f(\theta,\phi)$ when $r = a$, we should expect that

$$U(r,\theta,\phi) = \frac{a(a^2 - r^2)}{4\pi} \iint_\Omega \frac{f(\alpha,\beta)\,d\Omega}{(a^2 + r^2 - 2ar\cos\gamma)^{\frac{3}{2}}}. \tag{1}$$

This is called Poisson's integral.

It is readily verified that the expression on the right is harmonic when $r < a$. We have to show that it tends to $f(\theta,\phi)$ as $r \to a - 0$ under appropriate conditions on $f$. If $f$ is constant on $r = a$, $U$ is constant when $r \leqslant a$. Hence

$$U(r,\theta,\phi) - f(\theta,\phi) = \frac{a(a^2 - r^2)}{4\pi} \iint_\Omega \frac{f(\alpha,\beta) - f(\theta,\phi)}{(a^2 + r^2 - 2ar\cos\gamma)^{\frac{3}{2}}}\,d\Omega.$$

As we can rotate the axes in any way we please, it suffices to consider the limit when $\theta = \frac{1}{2}\pi, \phi = 0$. Then

$$U(r,\tfrac{1}{2}\pi,0) - f(\tfrac{1}{2}\pi,0) = \frac{a(a^2 - r^2)}{4\pi} \iint_\Omega \frac{f(\alpha,\beta) - f(\tfrac{1}{2}\pi,0)}{(a^2 + r^2 - 2ar\sin\alpha\cos\beta)^{\frac{3}{2}}}\,d\Omega.$$

We assume that $f(\theta,\phi)$ is continuous.

For every positive value of $\epsilon$, there exists a positive number $\delta$ such that

$$|f(\alpha,\beta) - f(\tfrac{1}{2}\pi,0)| < \epsilon,$$

when $-\delta < \beta < \delta, \frac{1}{2}\pi - \delta < \alpha < \frac{1}{2}\pi + \delta$. Call the part of $\Omega$ on which this holds $\Omega_0$, and the rest $\Omega_1$.

Firstly, we have

$$\left| \frac{1}{4\pi} a(a^2 - r^2) \iint_{\Omega_0} \frac{f(\alpha,\beta) - f(\tfrac{1}{2}\pi,0)}{(a^2 + r^2 - 2ar\sin\alpha\cos\beta)^{\frac{3}{2}}}\,d\Omega \right|$$

$$\leqslant \frac{1}{4\pi} a(a^2 - r^2)\,\epsilon \iint_{\Omega_0} \frac{d\Omega}{(a^2 + r^2 - 2ar\sin\alpha\cos\beta)^{\frac{3}{2}}}$$

$$\leqslant \frac{1}{4\pi} a(a^2 - r^2)\,\epsilon \iint_\Omega \frac{d\Omega}{(a^2 + r^2 - 2ar\sin\alpha\cos\beta)^{\frac{3}{2}}} < \epsilon.$$

Since $f(\alpha,\beta)$ is continuous, it is bounded on $\Omega$. Hence there exists a constant $M$ such that

$$|f(\alpha,\beta) - f(\tfrac{1}{2}\pi,0)| \leqslant M$$

on $\Omega$, and, in particular, on $\Omega_1$. On $\Omega_1$, $\sin\alpha\cos\beta \leqslant \cos\delta$, and so

$$a^2 + r^2 - 2ar\sin\alpha\cos\beta \geqslant a^2 + r^2 - 2ar\cos\delta.$$

Therefore

$$\left| \frac{1}{4\pi} a(a^2 - r^2) \iint_{\Omega_1} \frac{f(\alpha,\beta) - f(\tfrac{1}{2}\pi, 0)}{(a^2 + r^2 - 2ar\sin\alpha\cos\beta)^{\tfrac{3}{2}}} \, d\Omega \right|$$

$$\leqslant \frac{M}{4\pi} a(a^2 - r^2) \frac{1}{(a^2 + r^2 - 2ar\cos\delta)^{\tfrac{3}{2}}} \iint_{\Omega_1} d\Omega$$

$$< \frac{Ma(a^2 - r^2)}{(a^2 + r^2 - 2ar\cos\delta)^{\tfrac{3}{2}}},$$

which tends to zero as $r \to a - 0$. Hence with the same value of $\epsilon$, we can choose $r_1$ sufficiently near to $a$ so that

$$\left| \frac{1}{4\pi} a(a^2 - r^2) \iint_{\Omega_1} \frac{f(\alpha,\beta) - f(\tfrac{1}{2}\pi, 0)}{(a^2 + r^2 - 2ar\sin\alpha\cos\beta)^{\tfrac{3}{2}}} \, d\Omega \right| < \epsilon,$$

when $r_1 < r < a$. Therefore, when $r_1 < r < a$,

$$|U(r, \tfrac{1}{2}\pi, 0) - f(\tfrac{1}{2}\pi, 0)| < 2\epsilon.$$

As $\epsilon$ is arbitrary, $U(r, \tfrac{1}{2}\pi, 0)$ tends to $f(\tfrac{1}{2}\pi, 0)$ as $r \to a - 0$.

It follows that, when $f(\theta, \phi)$ is continuous, the problem of Dirichlet has a solution, necessarily unique, given by Poisson's integral.

The conditions on $f$ can be lightened. If $f$ is integrable, Poisson's integral represents a function harmonic in $r < a$, such that

$$U(r, \theta, \phi) \to f(\theta, \phi)$$

as $r \to a - 0$ at every point of continuity of $f$.

If $U(x, y, z)$ is harmonic in $r > a$, including the condition at infinity, and assumes the values $f(\theta, \phi)$ on $r = a$, then

$$\frac{a}{r} U\left( \frac{a^2 x}{r^2}, \frac{a^2 y}{r^2}, \frac{a^2 z}{r^2} \right)$$

is harmonic in $r < a$ and assumes the same boundary values on $r = a$. It follows that the solution of the external problem of Dirichlet for $r > a$ is

$$U(r, \theta, \phi) = \frac{1}{4\pi} a(r^2 - a^2) \iint_{\Omega} \frac{f(\alpha, \beta) \, d\Omega}{(a^2 + r^2 - 2ar\cos\gamma)^{\tfrac{3}{2}}},$$

where

$$\cos\gamma = \cos\theta\cos\alpha + \sin\theta\sin\alpha\cos(\phi - \beta).$$

## 11.8 The analytic character of harmonic functions

Let $U(x, y, z)$ be harmonic in a domain $D$. With any point of $D$ as origin, there exists a closed neighbourhood $r \leqslant a$ contained in $D$. $U$ is continuous in $D$, and so takes continuous values $f(\theta, \phi)$ on $r = a$

in polar coordinates. By Poisson's integral, if $(x, y, z)$ is a point with polar coordinates $(r, \theta, \phi)$ where $r < a$,

$$U(x, y, z) = \frac{1}{4\pi} a(a^2 - r^2) \iint_\Omega \frac{f(\alpha, \beta)\, d\Omega}{(a^2 + r^2 - 2ar\cos\gamma)^{\frac{3}{2}}}, \qquad (1)$$

where     $\cos\gamma = \cos\theta\cos\alpha + \sin\theta\sin\alpha\cos(\phi - \beta)$.

From the generating function

$$(1 - 2h\cos\gamma + h^2)^{-\frac{1}{2}} = \sum_0^\infty h^n P_n(\cos\gamma)$$

for the Legendre polynomials, it follows that

$$\frac{1 - h^2}{(1 - 2h\cos\gamma + h^2)^{\frac{3}{2}}} = \sum_0^\infty (2n + 1) h^n P_n(\cos\gamma). \qquad (2)$$

Hence     $U(x, y, z) = \dfrac{1}{4\pi} \iint_\Omega f(\alpha, \beta) \sum_0^\infty (2n + 1) \dfrac{r^n}{a^n} P_n(\cos\gamma)\, d\Omega.$ (3)

Now $|P_n(\cos\gamma)| \leqslant 1$. It follows that the infinite series under the sign of integration is absolutely convergent when $0 \leqslant r \leqslant b < a$, and that the convergence is uniform. We may therefore integrate term-by-term to obtain

$$U(x, y, z) = \frac{1}{4\pi} \sum_0^\infty (2n + 1) \frac{r^n}{a^n} \iint_\Omega f(\alpha, \beta) P_n(\cos\gamma)\, d\Omega. \qquad (4)$$

But $P_n(\cos\gamma)$ is a polynomial of degree $n$ in $\cos\gamma$. Hence

$$r^n \iint_\Omega f(\alpha, \beta) P_n(\cos\gamma)\, d\Omega$$

is a homogeneous polynomial of degree $n$ in $x, y$ and $z$ which satisfies Laplace's equation; it is a spherical harmonic. We have thus expressed $U(x, y, z)$ as a series of polynomial terms which converges uniformly and absolutely in $r \leqslant b$. Thus $U$ is analytic in any sphere contained in $D$; this proves the analytic character of $U$.

We can deduce from (4) the formula for the expansion in terms of the standard spherical harmonics by using the addition-theorem

$$P_n(\cos\gamma) =$$

$$P_n(\cos\theta) P_n(\cos\alpha) + 2 \sum_{m=1}^\infty \frac{(n-m)!}{(n+m)!} P_n^m(\cos\theta) P_n^m(\cos\alpha) \cos m(\phi - \beta).$$

It follows that

$$U(x, y, z) =$$

$$A_0 + \sum_1^\infty r^n \left\{ A_n P_n(\cos\theta) + \sum_{m=1}^n (A_n^m \cos m\phi + B_n^m \sin m\phi) P_n^m(\cos\theta) \right\},$$

where

$$A_n = \frac{2n+1}{2\pi a^n} \int_0^\pi d\alpha \int_0^{2\pi} d\beta \sin\alpha P_n(\cos\alpha)\, f(\alpha,\beta),$$

$$A_n^m = \frac{2n+1}{2\pi a^n}\frac{(n-m)!}{(n+m)!} \int_0^\pi d\alpha \int_0^{2\pi} d\beta \sin\alpha P_n^m(\cos\alpha)\cos m\beta\, f(\alpha,\beta),$$

$$B_n^m = \frac{2n+1}{2\pi a^n}\frac{(n-m)!}{(n+m)!} \int_0^\pi d\alpha \int_0^{2\pi} d\beta \sin\alpha P_n^m(\cos\alpha)\sin m\beta\, f(\alpha,\beta).$$

If $f$ is independent of $\beta$, this reduces to the expansion found in §11.3.

In a similar way we can use Poisson's formula for the solution of the exterior problem of Dirichlet with data on the sphere $r = a$. The solution is of the form

$$U(x,y,z) =$$

$$\frac{A_0 a}{r} + \sum_1^\infty \frac{a^{2n+1}}{r^{n+1}}\left\{ A_n P_n(\cos\theta) + \sum_{m=1}^\infty (A_n^m \cos m\phi + B_n^m \sin m\phi) P_n^m(\cos\theta) \right\}$$

with the same constants.

## 11.9   The linear equation of elliptic type

If the coefficients of the second derivatives in a linear equation of elliptic type are constants, we can find a non-singular linear transformation of the independent variables which turns the equation into

$$L(u) \equiv \nabla^2 u + 2au_x + 2bu_y + 2cu_z + du = F(x,y,z), \tag{1}$$

where the coefficients $a$, $b$, $c$, $d$ and $F$ are assumed to be analytic on a bounded domain $D$ whose frontier $\partial D$ consists of a finite number of non-intersecting regular closed surfaces. We denote by $(l,m,n)$ the direction cosines of the normal to $\partial D$ drawn out of $D$.

The operator $L^*$ adjoint to $L$ is defined by

$$L^*(v) \equiv \nabla^2 v - 2(av)_x - 2(bv)_y - 2(cv)_z + dv.$$

Then
$$vL(u) - uL^*(v) = \frac{\partial X}{\partial x} + \frac{\partial Y}{\partial y} + \frac{\partial Z}{\partial z},$$

where

$$X = vu_x - uv_x + 2auv, \quad Y = vu_y - uv_y + 2buv, \quad Z = vu_z - uv_z + 2cuv.$$

Green's theorem then gives

$$\iiint_D \{vL(u) - uL^*(v)\}\,dx\,dy\,dz = \iint_{\partial D} (lX + mY + nZ)\,dS,$$

a result which is true if $u$ and $v$ have bounded continuous second derivatives in $D$ and continuous first derivatives in $\bar{D}$. In particular, if $L(u) = F$, $L^*(v) = 0$, we obtain

$$\iiint_D vF\,dx\,dy\,dz = \iint_{\partial D} \left\{ v\frac{\partial u}{\partial N} - u\frac{\partial v}{\partial N} + 2(la+mb+nc)\,uv \right\} dS.$$
(2)

As Hadamard showed, the elementary solution of $L^*(v) = 0$ with singularity at $(x_0, y_0, z_0)$ is of the form

$$v = \frac{v_0(x,y,z)}{R},$$

where
$$R^2 = (x-x_0)^2 + (y-y_0)^2 + (z-z_0)^2$$

and $v_0$ is analytic and takes the value 1 at $(x_0, y_0, z_0)$. We can substitute this elementary function for $v$ in (2) provided that $(x_0, y_0, z_0)$ is not a point of $\bar{D}$. If $(x_0, y_0, z_0)$ is a point of $D$, we have

$$u(x_0,y_0,z_0) = \frac{1}{4\pi}\iint_{\partial D}\left\{ v\frac{\partial u}{\partial N} - u\frac{\partial v}{\partial N}\right.$$

$$\left. + 2(la+mb+nc)\,uv \right\} dS - \frac{1}{4\pi}\iiint_D vF\,dx\,dy\,dz. \quad (3)$$

This formula will not solve the problem of Dirichlet since it involves the boundary values of $u$ and $\partial u/\partial N$. If we could find a solution $v_1$ of $L^*(v) = 0$ with no singularities on $\bar{D}$ such that

$$v = \frac{v_0}{R} + v_1 \quad (4)$$

vanished on $\partial D$, formula (3) would not involve $\partial u/\partial N$, and so would solve the problem of Dirichlet. The function (4) would be the Green's function. Thus, if we could solve a particular case of the problem of Dirichlet for $L^*(v) = 0$, (3) would give a solution of the problem of Dirichlet for $L(u) = F$. But we should still have to show it is unique.

Using the argument of §10.4, we can show that: *If $d \leqslant 0$, the equation $L(u) = F$ has at most one solution which is continuously differentiable on $\bar{D}$, which has bounded continuous second derivatives on $D$ and which takes given continuous values on $\partial D$.*

## 11.10   The equation with constant coefficients

If the coefficients $a, b, c, d$ in

$$\nabla^2 U + 2aU_x + 2bU_y + 2cU_z + dU = \Phi(x,y,z)$$

are constants, the change of variable $U = u \exp(-ax - by - cz)$ gives

$$\nabla^2 u + \lambda u = F(x, y, z), \tag{1}$$

where

$$\lambda = d - a^2 - b^2 - c^2$$

is a constant. If $\lambda$ is positive, $\nabla^2 u + \lambda u = 0$ is called the reduced wave equation; it arises when we try to find 'monochromatic' wave functions in three-dimensional space.

Let $D$ be a domain bounded by one or more non-intersecting regular closed surfaces. If the problem of Dirichlet for (1) has two solutions $u_1$ and $u_2$, $u = u_1 - u_2$ satisfies

$$\nabla^2 u + \lambda u = 0$$

and vanishes on $\partial D$. But, by Green's transformation,

$$\iint_{\partial D} u \frac{\partial u}{\partial N} dS = \iiint_D \left\{ \frac{\partial}{\partial x}(u u_x) + \frac{\partial}{\partial y}(u u_y) + \frac{\partial}{\partial z}(u u_z) \right\} dx\,dy\,dz$$

$$= \iiint_D \{u_x^2 + u_y^2 + u_z^2 + u\nabla^2 u\}\,dx\,dy\,dz$$

$$= \iiint_D \{u_x^2 + u_y^2 + u_z^2 - \lambda u^2\}\,dx\,dy\,dz.$$

Hence

$$\iiint_D \{u_x^2 + u_y^2 + u_z^2\}\,dx\,dy\,dz = \lambda \iiint_D u^2\,dx\,dy\,dz.$$

If $\lambda \leqslant 0$, both sides of this equation are equal to zero. Hence $u_x, u_y, u_z$ vanish on $D$, and so $u$ is constant on $\bar{D}$. Therefore $u_1 = u_2$. Thus, if the problem of Dirichlet has a solution when $\lambda \leqslant 0$, it is unique.

If $\lambda > 0$, the solution may not be unique. For example, if $\lambda = k^2$ and $D$ is $r < a$, there is a solution

$$u = \frac{\sin kr}{r},$$

which vanishes on $r = a$ if $ka = n\pi$. Thus there are values of $k$ for which the problem of Dirichlet on $r \leqslant a$ does not have a unique solution.

## 11.11   The mean value theorem

*If $u$ is a solution of $\nabla^2 u + k^2 u = 0$ which has continuous second derivatives in a domain containing $(x - x_0)^2 + (y - y_0)^2 + (z - z_0)^2 \leqslant a^2$, the mean value of $u$ over $(x - x_0)^2 + (y - y_0)^2 + (z - z_0)^2 = R^2$, where $R \leqslant a$, is*

$$\frac{\sin kR}{kR} u(x_0, y_0, z_0).$$

This reduces to Gauss's mean value theorem when $k = 0$. We may take $(x_0, y_0, z_0)$ to be the origin, and use spherical polar coordinates $(r, \theta, \phi)$ so that $x = lr, y = mr, z = nr$, where

$$l = \sin\theta\cos\phi, \quad m = \sin\theta\sin\phi, \quad n = \cos\theta.$$

$\Omega$ denotes the whole solid angle at $O$; $d\Omega = \sin\theta\,d\theta\,d\phi$.

The mean value of $u$ over the sphere $\Sigma$, $x^2 + y^2 + z^2 = r^2$, where $r \leqslant a$ is

$$I(r) = \frac{1}{4\pi r^2}\iint_\Sigma u\,dS = \frac{1}{4\pi}\iint_\Omega u(lr, mr, nr)\,d\Omega.$$

Hence   $\dfrac{dI}{dr} = \dfrac{1}{4\pi}\displaystyle\iint_\Omega (lu_x + mu_y + nu_z)\,d\Omega = \dfrac{1}{4\pi r^2}\iint_\Sigma \dfrac{\partial u}{\partial N}\,dS,$

where $\partial/\partial N$ is differentiation along the outward normal. By Green's theorem

$$\frac{dI}{dr} = \frac{1}{4\pi r^2}\iiint \nabla^2 u(\xi, \eta, \zeta)\,d\xi\,d\eta\,d\zeta$$

$$= -\frac{k^2}{4\pi r^2}\iiint u(\xi, \eta, \zeta)\,d\xi\,d\eta\,d\zeta,$$

where integration is over $\xi^2 + \eta^2 + \zeta^2 \leqslant r^2$. Therefore

$$\frac{dI}{dr} = -\frac{k^2}{4\pi r^2}\int_0^r s^2\,ds\iint_\Omega u(ls, ms, ns)\,d\Omega$$

$$= -\frac{k^2}{r^2}\int_0^r s^2 I(s)\,ds.$$

Therefore   $\dfrac{d}{dr}r^2\dfrac{dI}{dr} = -k^2 r^2 I$

or   $r^2\dfrac{d^2 I}{dr^2} + 2r\dfrac{dI}{dr} + k^2 r^2 I = 0.$

Hence   $I = \dfrac{1}{r}(A\cos kr + B\sin kr).$

But $I$ tends to $u(0, 0, 0)$ as $r \to 0$, and so $A = 0$, and

$$I = \frac{\sin kr}{kr}u(0, 0, 0).$$

The corresponding result for $\nabla^2 u - k^2 u = 0$ is

$$I = \frac{\sinh kr}{kr}u(0, 0, 0).$$

From this it follows that a solution of $\nabla^2 u - k^2 u = 0$ which satisfies the usual differentiability conditions on a bounded domain $D$ cannot have a positive maximum or a negative minimum at any point of $D$. This implies the uniqueness theorem; for if $u$ vanished on $\partial D$, it would be identically zero on $\bar{D}$.

## 11.12 The solution of $\nabla^2 u - k^2 u = 0$ in polar coordinates

Suppose that we wish to solve the equation in the domain $r < a$ in polar coordinates, given that $u = f(\theta, \phi)$ when $r = a$. Since $f(\theta, \phi)$ can be expanded as a series

$$\sum_0^\infty S_n(\theta, \phi),$$

where $S_n(\theta, \phi)$ is a surface harmonic of order $n$, we try to find solutions of the form $R S_n(\theta, \phi)$ where $R$ depends on $r$ alone.

The differential equation in polar coordinates is

$$\frac{\partial^2 u}{\partial r^2} + \frac{2}{r} \frac{\partial u}{\partial r} - k^2 u + \frac{1}{r^2} \mathscr{L} u = 0,$$

where
$$\mathscr{L} u = \frac{\partial^2 u}{\partial \theta^2} + \cot \theta \frac{\partial u}{\partial \theta} + \frac{1}{\sin^2 \theta} \frac{\partial^2 u}{\partial \phi^2}.$$

Since $\mathscr{L} S_n = -n(n+1) S_n$, the corresponding function $R$ satisfies

$$\frac{d^2 R}{dr^2} + \frac{2}{r} \frac{dR}{dr} - \left\{ k^2 + \frac{n(n+1)}{r^2} \right\} R = 0.$$

If we put $R = S/\sqrt{r}$, this becomes

$$\frac{d^2 S}{dr^2} + \frac{1}{r} \frac{dS}{dr} - \left\{ k^2 + \frac{(n+\frac{1}{2})^2}{r^2} \right\} S = 0,$$

which is Bessel's equation for the functions of purely imaginary argument. It has the independent solutions†

$$I_{n+\frac{1}{2}}(kr), \quad I_{-n-\frac{1}{2}}(kr),$$

where
$$I_\nu(z) = (\tfrac{1}{2}z)^\nu \sum_{m=0}^\infty \frac{(\tfrac{1}{2}z)^{2m}}{m! \, \Gamma(\nu + m + 1)}.$$

Hence we have two solutions

$$\frac{I_{n+\frac{1}{2}}(kr)}{\sqrt{(kr)}} S_n(\theta, \phi), \quad \frac{I_{-n-\frac{1}{2}}}{\sqrt{(kr)}} S_n(\theta, \phi).$$

† See Watson, *A Treatise on the Theory of Bessel Functions*, 2nd edn (Cambridge, 1944), p. 77.

When $n = 0$, these solutions are constant multiples of

$$\frac{\sinh kr}{kr}, \quad \frac{\cosh kr}{kr}.$$

Now $I_\nu(z)/z^\nu$ is an integral function of the complex variable $z$ and has no real zeros. If $z$ is small, $I_{n+\frac{1}{2}}(z)/\sqrt{z}$ and $I_{-n-\frac{1}{2}}(z)/\sqrt{z}$ behave like multiples of $z^n$ and $z^{-n-1}$ respectively. As the solution is to be finite at the origin, it cannot contain the function $I_{-n-\frac{1}{2}}$. Hence if

$$f(\theta, \phi) = \sum_0^\infty S_n(\theta, \phi),$$

where $S_n$ is a surface harmonic of order $n$,

$$u = \sum_0^\infty \frac{I_{n+\frac{1}{2}}(kr)}{\sqrt{(kr)}} \frac{\sqrt{(ka)}}{I_{n+\frac{1}{2}}(ka)} S_n(\theta, \phi).$$

There is no restriction on $a$, since $I_{n+\frac{1}{2}}(ka)$ never vanishes.

If we wish to solve the external problem of Dirichlet with data on $r = a$, we have to use another solution of Bessel's equation since $I_{n+\frac{1}{2}}(kr)$ and $I_{-n-\frac{1}{2}}(kr)$ tend to infinity exponently as $r \to \infty$. But

$$K_{n+\frac{1}{2}}(kr) = (-1)^{n-1}\tfrac{1}{2}\pi\{I_{n+\frac{1}{2}}(kr) - I_{-n-\frac{1}{2}}(kr)\}$$

$$\sim \left(\frac{\pi}{2kr}\right)^{\frac{1}{2}} e^{-kr}$$

as $r \to \infty$. The required solution of the external problem is then

$$u = \sum_0^\infty \frac{K_{n+\frac{1}{2}}(kr)}{\sqrt{(kr)}} \frac{\sqrt{(ka)}}{K_{n+\frac{1}{2}}(ka)} S_n(\theta, \phi).$$

Again, there is no restriction on the value of $a$ since $K_\nu(z)$ has no positive zeros.

Lastly, if    $R^2 = (x - x_0)^2 + (y - y_0)^2 + (z - z_0)^2,$

the function      $\cosh kR/R$

is the elementary solution of $\nabla^2 u - k^2 u = 0$.

### 11.13   The solution of $\nabla^2 u + k^2 u = 0$ in polar coordinates

If

$$f(\theta, \phi) = \sum_0^\infty S_n(\theta, \phi),$$

the solution of the problem of Dirichlet of the equation $\nabla^2 u + k^2 u = 0$ for $r < a$, given that $u = f(\theta, \phi)$ on $r = a$ is

$$u = \sum_0^\infty \frac{J_{n+\frac{1}{2}}(kr)}{\sqrt{(kr)}} \frac{\sqrt{(ka)}}{J_{n+\frac{1}{2}}(ka)} S_n(\theta, \phi).$$

This can be obtained by repeating the argument of §11.2 or, more simply, by replacing $k$ by $ik$. This is satisfactory, subject to convergence considerations, provided that $ka$ is not a zero of $J_{n+\frac{1}{2}}(z)$.

The positive zeros of $J_\nu(z)$ form an increasing sequence; and, if the least positive zero is $j_\nu$, then $0 < j_\nu < j_{\nu+1}$. Since

$$J_{\frac{1}{2}}(z) = \sqrt{\left(\frac{2}{\pi z}\right)} \sin z,$$

the least positive zero of $J_{\frac{1}{2}}(z)$ is $\pi$. Therefore, if $ka < \pi$, $ka$ is not a zero of $J_{n+\frac{1}{2}}(z)$ for any positive integer $n$. The formal solution we have found thus holds if the radius of the sphere is sufficiently small. But if, say, $J_{n+\frac{1}{2}}(ka) = 0$, the equation has a solution

$$\frac{J_{n+\frac{1}{2}}(kr)}{\sqrt{(kr)}} S_n(\theta, \phi),$$

which vanishes on $r = a$; and the problem of Dirichlet for $r < a$ does not have a unique solution.

The particular solutions

$$\frac{J_{\frac{1}{2}}(kr)}{\sqrt{(kr)}}, \quad \frac{J_{-\frac{1}{2}}(kr)}{\sqrt{(kr)}},$$

do not depend on $\theta$ and $\phi$, and are proportional to

$$\frac{\sin kr}{r}, \quad \frac{\cos kr}{r}$$

respectively. If $\quad R^2 = (x - x_0)^2 + (y - y_0)^2 + (z - z_0)^2$,

the function $\qquad\qquad \dfrac{\cos kR}{R}$

is the elementary solution of $\nabla^2 u + k^2 u = 0$.

Sometimes it is more convenient to use the Hankel functions

$$H^{(1)}_{n+\frac{1}{2}}(z) = J_{n+\frac{1}{2}}(z) - (-1)^n i J_{-n-\frac{1}{2}}(z)$$

$$H^{(2)}_{n+\frac{1}{2}}(z) = J_{n+\frac{1}{2}}(z) + (-1)^n i J_{-n-\frac{1}{2}}(z),$$

$n$ being a positive integer or zero. These functions are of the form

$$\left(\frac{2}{\pi z}\right)^{\frac{1}{2}} \exp\{i(z - \tfrac{1}{2}(n+1)\pi)\} \left[1 + \sum_{p=1}^{n} \frac{(-1)^p (n+p)!}{p!(n-p)!(2iz)^p}\right],$$

$$\left(\frac{2}{\pi z}\right)^{\frac{1}{2}} \exp\{-i(z - \tfrac{1}{2}(n+1)\pi)\} \left[1 + \sum_{p=1}^{n} \frac{(n+p)!}{p!(n-p)!(2iz)^p}\right].$$

The solutions
$$\frac{H_{\frac{1}{2}}^{(1)}(kr)}{\sqrt{(kr)}}, \quad \frac{H_{\frac{1}{2}}^{(2)}(kr)}{\sqrt{(kr)}}$$

are multiples of
$$\frac{e^{ikr}}{r}, \quad \frac{e^{-ikr}}{r}.$$

## 11.14   Helmholtz's formula

In dealing with problems of wave propagation which vary simple-harmonically with the time, it is often convenient to use a complex wave function
$$U = u\exp(-ikct), \tag{1}$$
where $u$ does not depend on $t$, and to take real parts at the end. If
$$\nabla^2 U = \frac{1}{c^2}\frac{\partial^2 U}{\partial t^2},$$

$u$ is a complex solution of the reduced wave equation
$$\nabla^2 u + k^2 u = 0. \tag{2}$$

The only wave functions in which $u$ depends only on the distance $R$ from a fixed point are
$$\frac{\exp(ikR - ikct)}{R}, \quad \frac{\exp(-ikR - ikct)}{R};$$

the former represents expanding spherical waves, the latter contracting spherical waves. If we wish $u\exp(-ikct)$, where $u$ is a solution of (2), to represent expanding waves, we want $u$ to behave like $\exp(ikr)/r$, when $r$ is large. This is the origin of Sommerfeld's conditions, which appear later in this section.

We have seen that the reduced wave equation (2) has an elementary solution
$$\cos kR/R,$$
where $R$ is the distance from $(x_0, y_0, z_0)$ to $(x, y, z)$. We can, if we wish, use the complex elementary solution
$$\exp(ikR)/R.$$

We first state Helmholtz's formula, which is the analogue for the reduced wave equation of Green's equivalent layer formula in potential theory.

*Let $D$ be the domain bounded by a regular closed surface $S$. Let $u$ be a solution of $\nabla^2 u + k^2 u = 0$ which is continuously differentiable in $\overline{D}$, and has bounded continuous second derivatives in $D$. Then the value of*
$$\frac{1}{4\pi}\iint_S \left\{\frac{\exp(ikR)}{R}\frac{\partial u}{\partial N} - u\frac{\partial}{\partial N}\frac{\exp(ikR)}{R}\right\}dS,$$

*where*
$$R^2 = (x - x_0)^2 + (y - y_0)^2 + (z - z_0)^2$$

*is $u(x_0, y_0, z_0)$ or zero according as $(x_0, y_0, z_0)$ is inside or outside S.*

This is a straightforward consequence of Green's theorem. We note that, if $u$ is real, the value of

$$\frac{1}{4\pi} \iint_S \left\{ \frac{\cos kR}{R} \frac{\partial u}{\partial N} - u \frac{\partial}{\partial N} \frac{\cos kR}{R} \right\} dS$$

is $u(x_0, y_0, z_0)$ or zero according as $(x_0, y_0, z_0)$ is inside or outside S. But the value of

$$\frac{1}{4\pi} \iint_S \left\{ \frac{\sin kR}{R} \frac{\partial u}{\partial N} - u \frac{\partial}{\partial N} \frac{\sin kR}{R} \right\} dS$$

is zero everywhere, as we should expect; for $\sin kR/R$ is a solution of the reduced wave equation without any singularity.

*Let $\Delta$ be the unbounded domain exterior to a regular closed surface S Let $u$ be a solution of $\nabla^2 u + k^2 u = 0$ which is continuously differentiable in $\bar{\Delta}$ and has bounded continuous second derivatives in $\Delta$. Then if $\boldsymbol{n}$ is the unit normal vector to S drawn into $\Delta$, the value of*

$$\frac{1}{4\pi} \iint_S \left( \frac{\exp(ikR)}{R} \frac{\partial u}{\partial n} - u \frac{\partial}{\partial n} \frac{\exp(ikR)}{R} \right) dS$$

*is $-u(x_0, y_0, z_0)$ or zero according as $(x_0, y_0, z_0)$ lies outside or inside S, provided that $u$ satisfies Sommerfeld's conditions at infinity.*

Let $\Sigma$ be the sphere $R = a$, where $a$ is so large that $\Sigma$ contains S. Then the value of

$$\iint_S \left( \frac{\exp(ikR)}{R} \frac{\partial u}{\partial N} - u \frac{\partial}{\partial N} \frac{\exp(ikR)}{R} \right) dS$$

$$+ \int_\Sigma \left( \frac{\exp(ikR)}{R} \frac{\partial u}{\partial N} - u \frac{\partial}{\partial N} \frac{\exp(ikR)}{R} \right) d\Sigma$$

is $4\pi u(x_0, y_0, z_0)$ or zero according as $(x_0, y_0, z_0)$ lies outside or inside S. On S, $N$ is the unit normal vector drawn out of $\Delta$, and so is $-\boldsymbol{n}$. Hence to prove the result, we have to show that the integral over $\Sigma$ tends to zero as $a \to \infty$.

Now the double integral over $\Sigma$ is

$$a^2 \iint_\Omega \left[ \frac{\exp(ikR)}{R} \frac{\partial u}{\partial R} - u \frac{\exp(ikR)}{R} \left( ik - \frac{1}{R} \right) \right]_{R=a} d\Omega,$$

where $d\Omega$ is the element of solid angle. This can be written as

$$\exp(ika)\iint_\Omega \left[R\left(\frac{\partial u}{\partial R}-iku\right)+u\right]_{R=a} d\Omega.$$

Now
$$\iint_\Omega [u]_{R=a} d\Omega$$

tends to zero as $a\to\infty$ if $u\to 0$ as $R\to\infty$ uniformly with respect to the polar angles $\theta,\phi$; in particular, it tends to zero *if Ru is uniformly bounded as $R\to\infty$*, a condition which Sommerfeld called the *finiteness condition*.[†] The other integral tends to zero *if*

$$R\left(\frac{\partial u}{\partial R}-iku\right)$$

*tends to zero as $R\to\infty$, uniformly with respect to the polar angles,* a condition known as Sommerfeld's *radiation condition.* If these two conditions are satisfied, the result follows.

The two conditions lead to a result similar to Liouville's theorem in the theory of functions of a complex variable.

*Let $u$ be a solution of $\nabla^2 u + k^2 u = 0$ which has bounded continuous second derivatives everywhere and which satisfies Sommerfeld's conditions at infinity. Then $u$ is identically zero.*

For let $S$ be any regular closed surface. Then, if $(x_0,y_0,z_0)$ is any point inside $S$,

$$\iint_S \left(\frac{\exp(ikR)}{R}\frac{\partial u}{\partial N}-u\frac{\partial}{\partial N}\frac{\exp(ikR)}{R}\right)dS,$$

where $\partial/\partial N$ is differentiation along the normal out of the domain bounded by $S$, is equal to $4\pi u(x_0,y_0,z_0)$ by Helmholtz's result but is equal to zero by Sommerfeld's conditions. Hence $u$ is zero everywhere.

Sommerfeld's radiation condition implies that $u\exp(-ikct)$, or, rather, its real part, is the velocity potential of an expanding harmonic wave motion. The analogue of Liouville's theorem implies that such a wave motion must be due to sources at which the differentiability conditions do not hold.

### 11.15   The exterior problem of Dirichlet

Let $D$ be the unbounded domain outside a regular closed surface $S$. The exterior problem of Dirichlet for $\nabla^2 u + cu = 0$ is to find a solution which takes given values on $S$ and which has continuous first deriva-

---

† Sommerfeld, *Jahresberichte der D.M.V.* **21** (1912), 309–353. See also Magnus, *ibid.*, **52** (1943), 177–188 and Rellich, *ibid.*, **53** (1943), 57–65.

tives on $\bar{D}$, bounded continuous second derivatives on $D$. The boundary condition does not determine $u$ uniquely; conditions at infinity are needed. To show that a solution, if it exists, is unique, we have to show that, if $u$ is zero everywhere on $S$, $u$ vanishes everywhere on $D$.

We write the equation in the form $\Delta^2 u + k^2 u = 0$, but do not require $k$ to be real, so that the cases $c > 0$ and $c < 0$ are both covered. F. V. Atkinson† proved that, when $k$ is real, Sommerfeld's two conditions at infinity ensure uniqueness. But he also showed that the two conditions can be replaced by one condition which suffices to ensure uniqueness even when $k$ is complex.

Atkinson proves first two lemmas, using the notation of the first paragraph of this section.

**Lemma 1.** *Let $u$ be such that*

(i) $\nabla^2 u + k^2 u = 0$ *outside $S$,*

(ii) $u$ *is continuously differentiable on $\bar{D}$,*

(iii) $u$ *has bounded continuous second derivatives on $D$,*

(iv) $r \exp{(ikr)} \left\{ \left( ik - \dfrac{1}{r} \right) u - \dfrac{\partial u}{\partial r} \right\}$

*tends to zero as $r \to \infty$ uniformly with respect to the polar angles $\theta$ and $\phi$. Then $u$ can be expressed in the form*

$$u = \exp{(ikr)} \sum_1^\infty a_n r^{-n},$$

*where the coefficients $a_n$ are independent of $r$ and are continuous functions of direction. Moreover, there exists a constant $a$ such that the series is absolutely convergent when $r \geqslant a$ and uniformly convergent there with respect to $\theta$ and $\phi$.*

This is not an expansion in terms of the Hankel functions of order $n + \frac{1}{2}$ and surface harmonics of order $n$. A typical term $\exp{(ikr)} a_n r^{-n}$ does not satisfy the reduced wave equation.

If $P_0(x_0, y_0, z_0)$ lies outside $S$, it follows by the argument of §11.14 that, if $u$ satisfies condition (iv) of the lemma

$$u(x_0, y_0, z_0) = \frac{1}{4\pi} \iint_S \left( u \frac{\partial}{\partial n} \frac{\exp{(ikR)}}{R} - \frac{\exp{(ikR)}}{R} \frac{\partial u}{\partial n} \right) dS,$$

where $R$ is the distance from $P_0$ to the integration point $P(x, y, z)$ on $S$ and $\boldsymbol{n}$ is the unit normal vector drawn into $D$. If $OP_0 = r_0, OP = r$, then

$$R^2 = r_0^2 - 2r_0 r \cos{\psi} + r^2,$$

† *Phil. Mag.* **40** (1949), 645–651.

where $\psi$ is the angle $P_0 OP$. This gives

$$R = \frac{1}{\zeta}\sqrt{(1 - 2r\zeta\cos\psi + r^2\zeta^2)},$$

where $\zeta = 1/r_0$. If we suppose, more generally, that $\zeta$ is a complex variable, $R$ has a simple pole at $\zeta = 0$ and branch points at

$$\zeta = \exp(\pm i\psi)/r.$$

We then have

$$\frac{\exp(ikR)}{R} \bigg/ \frac{\exp(ikr_0)}{r_0} = (1 - 2r\zeta\cos\psi + r^2\zeta^2)^{-\frac{1}{2}}$$

$$\times \exp\left[\frac{ik}{\zeta}\{(1 - 2r\zeta\cos\psi + r^2\zeta^2)^{\frac{1}{2}} - 1\}\right].$$

If we take the branch of the square root which is equal to unity when $\zeta = 0$, the expression on the right of this equation is an analytic function of $\zeta$, regular when $|\zeta| < 1/r$. But $S$ lies everywhere at a finite distance from $O$, so that $r \leqslant \frac{1}{2}a$ say, on $S$. The analytic function is then certainly regular when $|\zeta| < 2/a$, and so can be expanded as a power series, which is absolutely convergent when $|\zeta| \leqslant 1/a$, the convergence being uniform with respect to $r$ and $\psi$.

It follows that

$$\left(\frac{\exp(ikr_0)}{r_0}\right)^{-1} \iint_S \frac{\exp(ikR)}{R} \frac{\partial u}{\partial n} dS$$

can be expanded as a power series

$$\sum_0^\infty A_n \zeta^n,$$

where the coefficients depend only on the direction of $OP_0$ and depend on it continuously. Hence, when $r_0 \geqslant a$,

$$\iint_S \frac{\exp(ikR)}{R} \frac{\partial u}{\partial n} dS = \exp(ikr_0)\sum_0^\infty \frac{A_n}{r_0^{n+1}}.$$

A similar argument applies to

$$\iint_S u \frac{\partial}{\partial n}\frac{\exp(ikr)}{r} dS,$$

and the result of the lemma follows. Note that it was not assumed that $k > 0$; the result holds for all real or complex values of $k$.

**Lemma 2.** *Let $u$ be such that*

(i) $\nabla^2 u + k^2 u = 0$ *outside $S$,*

(ii) $u$ *is continuously differentiable on $\bar{D}$,*

(iii) $u$ *has bounded continuous second derivatives on $D$.*

*If* im $k \geqslant 0$, *the conditions*

(A)    $r \exp{(ikr)} \left\{ \left( ik - \dfrac{1}{r} \right) u - \dfrac{\partial u}{\partial r} \right\} \to 0,$

(B)    $r^3 \exp{(-ikr)} \left\{ \left( ik - \dfrac{1}{r} \right) u - \dfrac{\partial u}{\partial r} \right\} = O(1),$

*as $r \to \infty$, are equivalent. They are also equivalent to*

(C)    $u \to 0, \quad r \left( iku - \dfrac{\partial u}{\partial r} \right) \to 0 \; as \; r \to \infty, \; and \, to$

(D)    $ru = O(1), \quad r^2 \left( iku - \dfrac{\partial u}{\partial r} \right) = O(1) \; as \; r \to \infty.$

Write $k = p + iq$, where $p$ and $q$ are real and $q \geqslant 0$. Suppose that $(B)$ holds. Then

$$\left| r \exp{(ikr)} \left\{ \left( ik - \frac{1}{r} \right) u - \frac{\partial u}{\partial r} \right\} \right|$$

$$= \left| r^3 \exp{(-ikr)} \left\{ \left( ik - \frac{1}{r} \right) u - \frac{\partial u}{\partial r} \right\} \frac{\exp{(2ikr)}}{r^2} \right| = O\left( \frac{\exp{(-2qr)}}{r^2} \right)$$

so that $(A)$ holds. If $(A)$ holds, we can apply the result of Lemma 1, to obtain

$$r^3 \exp{(-ikr)} \left\{ \left( ik - \frac{1}{r} \right) u - \frac{\partial u}{\partial r} \right\} = -r^2 \frac{\partial}{\partial r} \left( u \bigg/ \frac{\exp{(ikr)}}{r} \right)$$

$$= -r^2 \frac{\partial}{\partial r} \sum_1^\infty \frac{a_n}{r^{n-1}} = \sum_2^\infty \frac{(n-1) a_n}{r^{n-2}},$$

the series being absolutely and uniformly convergent when $r \geqslant a$. Hence $(B)$ holds.

$(C)$ implies $(A)$ since $|\exp{(ikr)}| = \exp{(-qr)} \leqslant 1$. If $(A)$ holds, then

$$u = \exp{(ikr)} \sum_1^\infty \frac{a_n}{r^n},$$

when $r \geqslant a$, and so $u \to 0$ as $r \to \infty$. Also

$$r \left( iku - \frac{\partial u}{\partial r} \right) = \exp{(ikr)} \sum_1^\infty \frac{n a_n}{r^n},$$

which tends to zero as $r \to \infty$. Thus $(A)$ implies $(C)$. Similarly we can show that $(D)$ implies $(A)$ and $(A)$ implies $(D)$. And thus $(C)$ and $(D)$ are also equivalent.

Thus when im $k \geqslant 0$, Sommerfeld's two conditions can be replaced by one condition $(A)$.

*The only function which satisfies the conditions of Lemma 1 and vanishes everywhere on $S$ is identically zero.* From this follows the uniqueness theorem for the external problem of Dirichlet. We assumed that $k = p + iq$, where $q \geqslant 0$. There are three cases to consider, viz. (i) $p \neq 0, q > 0$, (ii) $p = 0, q > 0$, (iii) $p \neq 0, q = 0$.

(i) Suppose that $p \neq 0, q > 0$, and that there is a solution which vanishes on $S$. Let $\Sigma$ be the sphere $r = b$, where $b > a$. Then $S$ lies inside $\Sigma$. Let $\Delta$ be the domain bounded externally by $\Sigma$, internally by $S$. If $\bar{u}$ is the complex conjugate of $u$,

$$\iiint_{\Delta} (u \nabla^2 \bar{u} - \bar{u} \nabla^2 u)\, dx\, dy\, dz = \iint_{\Sigma} \left( u \frac{\partial \bar{u}}{\partial N} - \bar{u} \frac{\partial u}{\partial N} \right) d\Sigma,$$

since $u$ and $\bar{u}$ vanish on $S$. The expression on the left is

$$(k^2 - \bar{k}^2) \iiint_{\Delta} u\bar{u}\, dx\, dy\, dz = 4ipq \iiint_{\Delta} |u^2|\, dx\, dy\, dz.$$

By Lemma 1,      $u = \exp(ipr - qr) \sum_{1}^{\infty} a_n r^{-n}$,

$$\bar{u} = \exp(-ipr - qr) \sum_{1}^{\infty} \bar{a}_n r^{-n},$$

when $r \geqslant a$, and so

$$u \frac{\partial \bar{u}}{\partial r} - \bar{u} \frac{\partial u}{\partial r} = O\left( \frac{\exp(-2qr)}{r^2} \right)$$

as $r \to \infty$. Therefore the integral over $\Sigma$ tends to zero as $b \to \infty$, and so, since $pq \neq 0$,

$$\iiint_{D} |u|^2 dx\, dy\, dz = 0.$$

Hence $u$ vanishes everywhere on $D$.

(ii) Suppose, next, that $p = 0, q > 0$ so that $k^2 = -q^2$. Let $u$ be a solution which satisfies the differentiability conditions of Lemma 1 and vanishes on $S$. Then

$$\iiint_{\Delta} (u \nabla^2 \bar{u} + u_x \bar{u}_x + u_y \bar{u}_y + u_z \bar{u}_z)\, dx\, dy\, dz = \iint_{\Sigma} u \frac{\partial \bar{u}}{\partial N}\, d\Sigma$$

because $u$ vanishes on $S$. Since

$$u\frac{\partial \bar{u}}{\partial r} = O\left(\frac{\exp(-2qr)}{r^2}\right)$$

as $r \to \infty$, the integral over $\Sigma$ tends to zero as $b \to \infty$. Therefore

$$\iiint_D (q^2|u|^2 + |\operatorname{grad} u|^2)\,dx\,dy\,dz = 0,$$

and so $u$ is again zero everywhere on $\bar{D}$.

(iii) Lastly suppose that $p \neq 0, q = 0$. If $u$ is a solution which vanishes on $S$ and satisfies the differentiability conditions,

$$\iiint_\Delta (u\nabla^2\bar{u} - \bar{u}\nabla^2 u)\,dx\,dy\,dz = \iint_\Sigma \left(u\frac{\partial \bar{u}}{\partial N} - \bar{u}\frac{\partial u}{\partial N}\right)d\Sigma.$$

The triple integral vanishes. By Lemma 1,

$$u\frac{\partial \bar{u}}{\partial r} - \bar{u}\frac{\partial u}{\partial r} = -2ip\sum_1^\infty \frac{a_n}{r^n}\sum_1^\infty \frac{\bar{a}_n}{r^n} - \frac{1}{r}\sum_1^\infty \frac{a_n}{r^n}\sum_1^\infty \frac{n\bar{a}_n}{r^2} - \frac{1}{r}\sum_1^\infty \frac{\bar{a}_n}{r^n}\sum_1^\infty \frac{na_n}{r^n}.$$

If we substitute this in

$$b^2\iint_\Omega \left(u\frac{\partial \bar{u}}{\partial r} - \bar{u}\frac{\partial u}{\partial r}\right)_{r=b} d\Omega = 0$$

and make $b \to \infty$, we find that $|a_1|^2 = 0$. Proceeding in this way, we find that every coefficient $a_n$ is identically zero, and $u$ vanishes everywhere on $\bar{D}$.

### Exercises

**1.** (i) Prove that, if $0 < h < 1, 0 < k < 1$,

$$\int_{-1}^1 \frac{d\mu}{\sqrt{(1-2h\mu+h^2)}\sqrt{(1-2k\mu+k^2)}} = \frac{1}{\sqrt{(hk)}}\log\frac{1+\sqrt{(hk)}}{1-\sqrt{(hk)}}.$$

(ii) The Legendre polynomials have the generating function

$$\frac{1}{\sqrt{(1-2h\mu+h^2)}} = \sum_0^\infty h^n P_n(\mu).$$

Deduce from (i) that $\displaystyle\int_{-1}^1 \{P_n(\mu)\}^2\,d\mu = \frac{2}{2n+1}$

$$\int_{-1}^1 P_m(\mu)P_n(\mu)\,d\mu = 0 \quad (m \neq n).$$

2. $u$ is harmonic in a domain containing $r \leqslant a$ in polar coordinates. On $r = a$, $u = f(\mu)$, where $\mu = \cos\theta$. Prove that

$$u = \sum_0^\infty a_n \frac{r^n}{a^n} P_n(\mu),$$

where $$a_n = \frac{2n+1}{2} \int_{-1}^1 f(\mu) P_n(\mu)\, d\mu.$$

3. Prove that $$\int_0^{2\pi} \frac{du}{z + ix\cos u + iy\sin u} = \frac{2\pi}{r},$$

where $z > 0$. Deduce that, if $r < 1$,

$$\int_0^{2\pi} \frac{du}{1 - z - ix\cos u - iy\sin u} = \frac{2\pi}{\sqrt{(1 - 2\mu r + r^2)}},$$

where $\mu = \cos\theta$. Hence show that

$$P_n(\cos\theta) = \frac{1}{2\pi} \int_0^{2\pi} (\cos\theta + i\sin\theta\cos t)^n\, dt$$

Deduce that $$|P_n(\cos\theta)| \leqslant 1.$$

4. If $$R_0^2 = (x - x_0)^2 + (y - y_0)^2 + (z - z_0)^2,$$
$$R_1^2 = (x - x_0)^2 + (y - y_0)^2 + (z + z_0)^2,$$

where $z_0 > 0$, prove that $$\frac{1}{R_0} - \frac{1}{R_1}$$

is the Green's function for the Laplace's equation in $z > 0$. Show that

$$\frac{1}{R_0} - \frac{1}{R_1} = \sum_1^\infty \frac{r_0^n}{r^{n+1}} \{P_n(\cos\alpha) - P_n(\cos\beta)\},$$

where $$\cos\alpha = \cos\theta\cos\theta_0 + \sin\theta\sin\theta_0\cos(\phi - \phi_0),$$
$$\cos\beta = -\cos\theta\cos\theta_0 + \sin\theta\sin\theta_0\cos(\phi - \phi_0),$$

$(r_0, \theta_0, \phi_0)$, $(r, \theta, \phi)$ being the polar coordinates of $(x_0, y_0, z_0)$ and $(x, y, z)$. Show that the series is absolutely convergent in $r \geqslant a$, for any $a > r_0$, and that the convergence is uniform with respect to $\theta$ and $\phi$.

If $u$ is harmonic in $z > 0$, prove that

$$u(x_0, y_0, z_0) = \frac{z_0}{2\pi} \iint_{-\infty}^\infty \frac{u(x, y, 0)\, dx\, dy}{\{(x - x_0)^2 + (y - y_0)^2 + z_0^2\}^{\frac{3}{2}}}$$

provided that $$\frac{\partial u}{\partial r} + \frac{2u}{r} \to 0$$

as $r \to \infty$ uniformly with respect to $\theta$ and $\phi$.

**5.** Show that, in spherical polar coordinates, Laplace's equation is

$$\frac{1}{r^2}\frac{\partial}{\partial r}\left(r^2\frac{\partial V}{\partial r}\right)+\frac{1}{r^2\sin\theta}\frac{\partial}{\partial\theta}\left(\sin\theta\frac{\partial V}{\partial\theta}\right)+\frac{1}{r^2\sin^2\theta}\frac{\partial^2 V}{\partial\phi^2}=0.$$

By the method of separation of variables, obtain the solution

$$(Ar^n+Br^{-n})\cos(m\phi+\epsilon)\,\Theta_n^m(\mu),$$

where $\mu=\cos\theta$, and $\Theta_n^m$ satisfies

$$(1-\mu^2)\frac{d^2\Theta}{d\mu^2}-2\mu\frac{d\Theta}{d\mu}+\left[n(n+1)-\frac{m^2}{1-\mu^2}\right]\Theta=0.$$

**6.** Show that, in cylindrical coordinates, Laplace's equation is

$$\frac{\partial^2 V}{\partial\rho^2}+\frac{1}{\rho}\frac{\partial V}{\partial\rho}+\frac{1}{\rho^2}\frac{\partial^2 V}{\partial\phi^2}+\frac{\partial^2 V}{\partial z^2}=0.$$

By the method of separation of variables, show that a solution is

$$(A\,e^{kz}+B\,e^{-kz})\,(CJ_m(k\rho)+DY_m(k\rho))\cos(m\phi+\epsilon),$$

where $A$, $B$, $C$, $D$ and $\epsilon$ are constants. If this solution is one-valued and has no singularity, prove that $m$ is an integer and that $D$ is zero.

**7.** Let $u(x,y,z)$ be a non-negative function, harmonic in a bounded domain $D$. Let

$$(x-x_0)^2+(y-y_0)^2+(z-z_0)^2\leqslant a^2$$

be a closed sphere contained in $D$. If $(x,y,z)$ is a point of the sphere at a distance $r<a$ from its centre, prove that

$$\frac{a(a-r)}{(a+r)^2}\,u(x_0,y_0,z_0)\leqslant u(x,y,z)\leqslant\frac{a(a+r)}{(a-r)^2}\,u(x_0,y_0,z_0).$$

Deduce that a non-negative function harmonic in every bounded domain is a constant.

**8.** Prove that Harnack's first and second theorems on convergence, proved in §8.14 and §8.16 for plane harmonic functions, also hold for sequences of harmonic functions in space of three dimensions.

**9.** If $\qquad R_0^2=(x-x_0)^2+(y-y_0)^2+(z-z_0)^2,$

prove that $\qquad\dfrac{e^{ikR_0}}{R_0}=\dfrac{e^{ikr}}{r}\left[1+\sum_{1}^{\infty}a_n\dfrac{r_0^n}{r^n}\right],$

where $a_n$ is a function of

$$\cos\alpha=\cos\theta\cos\theta_0+\sin\theta\sin\theta_0\cos(\phi-\phi_0)$$

alone, $(r_0,\theta_0,\phi_0)$ and $(r,\theta,\phi)$ being the polar coordinates of $(x_0,y_0,z_0)$ and $(x,y,z)$. Prove that $|a_n|\leqslant 1$ and that the series is absolutely convergent when $r\geqslant a$ for any $a>r_0$, and that the convergence is uniform.

If $k > 0$, $z_0 > 0$, show that

$$\frac{e^{ikR_0}}{R_0} - \frac{e^{ikR_1}}{R_1}$$

where $\qquad R_1^2 = (x - x_0)^2 + (y - y_0)^2 + (z + z_0)^2,$

is the Green's function of the equation $\nabla^2 u + k^2 u = 0$ for the half-space $z > 0$. Hence show that if $u$ is continuously differentiable on $z \geqslant 0$, has bounded continuous second derivatives on $z > 0$ and satisfies $\nabla^2 u + k^2 u = 0$, then

$$u(x_0, y_0, z_0) = \frac{z_0}{2\pi} \iint_{-\infty}^{\infty} \frac{e^{ikR}}{R^3} (1 - ikR) u(x, y, 0) \, dx \, dy,$$

where $\qquad R^2 = (x - x_0)^2 + (y - y_0)^2 + z_0^2,$

provided that $\qquad \dfrac{\partial u}{\partial r} + \dfrac{u}{r} (2 - ikr) \to 0$

as $r \to \infty$, uniformly with respect to $\theta$ and $\phi$.

# 12

# THE EQUATION OF HEAT

## 12.1  The equation of conduction of heat

When heat flows along an insulated uniform straight rod of thermal conductivity $K$, density $\rho$ and specific heat $c$, the temperature $u$ at time $t$ at a distance $x$ from a fixed point of the rod satisfies the equation

$$\frac{\partial^2 u}{\partial x^2} = \frac{\rho c}{K} \frac{\partial u}{\partial t}.$$

Since $K, \rho, c$ are constants, this can be written in the form

$$\frac{\partial^2 u}{\partial x^2} = \frac{\partial u}{\partial t}$$

by a change of time-scale. This is the simplest linear equation of parabolic type with two independent variables. It is called the equation of heat or the equation of diffusion. It has one family of characteristics, namely the lines $t = $ constant in the $xt$-plane.

The simplest problem is that of the infinite rod with a given initial temperature distribution $\quad u(x, 0) = f(x).$

On physical grounds, it is obvious that the temperature at any subsequent instant is uniquely determined. The problem is to find conditions satisfied by $f(x)$ so that this is true, and to find an explicit formula for $u$.

## 12.2  A formal solution of the equation of heat

If, in
$$\frac{\partial^2 u}{\partial x^2} = \frac{\partial u}{\partial t}, \tag{1}$$

we put $u = XT$ where $X$ and $T$ are functions of $x$ and $t$ respectively, we have
$$\frac{X''}{X} = \frac{\dot{T}}{T},$$

where dashes and dots denote differentiation with respect to $x$ and $t$. Hence
$$X'' = -a^2 X, \quad \dot{T} = -a^2 T,$$

where $a^2$ is the separation constant. Thus we have a solution

$$u = \exp\left(-a^2(t-t_0)\right)\cos a(x-x_0),$$

where $x_0$ and $t_0$ are constants.

In the physical problems of heat conduction, $u$ cannot increase indefinitely with $t$, so we assume that $a$ is real. A more general solution, valid when $t > t_0$, is

$$u = \int_{-\infty}^{\infty} \exp\left(-a^2(t-t_0)\right)\cos a(x-x_0)\,da$$

$$= \frac{\sqrt{\pi}}{\sqrt{(t-t_0)}}\exp\left\{-\frac{(x-x_0)^2}{4(t-t_0)}\right\}.$$

If $x \neq x_0$, this solution tends to zero as $t \to t_0 + 0$.

Other formal solutions can be obtained by integration. For example,

$$u = \frac{1}{2\sqrt{(\pi t)}}\int_{-\infty}^{\infty} f(\xi)\exp\left(-\tfrac14(x-\xi)^2/t\right)d\xi \tag{2}$$

is a solution valid when $t > 0$. If we put $\xi = x + 2\eta\sqrt{t}$ we obtain

$$u = \frac{1}{\sqrt{\pi}}\int_{-\infty}^{\infty} f(x + 2\eta\sqrt{t})\exp\left(-\eta^2\right)d\eta, \tag{3}$$

when $t > 0$. The limit of this as $t \to +0$ is $f(x)$. Hence (2) is the formal solution of the initial value problem for the infinite rod.

If $f(x)$ is zero when $x < a$ and when $x > b$ where $a < b$, the solution (2) becomes

$$u = \frac{1}{2\sqrt{(\pi t)}}\int_{a}^{b} f(\xi)\exp\left(-\tfrac14(x-\xi)^2/t\right)d\xi.$$

If, in addition, $f(x)$ is positive when $a < x < b$, $u(x,t)$ is positive when $t > 0$ for all values of $x$. The effect of an initial non-zero temperature distribution on a finite interval is immediately felt everywhere. This result is quite different from that for the equation of wave motions where an initial disturbance restricted to a finite interval is propagated with a finite velocity.

It is convenient to write

$$k(x,t) = \frac{1}{2\sqrt{(\pi t)}}\exp\left(-\tfrac14 x^2/t\right),$$

so that the formal solution (2) becomes

$$u(x,t) = \int_{-\infty}^{\infty} f(\xi)\,k(x-\xi,t)\,d\xi, \tag{4}$$

when $t > 0$. This result can be justified if we assume that $f(x)$ possesses a continuous second derivative and satisfies suitable conditions at infinity to ensure the uniform convergence of the integrals obtained from (3) by differentiation under the sign of integration. But the result holds under very much less restrictive conditions.

The solution (4) was obtained by integrating a multiple of

$$k(x-\xi, t-\tau)$$

along a path in the $\xi\tau$-plane. Another formal solution is

$$u(x,t) = \int_0^t \phi(\tau)\, k(x, t-\tau)\, d\tau, \tag{5}$$

where $t > 0$, the upper limit being $t$ since $k(x, t-\tau)$ is complex when $\tau > t$. This solution is an even function of $x$. Its value when $x = 0$ is

$$u(0,t) = \frac{1}{2\sqrt{\pi}} \int_0^t \phi(\tau) \frac{d\tau}{\sqrt{(t-\tau)}}.$$

This is Abel's integral equation for $\phi$. If $u(0,t)$ is continuous and vanishes when $t = 0$, the solution of the integral equation is

$$\phi(t) = \frac{2}{\sqrt{\pi}} \frac{d}{dt} \int_0^t u(0,\tau) \frac{d\tau}{\sqrt{(t-\tau)}}.$$

Thus (5) is a formal solution of the equation of heat in terms of the values taken by the solution when $x = 0$.

Yet another formal solution can be obtained by differentiating the expression on the right of (5) with respect to $x$. It is a multiple of

$$u(x,t) = \int_0^t \phi(\tau) \frac{x}{t-\tau} k(x, t-\tau)\, d\tau, \tag{6}$$

$t$ being positive. The expression (6) is an odd function of $x$. If we make the substitution $t-\tau = \frac{1}{4}x^2/u$ when $x > 0, t > 0$, (6) gives

$$u(x,t) = \frac{1}{\sqrt{\pi}} \int_{\frac{1}{4}x^2/t}^\infty \phi\left(t - \frac{x^2}{4u}\right) u^{-\frac{1}{2}} e^{-u}\, du.$$

The limit of this as $x \to +0$ is $\phi(t)$ when $t > 0$; but the limit as $x \to -0$ is $-\phi(t)$, since $u(x,t)$ is an odd function of $x$.

## 12.3 Use of integral transforms

Problems of heat conduction in an infinite or semi-infinite rod can often be solved by the use of integral transforms. As it is difficult to give this method a rigorous form, we content ourselves with illustrative examples.

To solve the initial value problem $u(x, 0) = f(x)$ for an infinite rod, we may use the complex Fourier transform

$$U(\alpha, t) = \frac{1}{\sqrt{(2\pi)}} \int_{-\infty}^{\infty} u(x, t)\, e^{i\alpha x} dx,$$

whose inverse is

$$u(x, t) = \frac{1}{\sqrt{(2\pi)}} \int_{-\infty}^{\infty} U(\alpha, t)\, e^{-i\alpha x} d\alpha.$$

Assuming that differentiation under the sign of integration is valid, we have

$$\frac{\partial U}{\partial t} = \frac{1}{\sqrt{(2\pi)}} \int_{-\infty}^{\infty} \frac{\partial u(x, t)}{\partial t}\, e^{i\alpha x} dx = \frac{1}{\sqrt{(2\pi)}} \int_{-\infty}^{\infty} \frac{\partial^2 u(x, t)}{\partial x^2}\, e^{i\alpha x} dx$$

$$= -\frac{\alpha^2}{\sqrt{(2\pi)}} \int_{-\infty}^{\infty} u(x, t)\, e^{i\alpha x} dx = -\alpha^2 U(\alpha, t),$$

provided that $u(x, t)$ behaves at $x = \pm\infty$ so that the integrated terms vanish. We then have

$$U(\alpha, t) = U(\alpha, 0) \exp(-\alpha^2 t) = \frac{1}{\sqrt{(2\pi)}} \exp(-\alpha^2 t) \int_{-\infty}^{\infty} f(\xi) \exp i\alpha\xi\, d\xi.$$

Using the formula for inverting the Fourier transform,

$$u(x, t) = \frac{1}{2\pi} \int_{-\infty}^{\infty} \exp(-\alpha^2 t - i\alpha x) \int_{-\infty}^{\infty} f(\xi) \exp(i\alpha\xi)\, d\xi\, d\alpha$$

$$= \frac{1}{2\pi} \int_{-\infty}^{\infty} f(\xi) \int_{-\infty}^{\infty} \exp(-\alpha^2 t - i\alpha(x - \xi))\, d\alpha\, d\xi$$

$$= \frac{1}{2\sqrt{(\pi t)}} \int_{-\infty}^{\infty} f(\xi) \exp(-\tfrac{1}{4}(x - \xi)^2/t)\, d\xi,$$

the result we obtained in §12.2.

To solve the problem for the semi-infinite rod $x > 0$ with data $u(x, 0) = f(x)$, $u(0, t) = \phi(t)$ when $t > 0$, it is more convenient to use the Fourier sine transform

$$U_s(\alpha, t) = \sqrt{\left(\frac{2}{\pi}\right)} \int_0^{\infty} u(x, t) \sin \alpha x\, dx,$$

whose inverse is

$$u(x, t) = \sqrt{\left(\frac{2}{\pi}\right)} \int_0^{\infty} U_s(\alpha, t) \sin \alpha x\, d\alpha.$$

Then

$$\frac{\partial U_s(\alpha, t)}{\partial t} = \sqrt{\left(\frac{2}{\pi}\right)} \int_0^{\infty} \frac{\partial u(x, t)}{\partial t} \sin \alpha x\, dx = \sqrt{\left(\frac{2}{\pi}\right)} \int_0^{\infty} \frac{\partial^2 u(x, t)}{\partial x^2} \sin \alpha x\, dx.$$

Integrating by parts and assuming that $u(x,t)$ behaves suitably at $x = +\infty$, we have

$$\frac{\partial U_s(\alpha, t)}{\partial t} = \sqrt{\left(\frac{2}{\pi}\right)}\, \alpha\phi(t) - \alpha^2 U_s(\alpha, t).$$

This gives

$$U_s(\alpha, t) = U_s(\alpha, 0)\exp\left(-\alpha^2 t\right) + \sqrt{\left(\frac{2}{\pi}\right)}\,\alpha \int_0^t \exp\left(-\alpha^2(t-\tau)\right)\phi(\tau)\,d\tau,$$

where $U_s(\alpha, 0)$ is the Fourier sine transform $F_s(\alpha)$ of $u(x, 0)$ (which is equal to $f(x)$.)

Using the formula for inverting the Fourier sine transform, we have

$$u(x,t) = \frac{2}{\pi}\int_0^\infty \sin\alpha x\exp\left(-\alpha^2 t\right)\int_0^\infty f(\xi)\sin\alpha\xi\,d\xi$$

$$+\frac{2}{\pi}\int_0^\infty \alpha\sin\alpha x\int_0^t \exp\left(-\alpha^2(t-\tau)\right)\phi(\tau)\,d\tau$$

$$=\frac{1}{\pi}\int_0^\infty f(\xi)\int_0^\infty \exp\left(-\alpha^2 t\right)\{\cos\alpha(x-\xi) - \cos\alpha(x+\xi)\}\,d\alpha\,d\xi$$

$$-\frac{2}{\pi}\frac{\partial}{\partial x}\int_0^t \phi(\tau)\int_0^t \exp\left(-\alpha^2(t-\tau)\right)\cos\alpha x\,d\alpha\,d\tau$$

$$=\int_0^\infty f(\xi)\{k(x-\xi, t) - k(x+\xi, t)\}\,d\xi - 2\frac{\partial}{\partial x}\int_0^t \phi(\tau)\,k(x, t-\tau)\,d\tau$$

$$=\int_0^\infty f(\xi)\{k(x-\xi, t) - k(x+\xi, t)\}\,d\xi + \int_0^t \phi(\tau)\frac{x}{t-\tau}k(x, t-\tau)\,d\tau.$$

The Laplace transform of an integrable function $\phi(t)$ is

$$\Phi(s) = \int_0^\infty \phi(t)\,e^{-st}\,dt.$$

If $|\phi(t)| < K e^{at}$, $\Phi(s)$ is an analytic function of the complex variable $s$, regular in $\operatorname{re} s > a$. The inverse of this Laplace transform is

$$\phi(t) = \frac{1}{2\pi i}\int_{c-\infty i}^{c+\infty i}\Phi(s)\,e^{st}\,ds,$$

where $c > a$. The theory of the Laplace transform lacks the symmetry of that of the Fourier transform. We illustrate the theory by considering the problem of heat conduction in a semi-infinite rod $x > 0$ under the conditions

$$u(x, 0) = 0\ (x > 0), \qquad u(0, t) = \phi(t)\ (t > 0).$$

If the Laplace transform of the solution $u(x, t)$ is

$$U(x, s) = \int_0^\infty u(x, t)\, e^{-st}\, dt,$$

we have
$$\frac{\partial^2 U}{\partial x^2} = \int_0^\infty \frac{\partial^2 u(x, t)}{\partial x^2}\, e^{-st}\, dt = \int_0^\infty \frac{\partial u}{\partial t}\, e^{-st}\, dt$$

$$= [u(x, t)\, e^{-st}]_0^\infty + s \int_0^\infty u(x, t)\, e^{st}\, dt.$$

Hence
$$\frac{\partial^2 U}{\partial x^2} - sU = 0$$

provided that the real part of $s$ is so large that $u(x, t)\, e^{-st}$ tends to zero as $t \to +\infty$.

Write $s = \sigma^2$ where the branch of $\sigma$ is taken which is positive when $s > 0$. Then
$$U(x, s) = A\, e^{\sigma x} + B\, e^{-\sigma x}.$$

Since $U(0, s)$ is the Laplace transform of $u(0, t)$, which is given to be $\phi(t)$, we have
$$A + B = \Phi(s).$$

We need another condition in order to determine $A$ and $B$. If we require $u(x, t)$ to be such that $U(x, s)$ is a bounded function of $x$, $A$ is zero, and
$$U(x, s) = \Phi(s)\, e^{-\sigma x},$$

which expresses $U(x, s)$ as the product of the Laplace transforms of

$$\phi(t), \quad \frac{x}{t}\, k(x, t).$$

That $e^{-\sigma x}$ is the Laplace transform of the second function can be found from tables of integral transforms.

It can be shown that, if $\Phi(s)$ and $\Psi(s)$ are the Laplace transforms of $\phi(t)$ and $\psi(t)$, then $\Phi(s)\, \Psi(s)$ is the Laplace transform of

$$\int_0^t \phi(u)\, \psi(t - u)\, du.$$

Hence the solution of the heat conduction problem is

$$u(x, t) = \int_0^t \phi(u)\, \frac{x}{t - u}\, k(x, t - u)\, du.$$

It is impossible to give in a short space an adequate account of integral transform methods: the reader should consult the books cited below.†

## 12.4 Use of Cauchy–Kowalewsky theorem

The equation
$$\frac{\partial^2 u}{\partial x^2} = \frac{\partial u}{\partial t} \tag{1}$$

has, by the Cauchy–Kowalewsky theorem, a unique analytic solution, regular in a neighbourhood of $(x_0, t_0)$, satisfying the conditions

$$u(x_0, t) = \phi(t), \quad u_x(x_0, t) = \psi(t), \tag{2}$$

provided that $\phi(t)$ and $\psi(t)$ are regular in a neighbourhood of $t_0$. If these conditions are satisfied at every point $(x_0, t_0)$ of a finite interval $\gamma$ of the initial line $x = x_0$, the problem has a unique solution regular near $\gamma$.

By a change of origin, we may take $x_0$ and $t_0$ to be zero. By a repeated differentiation of the differential equation, we can calculate all the derivatives of $u$ with respect to $x$, and obtain the Taylor series

$$u(x, t) = \phi(t) + \psi(t)\frac{x}{1!} + \phi_1(t)\frac{x^2}{2!} + \psi_1(t)\frac{x^3}{3!} + \cdots$$
$$+ \phi_n(t)\frac{x^{2n}}{(2n)!} + \psi_n(t)\frac{x^{2n+1}}{(2n+1)!} + \cdots \tag{3}$$

where $\phi_n$ and $\psi_n$ are the $n$th derivatives of $\phi$ and $\psi$.

Let
$$\phi(t) = \sum_0^\infty a_n t^n, \quad \psi(t) = \sum_0^\infty b_n t^n,$$

the series having radii of convergence $R_1$ and $R_2$, say. Let

$$R < \mathrm{Min}\,(R_1, R_2).$$

Since $\phi(t)$ and $\psi(t)$ are analytic functions of the *complex* variable $t$ regular in $|t| \leqslant R$, we have

$$|a_n| \leqslant \frac{M_1}{R^n}, \quad |b_n| \leqslant \frac{M_2}{R^n},$$

where $M_1$ and $M_2$ are the maxima of the moduli of $|\phi(t)|$ and $|\psi(t)|$ respectively on $|t| = R$. Therefore

$$|a_n| \leqslant \frac{M}{R^n}, \quad |b_n| \leqslant \frac{M}{R^n},$$

† E. C. Titchmarsh, *Introduction to the Theory of Fourier Integrals* 2nd edn (Oxford, 1948), pp. 281–8.
D. V. Widder, *The Laplace Transform* (Princeton, 1941).
H. S. Carslaw and J. C. Jaeger, *Operational Methods in Applied Mathematics*, 2nd edn (Oxford, 1948).
C. J. Tranter, *Integral Transforms in Mathematical Physics* (London, 1971).

where $M = \text{Max}\,(M_1, M_2)$. It follows that, when $|t| < R$, the function

$$\Phi(t) = \frac{MR}{R-t}$$

is a majorant for $\phi(t)$ and $\psi(t)$, and so

$$\frac{MR\,n!}{(R-t)^{n+1}}$$

is a majorant for $\phi_n(t)$ and for $\psi_n(t)$.

If we substitute in (3) the Taylor series for $\phi(t)$ and $\psi(t)$ and their derivatives, we obtain a formal double series $S$. The corresponding double series $S'$ obtained by replacing $\phi(t)$ and $\psi(t)$ by $\Phi(t)$ is a majorant of $S$. Typical terms in $S'$ are

$$M\frac{x^{2n}}{(2n)!}\frac{n!}{R^n}{}^{n+1}C_k\left(\frac{t}{R}\right)^k$$

and

$$M\frac{x^{2n+1}}{(2n+1)!}\frac{n!}{R^n}{}^{n+1}C_k\left(\frac{t}{R}\right)^k.$$

The double series $S'$ is therefore absolutely convergent for all values of $|x|$ and for all values of $|t|$ less than $R$. Hence the double series $S$ is absolutely convergent for all values of $|x|$ and for all $|t| < R$, and is uniformly convergent on any bounded closed subset. We can therefore differentiate $S$, and hence also the series (3), term-by-term. It follows that (3) satisfies the equation of heat under the given conditions when $|t| < R$. This verifies the result of the Cauchy–Kowalewsky theorem when

$$u(0,t) = \phi(t), \quad u_x(0,t) = \psi(t),$$

where $\phi(t)$ and $\psi(t)$ are analytic functions of $t$, regular when $|t| < R$.

This has an interesting consequence. When $t = 0$, $u(x,t)$ is equal to $F(x)$, where

$$F(x) = a_0 + b_0\frac{x}{1!} + 1!\,a_1\frac{x^2}{2!} + 1!\,b_1\frac{x^3}{3!} + 2!\,a_2\frac{x^4}{4!} + 2!\,b_2\frac{x^5}{5!} + \dots$$

The absolute value of the coefficient of $x^{2n}$ is

$$\left|\frac{n!}{(2n)!}a_n\right| \le \frac{n!}{(2n)!}\frac{M}{R^n}$$

and of $x^{2n+1}$ is

$$\left|\frac{n!}{(2n+1)!}b_n\right| \le \frac{n!}{(2n+1)!}\frac{M}{R^n}.$$

Hence $F(x)$, regarded as a function of a complex variable $x$, is an integral function.

Evidently
$$F(x) \ll M(1+x) \sum_0^\infty \frac{n!\,n!}{(2n)!} \frac{x^{2n}}{R^n n!}.$$

If
$$A_n = \frac{n!\,n!}{(2n)!} \frac{1}{(KR)^n}$$

we have
$$\frac{A_{n+1}}{A_n} = \frac{(n+1)^2}{(2n+1)(2n+2)} \frac{1}{KR} < \frac{1}{4KR}.$$

If we choose $K$ so that $4KR > 1$, $\{A_n\}$ is a decreasing sequence and so $A_n < 1$. Therefore
$$F(x) \ll M(1+x) \sum_0^\infty \frac{(Kx^2)^n}{n!} = M(1+x)\exp(Kx^2).$$

In particular, when $|x|$ is large,
$$|F(x)| < M\exp(2Kx^2),$$
since
$$|1+x| \leqslant 1+|x| < \exp(Kx^2).$$

Lastly, if we rearrange the double series $S$ as a power-series in $t$, we obtain
$$u(x,t) = F(x) + \sum_1^\infty \frac{t^n}{n!} F_{2n}(x),$$

where $F_n(x)$ is the $n$th derivative of $F(x)$. This is a solution of the equation of heat under the initial condition $u(x,0) = F(x)$.

## 12.5   An example due to Tikhonov

Tikhonov gave an example which showed that the initial value problem for the equation of heat
$$\frac{\partial^2 u}{\partial x^2} = \frac{\partial u}{\partial t}$$

does not have a unique solution unless $u$ satisfies a condition
$$|u(x,t)| < M\exp(ax^2).$$

Under suitable conditions,
$$u(x,t) = \sum_0^\infty \phi_k(t) \frac{x^{2k}}{(2k)!}, \tag{1}$$

where $\phi_k(t)$ is the $k$th derivative of $\phi(t)$, is a solution. In Tikhonov's example,
$$\phi(t) = \exp(-1/t^2).$$

By Cauchy's Theorem,

$$\phi_k(t) = \frac{k!}{2\pi i} \int_\Gamma \frac{\phi(z)}{(z-t)^{k+1}}\, dz,$$

where $\Gamma$ is a simple closed contour in the $z$-plane; the point $t$ is inside $\Gamma$, the origin is outside.

Take $\Gamma$ to be the circle $z = t + \frac{1}{2}t e^{i\theta}$. Then on $\Gamma$

$$\mathrm{re}\,\frac{1}{z^2} \geqslant \frac{4}{81t^2}$$

and so

$$|\phi_k(t)| \leqslant \frac{k!}{2\pi}\exp\left(-\frac{4}{81t^2}\right)\int_0^{2\pi}\left(\frac{2}{t}\right)^k d\theta = \frac{2^k k!}{t^k}\exp\left(-\frac{4}{81t^2}\right).$$

Hence, if $u$ is given by (1), we have

$$|u(x,t)| \leqslant \exp\left(-\frac{4}{81t^2}\right)\sum_0^\infty \frac{2^k k!}{(2k)!}\frac{x^{2k}}{t^k},$$

when $t > 0$. Therefore

$$|u(x,t)| \leqslant \exp\left(-\frac{4}{81t^2}\right)\sum_0^\infty \frac{1}{k!}\frac{x^{2k}}{t^k} = \exp\left(\frac{x^2}{t}-\frac{4}{81t^2}\right).$$

Therefore this solution tends to zero as $t \to +0$ uniformly on any finite interval of the $x$-axis. If we have found a solution of the initial value problem, another solution will be obtained if we add a multiple of Tikhonov's solution. But the resulting solution does not satisfy an inequality $|u(x,t)| \leqslant M\exp(ax^2)$ in $t \geqslant 0$.

## 12.6   The case of continuous initial data

In §12.1, we obtained the formal solution

$$u(x,t) = \frac{1}{2\sqrt{(\pi t)}}\int_{-\infty}^{\infty} f(\xi)\exp\left(-\tfrac{1}{4}(x-\xi)^2/t\right)d\xi, \tag{1}$$

which we expect to satisfy the initial condition $u(x,0) = f(x)$ for all $x$. We now consider the simplest case when $f(x)$ is continuous and satisfies the inequality
$$|f(x)| \leqslant M e^{ax^2} \tag{2}$$
for all $x$, where $a$ is a non-negative constant, and show that $u(x,t)$ tends to $f(x_0)$ as $(x,t)$ tends to $(x_0, +0)$ in any manner.

If $(x,t)$ lies in the rectangle $-\alpha \leqslant x \leqslant \alpha, 0 < \beta \leqslant t \leqslant \gamma$, we have

$$|f(\xi)\exp\left(-\tfrac{1}{4}(x-\xi)^2/t\right)| \leqslant M\exp\left\{a\xi^2 - \frac{1}{4\gamma}(x-\xi)^2\right\}$$

$$\leqslant M\exp\left\{-\left(\frac{1}{4\gamma}-a\right)\xi^2 + \frac{1}{2\gamma}\alpha|\xi|\right\}.$$

Hence, if $\gamma < 1/4a$, the integral (1) converges uniformly on the rectangle, and so $u(x,t)$ is continuous. Since $\alpha$ may be as large as we please, $u(x,t)$ is continuous on the strip $0 < t < 1/4a$. A similar argument shows that we may calculate the derivatives of $u(x,t)$ by differentiation under the sign of integration, and that the derivatives of all orders are continuous in the same strip. Since

$$\frac{1}{\sqrt{t}}\exp\left(-\tfrac{1}{4}(x-\xi)^2/t\right)$$

satisfies the equation of heat, (1) is a solution in the strip $0 < t < 1/4a$. And if (2) holds for every non-negative value of $a$, (1) is a solution in $t > 0$.

We next show that $u(x,t)$ tends to $f(x)$ as $t \to +0$, uniformly with respect to $x$ on any closed interval $-\alpha \leqslant x \leqslant \alpha$. We have

$$u(x,t)-f(x) = \frac{1}{2\sqrt{(\pi t)}}\int_{-\infty}^{\infty}\{f(\xi)-f(x)\}\exp\left(-\tfrac{1}{4}(x-\xi)^2/t\right)d\xi$$

$$= \frac{1}{2\sqrt{(\pi t)}}\int_{-\infty}^{\infty}\{f(x+\eta)-f(x)\}\exp\left(-\tfrac{1}{4}\eta^2/t\right)d\eta. \quad (3)$$

Since $f(x)$ is uniformly continuous, corresponding to any positive value of $\epsilon$, there exists a positive number $\delta$ such that, if $x$ and $x+\eta$ belong to $[-\alpha,\alpha]$, then

$$|f(x+\eta)-f(x)| < \epsilon,$$

whenever $|\eta| < \delta$. Hence

$$\left|\frac{1}{2\sqrt{(\pi t)}}\int_{-\delta}^{\delta}\{f(x+\eta)-f(x)\}\exp\left(-\tfrac{1}{4}\eta^2/t\right)d\eta\right|$$

$$< \frac{\epsilon}{2\sqrt{(\pi t)}}\int_{-\delta}^{\delta}\exp\left(-\tfrac{1}{4}\eta^2/t\right)d\eta < \epsilon$$

for all positive values of $t$. Next

$$\left|\frac{1}{2\sqrt{(\pi t)}}\int_{\delta}^{\infty}\{f(x+\eta)-f(x)\}\exp\left(-\tfrac{1}{4}\eta^2/t\right)d\eta\right|$$

$$\leqslant \frac{1}{2\sqrt{(\pi t)}}|f(x)|\int_{\delta}^{\infty}\exp\left(-\tfrac{1}{4}\eta^2/t\right)d\eta$$

$$+ \frac{1}{2\sqrt{(\pi t)}}\int_{\delta}^{\infty}|f(x+\eta)|\exp\left(-\tfrac{1}{4}\eta^2/t\right)d\eta.$$

The first term does not exceed

$$\frac{1}{2\sqrt{(\pi t)}} M e^{a\alpha^2} \int_\delta^\infty \exp\left(-\tfrac{1}{4}\eta^2/t\right) d\eta$$

$$= \frac{1}{2\sqrt\pi} M \exp\left(a\alpha^2\right) \int_{\delta/\sqrt t}^\infty \exp\left(-\tfrac{1}{4}\zeta^2\right) d\zeta < \epsilon,$$

if $\delta/\sqrt t$ is sufficiently large, that is, if $0 < t < \delta^2\tau_1$ where $\tau_1$ depends on $\alpha$ and $\epsilon$. The second term does not exceed

$$\frac{M}{2\sqrt{(\pi t)}} \int_\delta^\infty \exp\left\{a(x+\eta)^2 - \frac{\eta^2}{4t}\right\} d\eta$$

$$\leqslant \frac{M}{2\sqrt{(\pi t)}} \int_\delta^\infty \exp\left\{a(\eta+\alpha)^2 - \frac{\eta^2}{4t}\right\} d\eta$$

$$= \frac{M e^{a\alpha^2}}{2\sqrt\pi} \int_{\delta/\sqrt t}^\infty \exp\left\{-\tfrac{1}{4}\zeta^2(1-4at) + 2a\alpha\zeta\sqrt t\right\} d\zeta.$$

Hence, if $0 < t < 1/(8a)$, the second term does not exceed

$$\frac{M e^{a\alpha^2}}{2\sqrt\pi} \int_{\delta/\sqrt t}^\infty \exp\left\{-\tfrac{1}{8}\zeta^2 + \tfrac{1}{2}a\zeta\sqrt{(2a)}\right\} d\zeta$$

and this again is less than $\epsilon$ if $\delta/\sqrt t$ is sufficiently large, say if

$$0 < t < \delta^2\tau_2.$$

Similarly for the integral from $-\infty$ to $-\delta$.

We have thus proved that, for every positive value of $\epsilon$, there exists a positive number $t_0$ such that

$$|u(x,t)-f(x)| < 5\epsilon,$$

whenever $0 < t < t_0$, for every $x$ belonging to $[-\alpha,\alpha]$. Hence $u(x,t)$ tends to $f(x)$ as $t \to +0$, uniformly with respect to $x$.

Lastly, suppose that $x_0$ belongs to $[-\alpha,\alpha]$. Then, if $|x-x_0| < \delta$, $0 < t < t_0$,

$$|u(x,t)-f(x_0)| \leqslant |u(x,t)-f(x)| + |f(x)-f(x_0)| < 6\epsilon,$$

so that $u(x,t)$ tends to $f(x_0)$ as $(x,t)$ tends to $(x_0, +0)$ in any manner.

## 12.7 The existence and uniqueness theorem

A. N. Tikhonov† proved the existence and uniqueness theorem for the initial value problem of the equation of heat under quite general

† *Mat. sbornik* **42** (1935), 199–216. This paper is in Russian. The present account is based on D. V. Widder's paper in *Trans. Amer. Math. Soc.* **55** (1944), 85–95.

conditions. The formal solution found in §12.2 is

$$u(x,t) = \int_{-\infty}^{\infty} f(\xi)\,k(x-\xi,t)\,d\xi,$$

where $\qquad k(x,t) = u\tfrac{1}{2}\exp\left(-\tfrac{1}{4}x^2/t\right)/\sqrt{(\pi t)}.$

Let $D$ be the set of points in the $xt$-plane defined by $-R < x < R$, $0 < t \leqslant c$. The lines $x = \pm R, 0 \leqslant t \leqslant c$ and $t = 0, -R \leqslant x \leqslant R$ belong to the frontier of $D$, but do not belong to $D$; we denote them by $B$. Then $\bar{D}$, the closure of $D$, is the union of $D$ and $B$.

A function $u(x,t)$ with continuous first and second derivatives in a given region, which satisfies there the equation of heat, is said for brevity to belong to the set $\mathscr{H}$ in that region.

**Theorem 1.** *If $u(x,t)$ belongs to $\mathscr{H}$ in $D$ and if*

$$\liminf_{(x,t)\to(x_0,t_0)} u(x,t) \geqslant 0$$

*for every point $(x_0, t_0)$ of $B$, then $u(x,t) \geqslant 0$ in $D$.*

It is understood that $(x,t)$ approaches $(x_0,t_0)$ along a path in $D$.

If $(x_0, t_0)$ is any point of $B$, then, for every positive value of $\epsilon$, there exists a positive number $\delta_0$ such that $u(x,t) > -\epsilon$ whenever the point $(x,t)$ of $D$ lies in the open disc with centre $(x_0,t_0)$ and radius $\delta_0$. The family of such open discs, one for each point of $B$ is an infinite open covering of the bounded closed set $B$. By the Heine–Borel theorem, we can choose from this family a finite number of discs which also cover $B$. Hence there exists a positive number $\delta$, the smallest radius of this finite family of discs, such that $u(x,t) > -\epsilon$ whenever the point $(x,t)$ is at a distance less than $\delta$ from $B$.

If the theorem is false, there exists a point $(x_1, t_1)$ of $D$ at which $u$ is negative, so that

$$u(x_1, t_1) = -l < 0.$$

Let $\qquad v(x,t) = u(x,t) + \kappa(t-t_1)$

where $\kappa$ is a constant. Since $u$ satisfies the equation of heat, we have

$$\frac{\partial^2 v}{\partial x^2} = \frac{\partial v}{\partial t} - \kappa.$$

Choose $\kappa$ so that $0 < \kappa < l/t_1$, as we may since $t_1 > 0$. Let

$$0 < \epsilon < l - \kappa t_1,$$

and let $\delta$ be the corresponding positive number found in the preceding paragraph. Then, for all points $(x,t)$ at a distance less than $\delta$ from $B$,

$$v(x,t) > -\epsilon + \kappa(t-t_1) > -l + \kappa t > -l.$$

Since $v(x_1, t_1) = -l$, the minimum of $v(x, t)$ cannot exceed $-l$, and so is attained at a point $(x_2, t_2)$ of $D$. This is impossible. For at a minimum $\partial v/\partial t = 0$ (or possibly $\partial v/\partial t < 0$ if $t_2 = c$), and $\partial^2 v/\partial x^2 \geqslant 0$, yet

$$\frac{\partial^2 v}{\partial x^2} = \frac{\partial v}{\partial t} - \kappa,$$

where $\kappa > 0$. Hence the assumption that $u(x, t)$ is negative at a point of $D$ is false, and the theorem is proved.

**Theorem 2.** *If $u(x, t)$ belongs to $\mathscr{H}$ in the strip $0 < t \leqslant c$, if*

$$\lim_{(x, t) \to (x_0, +0)} u(x, t) = 0$$

*for all $x_0$, and if*  $\qquad \sup_{0 < t \leqslant c} u(x, t) = O(e^{ax^2})$

*as $x \to \pm \infty$, for some positive value of $a$, then $u(x, t)$ is identically zero in the strip.*

Note that we assume that $u(x, t)$ tends to zero as $(x, t)$ tends to $(x_0, +0)$ in any manner in $t > 0$.

Let

$$F(x) = \sup_{0 < t \leqslant c} |u(x, t)|.$$

Then there exists a constant $M$ such that

$$0 \leqslant F(x) \leqslant M e^{ax^2}$$

for all $x$. The function

$$U(x, t) = F(-R)\, k(x + R, t) + F(R)\, k(x - R, t)$$

belongs to $\mathscr{H}$ in $t > 0$, and, in particular, in the region $D$ defined by $-R < x < R, 0 < t \leqslant c$.

Now

$$U(R, t) \geqslant F(R)\, k(0, t) = \frac{F(R)}{2\sqrt{(\pi t)}}.$$

Hence, if $0 < t \leqslant c$,

$$|u(R, t)| \leqslant F(R) \leqslant F(R)\frac{\sqrt{c}}{\sqrt{t}} \leqslant 2U(R, t)\sqrt{(\pi c)}$$

or  $\qquad\qquad 2\sqrt{(\pi c)}\, U(R, t) \pm u(R, t) \geqslant 0.$

A similar result follows with $R$ replaced by $-R$.

Let

$$w_1(x, t) = 2\sqrt{(\pi c)}\, U(x, t) + u(x, t).$$

This function belongs to $\mathscr{H}$ in $0 < t \leqslant c$ and is non-negative when $x = \pm R, 0 < t \leqslant c$. If $(x, t) \to (x_0, +0)$ where $-R < x_0 < R$, then

$w_1(x,t)$ tends to zero. If $(x,t) \to (R,+0)$, then $F(-R)\,k(x+R,t)$ tends to zero and $F(R)\,k(x-R,t)$ is non-negative. Hence

$$\liminf_{(x,t)\to(R,+0)} U(x,t) \geqslant 0.$$

By hypothesis, $\quad \displaystyle\liminf_{(x,t)\to(R,+0)} u(x,t) = \lim u(x,t) = 0.$

Hence $\quad\quad\quad\quad \displaystyle\liminf_{(x,t)\to(R,+0)} w_1(x,t) \geqslant 0,$

with a similar result as $(x,t) \to (-R,+0)$. By Theorem 1, $w_1(x,t) \geqslant 0$ in $D$. In the same way, we can prove that

$$w_2(x,t) = 2\sqrt{(\pi c)}\,U(x,t) - u(x,t) \geqslant 0$$

in $D$. Therefore $\quad |u(x,t)| \leqslant 2\sqrt{(\pi c)}\,U(x,t)$
on $D$.

The function $u(x,t)$ does not depend on $R$. If we can show that $U(x,t)$ tends to zero as $R \to \infty$, we shall have $u(x,t) = 0$. Now

$$F(R)\,k(x-R,t) \leqslant \frac{M}{2\sqrt{(\pi t)}}\exp\left\{aR^2 - \frac{(R-x)^2}{4t}\right\}$$

so that $F(R)\,k(x-R,t)$ tends to zero as $R \to \infty$, provided that

$$0 < t < 1/(4a),$$

with a similar result for $F(-R)\,k(x+R,t)$. Hence if $c \leqslant 1/(4a)$, $u(x,t)$ is zero on the strip $0 < t \leqslant c$. If $c > 1/(4a)$, we can repeat the argument with $u(x,t+1/8a)$, and so forth as often as is necessary; $u(x,t)$ vanishes on $0 < t \leqslant c$ for any given value of $c$.

From this follows Tikhonov's uniqueness theorem:

*If $u_1(x,t)$ and $u_2(x,t)$ belong to $\mathscr{H}$ in $0 < t \leqslant c$, if both tend to $f(x_0)$ as $(x,t) \to (x_0,+0)$ for all values of $x_0$, and if*

$$\sup_{0<t\leqslant c} u_1(x,t) = O(e^{ax^2}), \quad \sup_{0<t\leqslant c} u_2(x,t) = O(e^{ax^2})$$

*as $x \to \pm\infty$ for some positive value of $a$, then $u_1(x,t)$ and $u_2(x,t)$ are identical in the strip.* For $u_1 - u_2$ satisfies the conditions of theorem 2.

**Theorem 3.** *If $f(x)\exp(-ax^2)$ is integrable in Lebesgue's sense, over $(-\infty,\infty)$ for some positive value of $a$, then, on the strip $0 < t < 1/(4a)$, the function*

$$u(x,t) = \frac{1}{2\sqrt{(\pi t)}}\int_{-\infty}^{\infty} f(\xi)\exp\left(-\tfrac{1}{4}(x-\xi)^2/t\right)d\xi$$

*is continuous, has continuous derivatives of all orders and satisfies the equation of heat. If $f(x)\exp(-ax^2)$ is integrable for every positive value of $a$, the conclusions hold in the half-plane $t > 0$.*

Since

$$k(x, t) = \frac{1}{2\sqrt{(\pi t)}}\exp\left(-\tfrac{1}{4}x^2/t\right)$$

satisfies the equation of heat, it suffices to prove that $u$ and its derivatives are continuous and that differentiation under the sign of integration is valid.

Write $f(x) = g(x)\,e^{ax^2}$, so that $g(x)$ is integrable over $(-\infty, \infty)$. On the rectangle $R$ defined by $-\alpha \leqslant x \leqslant \alpha, 0 < t_0 \leqslant t \leqslant t_1 < 1/(4a)$, we have

$$|f(\xi)\,k(x - \xi, t)|$$

$$= |g(\xi)|\frac{1}{2\sqrt{(\pi t)}}\exp\left\{a\xi^2 - \frac{1}{4t}(x - \xi)^2\right\}$$

$$\leqslant |g(\xi)|\frac{1}{2\sqrt{(\pi t_0)}}\exp\left\{a\xi^2 - \frac{1}{4t_1}(x - \xi)^2\right\}$$

$$\leqslant |g(\xi)|\frac{1}{2\sqrt{(\pi t_0)}}\exp\left\{a\xi^2 - \frac{1}{4t_1}(|x| - |\xi|)^2\right\}$$

$$= |g(\xi)|\frac{1}{2\sqrt{(\pi t_0)}}\exp\left\{-\left(\frac{1}{4t_1} - a\right)\xi^2 + \frac{1}{2t_1}|x|\,|\xi| - \frac{1}{4t_1}|x|^2\right\}$$

$$\leqslant |g(\xi)|\frac{1}{2\sqrt{(\pi t_0)}}\exp\left\{-\left(\frac{1}{4t_1} - a\right)\xi^2 + \frac{\alpha}{2t_1}|\xi|\right\}$$

If $0 < t_1 < 1/4a$,

$$\frac{\alpha}{2t_1}|\xi| - \left(\frac{1}{4t_1} - a\right)|\xi|^2 \leqslant \frac{\alpha^2}{4t_1(1 - 4at_1)}.$$

Therefore      $|f(\xi)\,k(x - \xi, t)| \leqslant C|g(\xi)|,$

where $C$ is a constant.

Since $g(\xi)$ is integrable, the integral defining $u(x, t)$ is uniformly convergent on $R$. Hence $u(x, t)$ is continuous on $R$ for all $\alpha$, $t_0$ and $t_1$; therefore $u(x, t)$ is continuous on $0 < t < 1/4a$.

Next, if $(x, t)$ and $(x + h, t)$ belong to $R$,

$$\frac{u(x + h, t) - u(x, t)}{h} = \int_{-\infty}^{\infty} g(\xi)\exp(a\xi^2)\frac{k(x + h - \xi, t) - k(x - \xi, t)}{h}\,d\xi.$$

By the mean-value theorem,

$$\frac{u(x + h, t) - u(x, t)}{h} = \int_{-\infty}^{\infty} g(\xi)\exp(a\xi^2)\,k_x(x + \theta h - \xi, t)\,d\xi,$$

where $0 < \theta < 1$. By an argument similar to that just used, there exists a constant $C_1$ such that

$$\left| \exp(a\xi^2) \, k_x(x + \theta h - \xi, t) \right| \leqslant C_1.$$

By Lebesgue's dominated convergence theorem, the last integral tends to

$$\int_{-\infty}^{\infty} g(\xi) \exp(a\xi^2) \, k_x(x - \xi, t) \, d\xi,$$

as $h \to 0$. Hence $\partial u/\partial x$ exists, and is given by

$$\frac{\partial u(x, t)}{\partial x} = \int_{-\infty}^{\infty} f(\xi) \, k_x(x - \xi, t) \, d\xi,$$

from which the continuity of $\partial u/\partial x$ follows. Similar methods apply to derivatives of all orders. We can therefore differentiate under the sign of integration; and as $k(x - \xi, t)$ satisfies the equation of heat, so also does $u(x, t)$.

**Theorem 4.** *If, in theorem 3, the limits $f(x_0 + 0)$ and $f(x_0 - 0)$ exist,*

$$\limsup_{(x, t) \to (x_0, +0)} |u(x, t)| \leqslant \max\{|f(x_0 + 0)|, |f(x_0 - 0)|\}.$$

*If $f(x)$ is continuous at $x_0$,*

$$\lim_{(x, t) \to (x_0 + 0)} u(x, t) = f(x_0).$$

Let $M$ be the greater of $|f(x_0 \pm 0)|$. Then, for every positive value of $\epsilon$, there exists a positive number $\delta$ such that $|f(x)| < M + \epsilon$ whenever $|x - x_0| < \delta$. We divide up the range of integration and write

$$u(x, t) = \int_{-\infty}^{x_0 - \delta} + \int_{x_0 - \delta}^{x_0 + \delta} + \int_{x_0 + \delta}^{\infty} f(\xi) \, k(x - \xi, t) \, d\xi = I_1 + I_2 + I_3,$$

and we shift the origin so that $x_0$ is zero. Then $|x| < \delta$.

The term $I_2$ is easy to handle, since

$$|I_2| < (M + \epsilon) \int_{-\delta}^{\delta} k(x - \xi, t) \, d\xi < (M + \epsilon) \int_{-\infty}^{\infty} k(x - \xi, t) \, d\xi = M + \epsilon.$$

Next, if $0 < 4at < 1$,

$$\exp(a\xi^2) \, k(x - \xi, t) = \frac{1}{2\sqrt{(\pi t)}} \exp\left(a\xi^2 - \tfrac{1}{4}(x - \xi)^2/t\right)$$

is a decreasing function of $\xi$ when $\xi > x/(1 - 4at)$. Hence if $|x| < \rho < \delta$, the function is decreasing when $\xi \geqslant \delta$, and so

$$\exp(a\xi^2) \, k(x - \xi, t) \leqslant \exp(a\delta^2) \, k(\delta - x, t) \leqslant \exp(a\delta^2) \, k(\delta - \rho, t).$$

Therefore

$$|I_3| \leqslant \exp(a\delta^2)\, k(\delta - \rho, t) \int_\delta^\infty \exp(-a\xi^2)\, |f(\xi)|\, d\xi.$$

Thus $I_3$ tends to zero as $t \to +0$. The integral $I_1$ does also. Hence

$$\limsup_{(x,t)\to(x_0,+0)} |u(x,t)| \leqslant M + \epsilon.$$

As $\epsilon$ is arbitrary,

$$\limsup_{(x,t)\to(x_0,+0)} |u(x,t)| \leqslant M = \max\{|f(x_0+0)|, |f(x_0-0)|\}.$$

If $f(x)$ is continuous at $x_0$, let

$$u_1(x,t) = u(x,t) - f(x_0).$$

Then 
$$u_1(x,t) = \int_{-\infty}^\infty \{f(\xi) - f(x_0)\}\, k(x - \xi, t)\, d\xi.$$

By the first part of the theorem,

$$\limsup_{(x,t)\to(x_0,+0)} |u_1(x,t)| \leqslant 0.$$

Hence, as $(x,t)$ tends to $(x_0, +0)$, $u_1(x,t)$ tends to zero and so $u(x,t)$ tends to $f(x_0)$.

A similar result, due to G. H. Hardy,[†] is that, *for fixed $x$, $u(x,t)$ tends to $f(x)$ as $t \to +0$ for almost all values of $x$ and that it tends to* $\frac{1}{2}\{f(x+0) + f(x-0)\}$ *whenever this expression has a meaning.*

## 12.8   The equation of heat in two and three dimensions

The function 
$$k(x, y; t) = \frac{1}{4\pi t} \exp\left(-\frac{x^2 + y^2}{4t}\right)$$

satisfies the equation 
$$\frac{\partial^2 u}{\partial x^2} + \frac{\partial^2 u}{\partial y^2} = \frac{\partial u}{\partial t}.$$

From this we can construct the formal solution

$$u(x,y,t) = \frac{1}{4\pi t} \iint_{-\infty}^\infty f(\xi, \eta) \exp\left\{-\frac{(x-\xi)^2 + (y-\eta)^2}{4t}\right\} d\xi\, d\eta.$$

If we put 
$$\xi = x + 2X\sqrt{t}, \quad \eta = y + 2Y\sqrt{t},$$

we obtain

$$u(x,y,t) = \frac{1}{\pi} \int_{-\infty}^\infty f(x + 2X\sqrt{t}, y + 2Y\sqrt{t}) \exp(-X^2 - Y^2)\, dX\, dY.$$

As $t \to +0$, this solution tends to $f(x,y)$ under suitable conditions.

† *Mess. Math.*, **46** (1916), 43–48.

Similarly, a formal solution of the initial value problem for

$$\frac{\partial^2 u}{\partial x^2} + \frac{\partial^2 u}{\partial y^2} + \frac{\partial^2 u}{\partial z^2} = \frac{\partial u}{\partial t}$$

is $\quad u(x, y, z, t) = \dfrac{1}{8(\pi t)^{\frac{3}{2}}} \displaystyle\iiint_{-\infty}^{\infty} f(\xi, \eta, \zeta) \exp\left(-\tfrac{1}{4}R^2/t\right) d\xi\, d\eta\, d\zeta,$

where $\quad R^2 = (x-\xi)^2 + (y-\eta)^2 + (z-\zeta)^2.$

## 12.9  Boundary conditions

In the problem of heat conduction in a finite rod, there are, in addition to the initial condition, boundary conditions at the end points of the rod. Similar problems arise in the theory of heat conduction in the plane or in space.

Suppose that we have a conducting solid bounded by a closed surface $S$. The temperature $u$ satisfies

$$\frac{\partial^2 u}{\partial x^2} + \frac{\partial^2 u}{\partial y^2} + \frac{\partial^2 u}{\partial z^2} = \frac{\partial u}{\partial t}$$

and is given initially everywhere inside $S$. There are three possible types of boundary condition on $S$.

(i) The temperature may be prescribed on $S$ for all time.

(ii) There may be no flow of heat across $S$ so that $\partial u/\partial N$ vanishes on $S$.

(iii) If the flux of heat across $S$ is proportional to the difference between the temperature at the surface and the temperature $u_0$ of the surrounding medium, it is equal to $H(u_0 - u)$ where $H$ is a positive constant. The boundary condition is then

$$K\frac{\partial u}{\partial N} = H(u_0 - u),$$

where $\partial/\partial N$ is differentiation along the outward normal, and $K$ is a positive constant. We write this as

$$\frac{\partial u}{\partial N} + hu = hu_0,$$

where $h$ is a positive constant.

If the solid is bounded externally by a closed surface $S_1$, internally by a closed surface $S_2$, we could have different types of boundary condition on $S_1$ and $S_2$.

## 12.10   The finite rod

In the case of an insulated uniform rod of finite length $a$, we have to show that

$$\frac{\partial^2 u}{\partial x^2} = \frac{\partial u}{\partial t}$$

has a unique solution which satisfies the initial condition $u = f(x)$, when $t = 0, 0 < x < a$, and also satisfies conditions at the ends.

There are the four types of condition at the ends.

(i) The temperature is given at the ends by

$$u(0, t) = g_0(t), \quad u(a, t) = g_1(t),$$

when $t > 0$. We do not require $u(x, t)$ to tend to a limit as $(x, t)$ tends to $(0, +0)$ in any manner or as $(x, t)$ tends to $(a, +0)$. For example, we might have the problem of a finite rod at a unique temperature whose end points are suddenly cooled to zero. The conditions are then

$$u(x, 0) = 1 \quad (0 < x < a),$$

$$u(0, t) = 0, \quad u(a, t) = 0 \quad (t > 0).$$

This is the idealisation of a real problem in which the cooling of the ends of the rod takes place in a very short time.

(ii) $\dfrac{\partial u}{\partial x} - h_0 u = g_0(t) \, (x = 0, t > 0), \quad \dfrac{\partial u}{\partial x} + h_1 u = g_1(t) \, (x = a, t > 0),$

where $h_0$ and $h_1$ are non-negative constants.

(iii) $u = g_0(t) \, (x = 0, t > 0), \quad \dfrac{\partial u}{\partial x} + h_1 u = g_1(t) \, (x = a, t > 0);$

(iv) $\dfrac{\partial u}{\partial x} - h_0 u = g_0(t) \, (x = 0, t > 0), \quad u = g_1(t) \, (x = a, t > 0).$

To prove uniqueness, we have to show that, if $f$, $g_0$ and $g_1$ are all zero, $u(x, t)$ is identically zero for all $t > 0$.

Let $D$ be the rectangle $0 < x < a, 0 < t < b$, and let $\Gamma$ be its boundary. In the relation

$$\iint_D (\phi_{xx} - \phi_t) \, dx \, dt = \int_\Gamma (l\phi_x - m\phi) \, ds,$$

where $(l, m)$ are the direction cosines of the outward normal, put $\phi = \frac{1}{2}u^2$, where $u$ satisfies the equation of heat. Then

$$\iint_D u_x^2 \, dx \, dt = \int_\Gamma (luu_x - \tfrac{1}{2}mu^2) \, ds.$$

If $f, g_0$ and $g_1$ are all zero, this equation becomes

(i) $$\iint_D u_x^2\,dx\,dt + \tfrac{1}{2}\int_0^a \{u(x,b)\}^2\,dx = 0,$$

(ii) $$\iint_D u_x^2\,dx\,dt + \tfrac{1}{2}\int_0^a \{u(x,b)\}^2\,dx$$
$$+ h_0\int_0^b \{u(0,t)\}^2\,dt + h_1\int_0^b \{u(a,t)\}^2\,dt = 0,$$

(iii) $$\iint_D u_x^2\,dx\,dt + \tfrac{1}{2}\int_0^a \{u(x,b)\}^2\,dx + h_1\int_0^b \{u(a,t)\}^2\,dt = 0,$$

(iv) $$\iint_D u_x^2\,dx\,dt + \tfrac{1}{2}\int_0^a \{u(x,b)\}^2\,dx + h_0\int_0^b \{u(0,t)\}^2\,dt = 0,$$

in the four cases. Since $h_0$ and $h_1$ are not negative, we have $u(x,b) = 0$ when $0 < x < a$, for every positive value of $b$. Hence $u(x,t)$ vanishes when $0 < x < a, t > 0$, which proves the uniqueness theorem.

## 12.11    The semi-infinite rod

In the problem of the semi-infinite rod, there is an initial condition, an end condition and an order condition at infinity. The initial condition we take to be $$u(x,0) = f(x) \quad (x > 0),$$

where $f(x)\,e^{-ax^2}$ is integrable over $(0,\infty)$ for every positive value of $a$. There are various end conditions. We start by considering the cases when either

(i) $u = \phi(t) \quad (x = 0, t > 0),$

or        (ii) $u_x = \psi(t) \quad (x = 0, t > 0),$

where $\phi(t)$ and $\psi(t)$ are integrable over any finite interval.

The first problem can be split into two simpler problems $(a)$ when $\phi(t)$ is identically zero, $(b)$ when $f(x)$ is identically zero. The required solution is the sum of the solutions of the two simpler problems.

The solution of problem $(a)$ is an odd function of $x$. To solve it, we extend the initial data to $x < 0$ by introducing a new function

$$F(x) = f(x)\,(x > 0), \quad F(x) = -f(-x)\,(x < 0).$$

We can assign to $F(0)$ any value we please. The problem with this initial function has a unique solution

$$u_1(x,t) = \int_{-\infty}^{\infty} F(\xi)\,k(x-\xi,t)\,d\xi.$$

This is an odd function of $x$, continuous when $t > 0$, and so $u_1(0, t) = 0$ there. At every point of continuity of $F(x)$, $u_1(x, t) \to F(x_0)$ as $(x, t) \to (x_0, +0)$ in any manner. The solution of problem $(b)$ has already been found; it is

$$u_2(x, t) = \int_0^t \phi(\tau) \frac{x}{t - \tau} k(x, t - \tau) \, d\tau.$$

Hence the solution of problem (i) is

$$u(x, t) = \int_0^\infty f(\xi) \{k(x - \xi, t) - k(x + \xi, t)\} \, d\xi + \int_0^t \phi(\tau) \frac{x}{t - \tau} k(x, t - \tau) \, d\tau.$$

The solution of problem (ii) can also be split into two simpler problems, $(c)$ when $\psi(t)$ is identically zero, $(d)$ when $f(x)$ is identically zero. To solve problem $(c)$, we introduce a new function

$$G(x) = f(x) \, (x > 0), \quad G(x) = f(-x) \, (x < 0).$$

The solution is    $u_1(x, t) = \int_{-\infty}^\infty G(\xi) \, k(x - \xi, t) \, d\xi.$

At any point of continuity $x_0$ of $G(x)$, this tends to $G(x_0)$ as $t \to +0$. In $t > 0$, $u_1(x, t)$ is continuous and continuously differentiable. Hence

$$\frac{\partial u_1(x, t)}{\partial x} = -\int_{-\infty}^\infty G(\xi) \frac{x - \xi}{2t} k(x - \xi, t) \, d\xi.$$

When $x = 0, t > 0$, this is equal to

$$\int_{-\infty}^\infty G(\xi) \frac{\xi}{2t} k(\xi, t) \, d\xi,$$

which vanishes since $G(\xi)$ is an even function. The solution of problem $(d)$ is

$$u_2(x, t) = -2 \int_0^t k(x, t - \tau) \, \psi(\tau) \, d\tau.$$

This vanishes when $t = 0$. Also

$$\frac{\partial u_2(x, t)}{\partial x} = \int_0^t \psi(\tau) \frac{x}{t - \tau} k(x, t - \tau) \, d\tau,$$

and this tends to $\psi(t_0)$ as $(x, t)$ tends to $(+0, t_0)$ in any manner when $t_0 > 0$. The solution of problem (ii) is

$$u(x, t) = \int_0^\infty f(\xi) \{k(x - \xi, t) + k(x + \xi, t)\} \, d\xi - 2 \int_0^t \psi(\tau) \, k(x, t - \tau) \, d\tau.$$

When the end condition for the semi-infinite rod was $u = 0$ or $u_x = 0$ when $x = 0$, we started with the solution

$$u(x,t) = \int_0^\infty f(\xi)\,k(x-\xi,t)\,d\xi + \int_0^\infty g(\xi)\,k(x+\xi,t)\,d\xi$$

of the equation of heat and chose the function $g(\xi)$ appropriately. The same method can be applied when the end condition is $u_x - hu = 0$ where $h > 0$.

When $x = 0$,

$$u_x = \int_0^\infty f(\xi)\frac{\xi}{2t}k(\xi,t)\,d\xi - \int_0^\infty g(\xi)\frac{\xi}{2t}k(\xi,t)\,d\xi$$

$$= -\int_0^\infty \{f(\xi)-g(\xi)\}\frac{\partial}{\partial\xi}k(\xi,t)\,d\xi.$$

This we can integrate by parts if $f(\xi)$ and $g(\xi)$ are absolutely continuous. For then the derivatives $f'$ and $g'$ exist almost everywhere. We then have

$$u_x(0,t) = -[\{f(\xi)-g(\xi)\}k(\xi,t)]_0^\infty + \int_0^\infty \{f'(\xi)-g'(\xi)\}k(\xi,t)\,d\xi$$

$$= \int_0^\infty \{f'(\xi)-g'(\xi)\}k(\xi,t)\,d\xi$$

provided that the terms at the limits vanish. We can ensure this by choosing $g(\xi)$ so that $g(0) = f(0)$ and

$$\lim_{\xi\to\infty} \{f(\xi)-g(\xi)\}k(\xi,t) = 0.$$

The end condition is then

$$\int_0^\infty \{f'(\xi)-hf(\xi)-g'(\xi)-hg(\xi)\}k(\xi,t)\,d\xi = 0,$$

for $t > 0$. This is satisfied if $g(\xi)$ satisfies the differential equation

$$g' + hg = f' - hf.$$

The solution is

$$g(\xi)e^{h\xi} - g(0) = \int_0^\xi \{f'(\eta)-hf(\eta)\}e^{h\eta}\,d\eta$$

$$= f(\xi)e^{h\xi} - f(0) - 2h\int_0^\xi f(\eta)e^{h\eta}\,d\eta.$$

Since $g(0) = f(0)$,

$$f(\xi)-g(\xi) = 2he^{-h\xi}\int_0^\xi f(\eta)e^{h\eta}\,d\eta.$$

If       $|f(\xi)| < Me^{a\xi^2}$

for every positive value of $a$,

$$|f(\xi) - g(\xi)| \, k(\xi, t) < \frac{Mh}{\sqrt{(\pi t)}} \exp\left(-\xi^2/4t - h\xi\right) \int_0^\xi \exp\left(a\eta^2 + h\eta\right) d\eta$$

$$< \frac{Mh}{\sqrt{(\pi t)}} \, \xi \exp\left(-\xi^2/4t + a\xi^2\right),$$

which tends to zero as $\xi \to \infty$ when $t < 1/4a$. As $a$ can be as small as we please, $|f(\xi) - g(\xi)| \, k(\xi, t)$ tends to zero as $\xi \to \infty$ for any positive value of $t$.

The solution of the equation of heat we are trying to obtain is thus

$$u(x, t) = \int_0^\infty f(\xi) \left\{ k(x - \xi, t) + k(x + \xi, t) \right\} d\xi$$

$$- 2h \int_0^\infty k(x + \xi, t) \, e^{-h\xi} \int_0^\xi e^{h\eta} f(\eta) \, d\eta \, d\xi.$$

Inverting the order of integration

$$u(x, t) = \int_0^\infty f(\xi) \left\{ k(x - \xi, t) + k(x + \xi, t) \right\} d\xi - 2h \int_0^\infty f(\eta) \, K(x + \eta, t) \, d\eta,$$

where $$K(x, t) = \int_0^\infty k(x + \xi, t) \, e^{-h\xi} d\xi.$$

$K(x, t)$ can be expressed in terms of known functions. For

$$K(x, t) = \frac{1}{2\sqrt{(\pi t)}} \int_0^\infty \exp\left\{ -\frac{(x + \xi)^2}{4t} - h\xi \right\} d\xi$$

$$= \frac{1}{2\sqrt{(\pi t)}} \, e^{hx} \int_x^\infty \exp\left\{ -\frac{\eta^2}{4t} - h\eta \right\} d\eta$$

$$= \frac{1}{2\sqrt{(\pi t)}} \exp\left\{ hx + h^2t \right\} \int_x^\infty \exp\left\{ -\frac{(\eta + 2ht)^2}{4t} \right\} d\eta$$

$$= \frac{1}{\sqrt{\pi}} \exp\left\{ hx + h^2t \right\} \int_y^\infty \exp\left\{ -\tau^2 \right\} d\eta,$$

where $$y = \frac{x}{2\sqrt{t}} + h\sqrt{t}.$$

Hence $$K(x, t) = \frac{1}{\sqrt{\pi}} \exp\left\{ hx + h^2t \right\} \mathrm{Erfc}\left( \frac{x}{2\sqrt{t}} + h\sqrt{t} \right),$$

where $\mathrm{Erfc}\, x$ is the Error Function.

## 12.12 The finite rod again

In the problem of heat conduction in a finite rod we have to solve the equation of heat in $0 < x < a, t > 0$, given certain end conditions and an initial condition $u(x, 0) = f(x)$, where $f(x)$ is bounded and integrable. Suppose that the temperature is assigned at the ends, so that

$$u(0, t) = \phi(t), \quad u(a, t) = \psi(t),$$

when $t > 0$, where $\phi$ and $\psi$ are bounded and integrable over any finite interval. This problem can be split up into three problems: in problem (a), $\phi$ and $\psi$ are zero; in problem (b) $f$ and $\psi$ are zero; in problem (c), $f$ and $\phi$ are zero.

We solve problem (a) by extending the initial data, just as we did in the case of the semi-infinite rod. Since $u(0, t) = 0$, let $u(x, t)$ be an odd function of $x$; since $u(a, t) = 0$, let $u(a+x, t)$ be an odd function of $x$. We consider then the infinite rod with

$$F(x) = f(x) \quad (0 < x < a),$$

$$F(-x) = -F(x)$$

$$F(a+x) = -F(a-x).$$

The second and third equations give

$$F(a+x) = F(x-a)$$

and so $$F(x) = F(x+2a).$$

The extended initial data satisfy

$$F(x) = f(x)\,(0 < x < a), \quad F(-x) = -F(x), \quad F(x+2a) = F(x).$$

Since $f(x)$ is bounded, $F(x)$ certainly satisfies the condition

$$|F(x)| < M\,e^{cx^2}$$

for every positive value of $c$, on which the theory of the infinite rod depended. The solution of problem (a) is then

$$u_1(x, t) = \int_{-\infty}^{\infty} k(x-\xi, t)\,F(\xi)\,d\xi.$$

This function is continuous in $t > 0$ for all values of $x$, has there continuous derivatives of all orders and satisfies the equation of heat. As $(x, t)$ tends to $(x_0, +0)$ in any manner, $u_1(x, t)$ tends to $F(x_0)$ at every point of continuity, and so does so almost everywhere. In particular it tends to $f(x_0)$ almost everywhere on $0 < x < a$.

When $t > 0$,
$$u_1(0, t) = \int_{-\infty}^{\infty} k(\xi, t) F(\xi) d\xi = 0$$

since $F$ is an odd function. Also

$$u_1(a, t) = \int_{-\infty}^{\infty} k(a - \xi, t) F(\xi) d\xi$$

$$= \int_{-\infty}^{\infty} k(\eta, t) F(a - \eta) d\eta = 0$$

since $F(a - \eta)$ is an odd function.

This solution can be written as

$$u_1(x, t) = \sum_{-\infty}^{\infty} \int_{(2n-1)a}^{(2n+1)a} F(\xi) k(x - \xi, t) d\xi$$

$$= \sum_{-\infty}^{\infty} \int_{-a}^{a} F(\xi) k(x - \xi - 2na, t) d\xi$$

$$= \sum_{-\infty}^{\infty} \int_{0}^{a} f(\xi) \{k(x - \xi - 2na, t) - k(x + \xi - 2na, t)\} d\xi$$

$$= \int_{0}^{a} f(\xi) K(x, \xi, t) d\xi,$$

where

$K(x, \xi, t)$
$$= \frac{1}{2\sqrt{(\pi t)}} \sum_{-\infty}^{\infty} (\exp\{-\tfrac{1}{4}(x - \xi - 2na)^2/t\} - \exp\{-\tfrac{1}{4}(x + \xi - 2na)^2/t\}).$$

Inversion of the order of integration and summation is readily justified. The function $K$ can be expressed in terms of Jacobi's theta function†

$$\vartheta_3(z \,|\, \tau) = \sum_{-\infty}^{\infty} \exp(\pi i \tau n^2 - 2niz)$$

where im $\tau > 0$. In this notation,

$$\sum_{-\infty}^{\infty} \exp\left(-\frac{(x - 2na)^2}{4t}\right) = \exp\{-\tfrac{1}{4}x^2/t\} \vartheta_3\left(\frac{iax}{2t} \,\Big|\, \frac{ia^2}{\pi t}\right)$$

and this is equal to
$$\frac{\sqrt{(\pi t)}}{a} \vartheta_3\left(\frac{\pi x}{2a} \,\Big|\, \frac{i\pi t}{a^2}\right)$$

by Jacobi's imaginary transformation. Hence

$$K(x, \xi, t) = \frac{1}{2a}\left\{\vartheta_3\left(\frac{\pi(x - \xi)}{2a} \,\Big|\, \frac{i\pi t}{a^2}\right) - \vartheta_3\left(\frac{\pi(x + \xi)}{2a} \,\Big|\, \frac{i\pi t}{a^2}\right)\right\}.$$

† See Whittaker and Watson, *Modern Analysis*, pp. 464 and 474.

The last formula gives

$$K(x, \xi, t) = \frac{2}{a} \sum_{1}^{\infty} \exp\left(-n^2\pi^2 t/a^2\right) \sin\frac{n\pi x}{a} \sin\frac{n\pi\xi}{\alpha}.$$

Hence we have the Fourier solution

$$u_1(x, t) = \sum_{1}^{\infty} b_n \exp\left(-n^2\pi^2 t/a^2\right) \sin\frac{n\pi x}{a},$$

where
$$b_n = \frac{2}{a} \int_0^a f(\xi) \sin\frac{n\pi\xi}{a} \, d\xi,$$

which we discuss later.

The second problem is that in which $u$ vanishes when $0 < x < a, t = 0$ and when $x = a, t > 0$, but $u = \phi(t)$ when $x = 0, t > 0$. We have seen that

$$u_0(x, t) = \int_0^t \frac{x}{t-\tau} \phi(\tau) \, k(x, t-\tau) \, d\tau$$

is a solution of the equation of heat which is continuously differentiable as often as we please in $x > 0, t > 0$ and in $x < 0, t > 0$. It vanishes when $t = 0$ and is an odd function of $x$. It is discontinuous across $x = 0, t > 0$; for it tends to $\phi(t_0)$ at every point of continuity of $\phi(t)$ as $(x, t)$ tends to $(+0, t_0)$, but it tends to $-\phi(t_0)$ as $(x, t) \to (-0, t_0)$.

Consider the function

$$u_2(x, t) = \sum_{-\infty}^{\infty} u_0(x + 2na, t).$$

Since $\phi(t)$ is bounded, say $|\phi(t)| < M$, we have, when

$$0 \leqslant x \leqslant a, \quad 0 < t_0 \leqslant t \leqslant t_1, \quad n > 0,$$

$$|u_0(x + 2na, t)| \leqslant (2n+1) M a \int_0^t k(2na, \tau) \frac{d\tau}{\tau}$$

$$= \frac{(2n+1) M a}{2\sqrt{\pi}} \int_0^t \exp\left(-n^2 a^2/\tau\right) \frac{d\tau}{\tau^{\frac{3}{2}}}$$

$$< \frac{(2n+1) M a \sqrt{t}}{2\sqrt{\pi}} \int_0^t \exp\left(-n^2 a^2/\tau\right) \frac{d\tau}{\tau^2}$$

$$= \frac{(2n+1) M \sqrt{t}}{2n^2 a \sqrt{\pi}} \exp\left(-n^2 a^2/t\right)$$

$$\leqslant \frac{(2n+1) M \sqrt{t_1}}{2n^2 a \sqrt{\pi}} \exp\left(-n^2 a^2/t_0\right),$$

and similarly for $|u_0(x - 2na, t)|$. Hence the series defining $u_2(x, t)$ is uniformly and absolutely convergent when $0 \leqslant x \leqslant a, 0 < t_0 \leqslant t \leqslant t_1$.

The series obtained by term-by-term differentiation can be discussed in the same way. It follows that $u_2(x, t)$ satisfies the equation of heat in $0 < x \leqslant a, t > 0$.

Write
$$u_2(x, t) = u_0(x, t) + \sum_{-\infty}^{\infty}{}' u_0(x + 2na, t),$$

where the term $n = 0$ is omitted in $\Sigma'$. Hence at every point of continuity of $\phi(t)$ and so almost everywhere, we have

$$u_2(+0, t) = \phi(t) + \sum_{-\infty}^{\infty}{}' u_0(2na, t) = \phi(t),$$

since $u_0(x, t)$ is an odd function of $x$. Also

$$u_2(a, t) = \sum_{-\infty}^{\infty} u_0((2n+1)a, t) = 0.$$

Thus $u_2(x, t)$ does satisfy the prescribed initial and end conditions.

From this form of the solution, we can deduce the solution by Fourier series. For

$$u_2(x, t) = -2 \frac{\partial}{\partial x} \sum_{-\infty}^{\infty} \int_0^t \phi(\tau) k(x + 2na, t - \tau) d\tau$$

$$= -\frac{\partial}{\partial x} \sum_{-\infty}^{\infty} \int_0^t \phi(t) \exp\left(-\tfrac{1}{4}(x + 2na)^2/(t - \tau)\right) \frac{d\tau}{\sqrt{\pi}\sqrt{(t - \tau)}}$$

$$= -\frac{\partial}{\partial x} \int_0^t \phi(t) \vartheta_3\left(\frac{\pi x}{2a} \bigg| \frac{i\pi(t - \tau)}{a^2}\right) \frac{d\tau}{a}$$

$$= -\frac{\partial}{\partial x} \int_0^t \phi(\tau) \left[1 + 2 \sum_1^{\infty} \exp\left(-n^2\pi^2(t - \tau)/a^2\right) \cos\frac{n\pi x}{a}\right] \frac{d\tau}{a},$$

and so

$$u_2(x, t) = \frac{2\pi}{a^2} \sum_1^{\infty} n \sin\frac{n\pi x}{a} \exp\left(-n^2\pi^2 t/a^2\right) \int_0^t \phi(\tau) \exp\left(n^2\pi^2\tau/a^2\right) d\tau.$$

The solution of the problem when the initial and end conditions are
$$u(x, 0) = 0, \quad u(0, t) = 0, \quad u(a, t) = \psi(t)$$

can be deduced by replacing $x$ by $a - x$ and $\phi$ by $\psi$.

## 12.13    The use of Fourier series

The problem of the finite rod when the temperature is given initially and the ends of the rod are kept at the same constant temperature can also be solved by using Fourier series. We have to find the solution of

$$\frac{\partial^2 u}{\partial x^2} = \frac{\partial u}{\partial t}$$

given that $\quad u(x, +0) = f(x) \quad (0 < x < a)$,

$$u(+0, t) = u(a-0, t) = 0 \quad (t > 0).$$

The particular solution

$$\exp(-n^2\pi^2 t/a^2)\sin\frac{n\pi x}{a}$$

vanishes when $x = 0$ and $x = a$. If the Fourier half-range sine series for $f(x)$ is

$$\sum_{1}^{\infty} b_n \sin\frac{n\pi x}{a},$$

where $\qquad b_n = \frac{2}{a}\int_0^a f(x)\sin\frac{n\pi x}{a}\,dx,$

we should expect

$$u(x,t) = \sum_{1}^{\infty} b_n \exp(-n^2\pi^2 t/a^2)\sin\frac{n\pi x}{a} \qquad (1)$$

would be the required solution.

We assume that $f(x)$ is integrable in Lebesgue's sense. Then $\{b_n\}$ is a null-sequence, and so is bounded; there exists a constant $K$ such that $|b_n| < K$ for all values of $n$. Hence, if $0 \leqslant x \leqslant a, 0 < t_0 \leqslant t \leqslant t_1$, we have

$$\left| b_n \exp(-n^2\pi^2 t/a^2)\sin\frac{n\pi x}{a} \right| < K\exp(-n^2\pi^2 t_0/a).$$

The series (1) therefore converges uniformly and absolutely in the rectangle, and so its sum is continuous there. In particular $u(0,t)$ and $u(a,t)$ vanish when $t \geqslant t_0 > 0$. The series obtained from (1) by differentiation under the sign of summation are also uniformly and absolutely convergent. Therefore (1) defines a solution of the equation of heat which satisfies the end conditions.

The series $\sum_{1}^{\infty} b_n \sin n\pi x/a$ is not necessarily convergent, but it is summable $(C, 1)$† almost everywhere; and its Cesàro sum is $f(x)$ at every point of continuity, again almost everywhere. To complete the proof we use a theorem due to Bromwich,‡ that if $\sum_{0}^{\infty} a_k$ is summable $(C, 1)$ with sum $s$, then

$$\lim_{\theta \to +0} \sum_{0}^{\infty} a_n \exp(-\theta\lambda_n) = s$$

---

† The terms summable $(C, 1)$ and Cesàro sum are defined in note 8 of the Appendix.
‡ Bromwich, *An Introduction to Infinite Series*, 2nd edn (1926), p. 429.

if $\{\lambda_n\}$ is an increasing sequence of postitive integers such that, for all $n$,

$$\lambda_n - 2\lambda_{n+1} + \lambda_{n+2} > 0,$$

$$n(\lambda_n - \lambda_{n-1}) < K\lambda_n,$$

where $K$ is a constant. In the present case $\theta = \pi^2 t/a^2$, and $\lambda_n = n^2$, so that Bromwich's conditions are satisfied. Hence the series in (1) tends to $f(x)$ as $t \to +0$ whenever $f(x)$ is continuous.

The problem of solving the equation of heat in $0 \leqslant x \leqslant a, t > 0$, when $u(x, 0) = 0, u(0, t) = \phi(t), u(a, t) = 0$ can be solved by a modification of the Fourier method. Assume that

$$u(x, t) = \sum_1^\infty b_n(t) \sin \frac{n\pi x}{a},$$

where
$$b_n(t) = \frac{2}{a} \int_0^a u(x, t) \sin \frac{n\pi x}{a} \, dx.$$

Since $u(x, t)$ vanishes when $t = 0, b_n(0) = 0$. We have

$$\frac{db_n(t)}{dt} = \frac{2}{a} \int_0^a \frac{\partial u(x, t)}{\partial t} \sin \frac{n\pi x}{a} \, dx$$

$$= \frac{2}{a} \int_0^a \frac{\partial^2 u(x, t)}{\partial x^2} \sin \frac{n\pi x}{a} \, dx$$

$$= \frac{2}{a} \left[ \frac{\partial u}{\partial x} \sin \frac{n\pi x}{a} - \frac{n\pi}{a} u \cos \frac{n\pi x}{a} \right]_0^a - \frac{2n^2\pi^2}{a^3} \int_0^a u(x, t) \sin \frac{n\pi x}{a} \, dx$$

$$= \frac{2n\pi}{a^2} \phi(t) - \frac{n^2\pi^2}{a^2} b_n(t).$$

From
$$\frac{db_n}{dt} + \frac{n^2\pi^2}{a^2} b_n = \frac{2n\pi}{a^2} \phi(t)$$

we have
$$b_n = \frac{2n\pi}{a^2} \int_0^t \phi(\tau) \exp\left(-n^2\pi^2(t-\tau)/a^2\right) d\tau.$$

Thus
$$u(x, t) = \frac{2\pi}{a^2} \sum_1^\infty n \sin \frac{n\pi x}{a} \int_0^t \phi(\tau) \exp\left(-n^2\pi^2(t-\tau)/a^2\right) d\tau.$$

This formal argument gives the answer found in §12.12, but it would be difficult to give a rigorous proof.

Note that $u(0, t) = 0$. The reason for this is that $u(x, t)$ is an odd function of $x$. $u(x, t)$ tends to $\phi(t)$ as $x \to +0$, but it tends to $-\phi(t)$ as $x \to -0$. The sum of the series when $x = 0$ is $\frac{1}{2}\{u(+0, t) + u(-0, t)\}$, which is zero.

If we try to solve the equation of heat for the finite rod when the conditions are
$$u(x, 0) = f(x) \quad (0 < x < a),$$

$$u_x(0, t) - hu(0, t) = 0, \quad u_x(a, t) + Hu(a, t) = 0 \quad (t > 0),$$

where $h$ and $H$ are positive constants, we might start with the solution

$$u = \exp(-k^2 t)(A \cos kx + B \sin kx).$$

In order to satisfy the end conditions, we must have

$$(k^2 - hH) \sin ka = k(h + H) \cos ka.$$

This equation in $k$ has no complex roots. Its real roots occur in pairs $\pm k_1, \pm k_2, \ldots$ where $\{k_n\}$ is a strictly increasing sequence of positive numbers. When $n$ is large,

$$k_n = \frac{n\pi}{a} + \frac{a(h + H)}{n\pi} + O\left(\frac{1}{n^2}\right).$$

Since $ak_n/\pi$ is not an integer, the Fourier series method is not applicable. We have to use instead an expansion as a series of Sturm–Liouville functions. The theory is outside the scope of this book.

## Exercises

**1.** If $U(\alpha, t)$ is the Fourier transform of a solution $u(x, t)$ of

$$\frac{\partial^2 u}{\partial x^2} = \frac{\partial u}{\partial t} + xu,$$

prove that
$$\frac{\partial U}{\partial t} + i \frac{\partial U}{\partial \alpha} = -\alpha^2 U,$$

and hence that      $U(\alpha, t) = F(\alpha + it) \exp(-\tfrac{1}{3} i\alpha^3).$

Deduce that, if $u(x, 0) = f(x)$ for all values of $x$

$$u(x, t) = \frac{1}{2\sqrt{(\pi t)}} \exp(\tfrac{1}{3} t^3 - xt) \int_{-\infty}^{\infty} f(\xi) \exp(-\tfrac{1}{4}(x - \xi - t^2)^2 / t) \, d\xi.$$

**2.** $u(x, t)$ satisfies the equation of heat in $x > 0, t > 0$ under the conditions

$$u(x, 0) = f(x), \quad u_x(0, t) = \psi(t).$$

The Fourier cosine transform of $u(x, t)$ is

$$U_c(\alpha, t) = \sqrt{\left(\frac{2}{\pi}\right)} \int_0^{\infty} u(x, t) \cos \alpha x \, dx,$$

with inverse $\qquad u(x,t) = \sqrt{\left(\dfrac{2}{\pi}\right)} \displaystyle\int_{-\infty}^{\infty} U_c(\alpha,t) \cos \alpha x \, d\alpha.$

Prove that $\qquad \dfrac{\partial U_c}{\partial t} = \sqrt{\left(\dfrac{2}{\pi}\right)} \psi(t) - \alpha^2 U_c.$

Deduce that

$$U_c(\alpha,t) = F_c(\alpha) e^{-\alpha^2 t} - \sqrt{\left(\dfrac{2}{\pi}\right)} \int_0^t \psi(\tau) e^{-\alpha^2 (t-\tau)} d\tau,$$

where $F_c(\alpha)$ is the Fourier cosine transform of $f(x)$, and hence that

$$u(x,t) = \int_0^\infty f(\xi)\{k(x-\xi,t)+k(x+\xi,t)\}\,d\xi - 2\int_0^t \psi(\tau)\,k(x,t-\tau)\,d\tau.$$

**3.** Prove that, if $a$ and $b$ are positive,

$$\int_{-\infty}^{\infty} k(x-\xi,a)\,k(\xi,b)\,d\xi = k(x,a+b).$$

Hence show that, if the operator $\mathscr{G}_x^t$ is defined by

$$\mathscr{G}_x^t[f] = \int_{-\infty}^{\infty} f(\xi)\,k(x-\xi,t)\,d\xi \quad (t>0),$$

then $\qquad \mathscr{G}_x^{t_1}[\mathscr{G}_x^{t_2}[f]] = \mathscr{G}_x^{t_1+t_2}[f],$

if $t_1$ and $t_2$ are positive.

**4.** Find the solution of $\qquad \dfrac{\partial^2 u}{\partial x^2} = \dfrac{\partial u}{\partial t}$

in $0 < x < \pi,\, t > 0$ such that

$$u(x,0) = 0 \ (0 < x < \pi), \quad u(0,t) = 1-e^{-t} \ (t>0), \quad u(\pi,t) = 0 \ (t>0).$$

Prove that this solution tends to $(\pi-x)/\pi$ as $t \to +\infty$. Why would you expect this result?

**5.** $u(x,t)$ is the solution of the equation of heat which satisfies the conditions
$$u = 0 \ (t=0, x \geqslant 0), \quad u_x - hu = \phi(t) \ (x=0, t \geqslant 0),$$
where $h$ is a positive constant and $\phi(t)$ is continuous. Prove that, when $x > 0$ and $t > 0$
$$\frac{\partial u}{\partial x} - hu = \int_0^t \phi(\tau)\,\frac{x}{t-\tau}\,k(x,t-\tau)\,d\tau.$$
Hence show that

$$u(x,t) = -\int_0^\infty e^{-h\xi} \int_0^t \frac{x+\xi}{\tau}\,\phi(t-\tau)\,k(x+\xi,\tau)\,d\xi\,d\tau.$$

**6.** $u(x,t)$ satisfies the equation of heat in $x > 0, t > 0$, under the conditions
$$u(x,0) = 0 \ (x>0), \quad u(0,t) = \phi(t) \ (t>0).$$

Prove that $$u(x, t) = \frac{2}{\sqrt{\pi}} \int_X^\infty \phi\left(t - \frac{x^2}{4\xi^2}\right) e^{-\xi^2} d\xi,$$

where $X = \tfrac{1}{2}x/\sqrt{t}$.

**7.** Prove that the solution of the equation of heat in $t \geqslant 0$ which satisfies the conditions
$$u(x, 0) = 1 \ (x > 0), \quad u(x, 0) = -1 \ (x < 0)$$

is $$u(x, t) = \frac{2}{\sqrt{\pi}}\left[1 - \mathrm{Erfc}\,\frac{x}{2\sqrt{t}}\right].$$

**8.** Prove that the solution of the equation of heat in $x \geqslant 0, t \geqslant 0$ which satisfies the conditions

$$u(x, 0) = 0 \ (x > 0), \quad u(0, t) = 1 \ (0 < t < T), \quad u(0, t) = 0 \ (t > T)$$

is $$u(x, t) = \frac{2}{\sqrt{\pi}}\,\mathrm{Erfc}\,\frac{x}{2\sqrt{t}} \quad (0 < t \leqslant T),$$

$$= \frac{2}{\sqrt{\pi}}\left[\mathrm{Erfc}\,\frac{x}{2\sqrt{t}} - \mathrm{Erfc}\,\frac{x}{2\sqrt{(t-T)}}\right] \quad (t > T).$$

Verify that the solution is continuous and continuously differentiable.

**9.** $u(x, t)$ satisfies the equation of heat in $0 \leqslant x \leqslant 0, t \geqslant 0$, under the conditions

$$u(x, 0) = f(x) \ (0 < x < a), \quad u_x(0, t) = u_x(a, t) = 0 \ (t > 0),$$

where $f(x)$ is integrable. Prove that

$$u(x, t) = \tfrac{1}{2}a_0 + \sum_1^\infty a_n e^{-n^2\pi^2 t/a^2} \cos\frac{n\pi x}{a},$$

where $$\tfrac{1}{2}a_0 + \sum_1^\infty a_n \cos\frac{n\pi x}{a}$$

is the Fourier half-range cosine series for $f(x)$.

**10.** Show that the solution of the equation of heat in $x > 0, t > 0$ such that
$$u(x, 0) = f(x) \ (x > 0), \quad u_x(0, t) - hu(0, t) = 0 \ (t > 0)$$

is $$u(x, t) = \int_0^\infty \{f(\xi)\,k(x - \xi, t) + g(\xi)\,k(x + \xi, t)\}\,d\xi,$$

where $$g(x) = f(x) - 2he^{-hx}\int_0^x f(\xi)\,e^{h\xi}\,d\xi.$$

# APPENDIX

## Note 1. Analytic functions

The function $f(x,y)$ of two real variables is said to be an analytic function regular in a neighbourhood of $(x_0, y_0)$ if it can be expanded as a double series

$$\sum_{m,n=0}^{\infty} a_{mn}(x-x_0)^m (y-y_0)^n$$

absolutely convergent on a disc $(x-x_0)^2 + (y-y_0)^2 < R_0^2$.

If $(x_0 + R_1, y_0 + R_1)$ is a point of this disc, $\{a_{mn}R_1^{m+n}\}$ is a bounded double sequence. Hence if $R < R_1$ the double series is uniformly and absolutely convergent on the square $|x-x_0| \leqslant R$, $|y-y_0| \leqslant R$. The series can therefore be differential term-by-term on that closed square as often as we please. The definition can be extended in the obvious way to any number of variables.

## Note 2. Dominant functions

Suppose that, by some formal process, we have obtained a double series

$$\sum_{m,n=0}^{\infty} a_{mn}(x-x_0)^m (y-y_0)^n$$

and that there exists a double sequence of *positive* numbers $\{A_{mn}\}$ such that the double series

$$\sum_{m,n=0}^{\infty} A_{mn}(x-x_0)^m (y-y_0)^n$$

is convergent absolutely and uniformly when $|x-x_0| \leqslant R$, $|y-y_0| \leqslant R$ and that $|a_{mn}| < A_{mn}$ for all $m$ and $n$. Then, by the comparison test,

$$\sum_{m,n=0}^{\infty} a_{mn}(x-x_0)^m (y-y_0)^n$$

is also uniformly and absolutely convergent when $|x-x_0| \leqslant R$, $|y-y_0| \leqslant R$. If we denote the sums of the two series by $f$ and $F$, we say that $F$ is a dominant (or majorant) function of $f$, and we write $f \ll F$. If, on the same square, $f \ll F$, $g \ll G$, then

$$f+g \ll F+G, \quad fg \ll FG.$$

[ 271 ]

Also                    $\dfrac{\partial f}{\partial x} \ll \dfrac{\partial F}{\partial x}, \quad \dfrac{\partial f}{\partial y} \ll \dfrac{\partial F}{\partial y};$

with similar expressions for derivatives of all orders. The definition can obviously be extended to any number of variables.

If $f(x, y)$ is an analytic function, regular in a neighbourhood of some point, the origin say, it is expansible as a double series

$$\Sigma a_{mn} x^m y^n$$

uniformly and absolutely convergent on a square $|x| \leqslant R, |y| \leqslant R$. The double sequence $\{a_{mn} R^{m+n}\}$ is therefore bounded, $|a_{mn} R^{m+n}| < M$. It follows that

$$f \ll \sum_{mn} M \frac{x^m y^n}{R^{m+n}} = M \bigg/ \left(1 - \frac{x}{R}\right)\left(1 - \frac{y}{R}\right).$$

Also         $f \ll \sum_{mn} M \dfrac{(m+n)!}{m!\, n!} \dfrac{x^m y^n}{R^{m+n}} = M \bigg/ \left(1 - \dfrac{x}{R} - \dfrac{y}{R}\right).$

## Note 3. Regular arcs

A set of points in the plane defined by parametric equations

$$x = f(t), \quad y = g(t) \quad (a \leqslant t \leqslant b),$$

where $f(t)$ and $g(t)$ are continuous functions, is called an arc. If no point of the arc corresponds to two different values of $t$, the arc is said to be simple. If, in addition, $f(t)$ and $g(t)$ are continuously differentiable, the derivatives being one-sided at the end points, we shall call such a simple arc a *regular arc*. A regular arc has an arc length defined in the usual way. The definition also holds with the obvious changes for arcs in space.

## Note 4. Regular closed curves

A set of points in the plane defined by parametric equations

$$x = f(t), \quad y = g(t) \quad (a \leqslant t \leqslant b),$$

when $f(t)$ and $g(t)$ are continuous, is called a simple closed curve if no point corresponds to two different values of $t$ except that

$$f(a) = f(b), \quad g(a) = g(b).$$

If the simple closed curve is a chain of a finite number of regular arcs we shall call it a *regular closed curve*. A regular closed curve may have a finite number of corners; it has a piece-wise continuously-turning tangent. A regular closed curve in the plane divides the plane into two domains, a bounded interior domain and an unbounded exterior domain.

## Note 5. Green's theorem in the plane

Let $D$ be the domain bounded by a regular closed curve $C$. If

  (i)  $u$ and $v$ are continuous in $\bar{D}$, the closure of $D$,

  (ii)  $u_x$ and $v_y$ exist and are bounded in $D$,

  (iii)  the double integrals of $u_x$ and $v_y$ over $D$ exist,

then

$$\iint_D (u_x + v_y)\, dx\, dy = \int_C (lu + mv)\, ds,$$

where $(l, m)$ are the direction cosines of the normal to $C$ drawn out of $D$. The fact that the normal to $C$ may suddenly change direction at a finite number of points of $C$ does not affect the truth of the result. And the result is evidently true if $u_x$ and $v_y$ are continuous in $\bar{D}$.

There is an alternative form of the result corresponding to Stokes's theorem in space, namely that

$$\int_C P\, dx + Q\, dy = \iint_D (Q_x - P_y)\, dx\, dy,$$

where integration over $C$ is in the positive sense.

If $D$ is bounded externally by a regular closed curve $C_1$ and internally by a regular closed curve $C_2$,

$$\iint_D (u_x + v_y)\, dx\, dy = \int_{C_1} (lu + mv)\, ds + \int_{C_2} (lu + mv)\, ds,$$

where $(l, m)$ are the direction cosines of the normal to $C_1$ or $C_2$, again drawn out of $D$. The Stokes form is, in this case,

$$\iint_D (Q_x - P_y)\, dx\, dy = \int_{C_1} P\, dx + Q\, dy - \int_{C_2} P\, dx + Q\, dy,$$

where integration over $C_1$ and over $C_2$ are both in the positive sense.

## Note 6. Surfaces

A simple closed surface in three-dimensional Euclidean space is one which is topologically equivalent to a sphere. Two surfaces are topologically equivalent if each can be transformed into the other by continuous deformation. A simple closed surface is bounded and divides the whole space into a bounded interior domain and an unbounded exterior domain.

A simple closed curve drawn on a simple closed surface divides the surface into two portions, which we call caps. A cap has parametric equations $r = r(\lambda, \mu)$ in vector notation, where each component is a continuous function of $\lambda$ and $\mu$. If $r(\lambda, \mu)$ is differentiable, the vectors $r_\lambda$ and $r_\mu$ are tangent to the cap, and

$$N = (r_\lambda \times r_\mu)/|r_\lambda \times r_\mu|$$

is a unit vector normal to the cap at the point of parameters $\lambda$ and $\mu$. If $r_\lambda$ and $r_\mu$ are continuous, the direction of $N$ varies continuously as $(\lambda, \mu)$ moves on the cap. We then call the cap a regular cap.

If a simple closed surface is formed of a finite number of regular caps, we call it a regular closed surface.

### Note 7. Green's theorem in space

Let $D$ be a domain bounded by a regular closed surface $S$. If

(i) $u, v$ and $w$ are continuous in $\bar{D}$, the closure of $D$,

(ii) $u_x, v_y$ and $w_z$ exist and are bounded in $D$,

(iii) the integrals of $u_x, v_y$ and $w_z$ over $D$ exist,

then

$$\iiint_D (u_x + v_y + w_z)\, dx\, dy\, dz = \iint_S (lu + mv + nw)\, dS,$$

where $(l, m, n)$ are the direction cosines of the normal to $S$ drawn out of $D$.

More generally, if $D$ is bounded internally by a regular closed surface $S_0$ and externally by a regular closed surface $S_1$, the triple integral is then equal to

$$\iint_{S_0} (lu + mv + nw)\, dS + \iint_{S_1} (lu + mv + nw)\, dS,$$

where $(l, m, n)$ are still the direction cosines of the normal drawn out of $D$.

### Note 8. Summability

The infinite series $\sum_1^\infty a_n$ is said to be convergent with sum $s$ if the sequence $\{s_n\}$ defined by

$$s_n = a_0 + a_1 + a_2 + \dots + a_{n-1}$$

tends to the limit $s$ as $n \to \infty$.

If the series is not convergent, it may be summable in some other sense. If

$$\sigma_n = \frac{s_0 + s_1 + s_2 + \dots + s_{n-1}}{n},$$

it may happen that the sequence $\{\sigma_n\}$ tends to the limit $\sigma$ as $n \to \infty$. If this is the case, the series $\sum\limits_{0}^{\infty} a_n$ is said to be summable by Cesàro's mean, or summable $(C, 1)$, with Cesàro sum $\sigma$. If $\sum\limits_{0}^{\infty} a_n$ is convergent with sum $s$, it is summable $(C, 1)$, and $\sigma = s$. But a series summable $(C, 1)$ is not necessarily convergent.

If the series $\sum\limits_{0}^{\infty} a_n x^n$ has radius of convergence unity and has sum $f(x)$ when $|x| < 1$, the series $\sum\limits_{1}^{\infty} a_n$ is said to be summable in Abel's sense, or summable $(A)$, if $f(x)$ tends to a finite limit $S$ as $x \to 1-0$; $S$ is called the Abel sum. If $\sum\limits_{0}^{\infty} a_n$ is convergent with sum $s$, the Abel sum is also $s$, by Abel's theorem on the continuity of power series. If $\sum\limits_{0}^{\infty} a_n$ is summable $(C, 1)$, with sum $\sigma$, its Abel sum is also $\sigma$.

## Note 9. Fourier series

If $f(\theta)$ is integrable in Lebesgue's sense over $0 \leqslant \theta \leqslant 2\pi$, it has a Fourier series

$$\tfrac{1}{2}a_0 + \sum_{1}^{\infty} (a_n \cos n\theta + b_n \sin n\theta),$$

where $\quad a_n = \dfrac{1}{\pi} \displaystyle\int_0^{2\pi} f(\theta) \cos n\theta \, d\theta, \quad b_n = \dfrac{1}{\pi} \displaystyle\int_0^{2\pi} f(\theta) \sin n\theta \, d\theta.$

The sequences $\{a_n\}$ and $\{b_n\}$ are null-sequences, but this does not imply that the Fourier series is convergent. But the Fourier series is summable $(C, 1)$ to the sum

$$\tfrac{1}{2}\{f(\theta+0) + f(\theta-0)\}$$

for every value of $\theta$ for which this has a meaning. If we assume that $f(\theta)$ is periodic of period $2\pi$, the sum $(C, 1)$ when $\theta = 0$ is

$$\tfrac{1}{2}\{f(+0) + f(2\pi-0)\}.$$

In particular, the series is summable $(C, 1)$ to the sum $f(\theta)$ at every point where $f(\theta)$ is continuous; it is summable $(C, 1)$ to $f(\theta)$ for almost all $\theta$.

Since $\{a_n\}$ and $\{b_n\}$ are null-sequences, the power series

$$\tfrac{1}{2}a_0 + \sum_{1}^{\infty} (a_n \cos n\theta + b_n \sin n\theta) x^n$$

has radius of convergence unity. Since the series with $x = 1$ is summable $(C, 1)$ with sum $f(\theta)$ at every point where $f$ is continuous, it follows that

$$\lim_{x \to 1-0} \left[ \tfrac{1}{2}a_0 + \sum_1^\infty (a_n \cos n\theta + b_n \sin n\theta) x^n \right] = f(\theta)$$

at every point where $f$ is continuous. The Abel sum of the Fourier series is $f(\theta)$ for almost all $\theta$.

# BOOKS FOR FURTHER READING

Bateman, H. *Partial Differential Equations of Mathematical Physics* (Cambridge, 1932).

Courant, R. and Hilbert, D. *Methods of Mathematical Physics* (New York, 1 (1953), 2 (1962)). This is the revised translation of *Methoden der Mathematischen Physik* (Berlin, 1 (1924), 2 (1937)).

Duff, G. F. D. *Partial Differential Equations* (Toronto, 1956).

Epstein, B. *Partial Differential Equations* (New York, 1962).

Garabedian, P. R. *Partial Differential Equations* (New York, 1964).

Goursat, E. *Course in Mathematical Analysis*, 3 parts 1 and 2 (New York, 1964). This is the translation of tome 3 of *Cours d'Analyse Mathématique* (Paris, 1923).

Hadamard, J. *Lectures on Cauchy's Problem in Linear Partial Differential Equations* (Yale, 1923; reprinted New York, 1952).

Hellwig, G. *Partial Differential Equations, an Introduction* (Waltham, Mass., 1964).

Jeffreys, H. and B. S. *Methods of Mathematical Physics*, 3rd edn (Cambridge, 1956).

Kellogg, O. D. *Foundations of Potential Theory* (Berlin, 1923; reprinted, New York, 1953).

Petrovsky, I. G. *Lectures on Partial Differential Equations* (New York and London, 1955).

Sauer, R. *Anfangswertprobleme bei Partiellen Differentialgleichungen* (Berlin, 1952).

Sneddon, I. N. *Mixed Boundary Value Problems in Potential Theory* (Amsterdam, 1966).

Steinberg, W. J. and Smith, T. L. *The Theory of Potential and Spherical Harmonics* (Toronto, 1944).

# INDEX